International Conference on Synthesis and Consolidation of Powder Materials (SCPM-2018)

AA006195

IOP Conference Series: Materials Science and Engineering
Volume 558

Chernogolovka, Russia
23 – 26 October 2018

ISBN: 978-1-5108-9092-3
ISSN: 1757-8981

Printed from e-media with permission by:

Curran Associates, Inc.
57 Morehouse Lane
Red Hook, NY 12571

Some format issues inherent in the e-media version may also appear in this print version.

Printed by Curran Associates, Inc. (2019)

For permission requests, please contact the Institute of Physics at the address below.

Institute of Physics
Dirac House, Temple Back
Bristol BS1 6BE UK

Phone: 44 1 17 929 7481
Fax: 44 1 17 920 0979

techtracking@iop.org

Additional copies of this publication are available from:

Curran Associates, Inc.
57 Morehouse Lane
Red Hook, NY 12571 USA
Phone: 845-758-0400
Fax: 845-758-2633
Email: curran@proceedings.com
Web: www.proceedings.com

Table of contents

Volume 558

Preface

Papers

IOP Conf. Series: Materials Science and Engineering **558** (2019) 011001 doi:10.1088/1757-899X/558/1/011001

PREFACE

International Conference on Synthesis and Consolidation of Powder Materials (SCPM-2018)

The International Conference on Synthesis and Consolidation of Powder Materials (SCPM-2018) held in Chernogolovka on October 23–26, 2018 was organized by Merzhanov Institute of Structural Macrokinetics and Materials Science, Russian Academy of Sciences under the auspices of the Russian Foundation for Basic Research. About 160 oral and poster presentations were presented by participants from Belorussia, France, Georgia, Germany, Latvia, Turkey, USA, and Russia.

Modern methods for high-temperature synthesis and consolidation of advanced powder materials are characterized by such features as extreme temperature–time conditions, short duration, and impact of electric field. Both processes, synthesis and consolidation, can be performed in one stage. SCPM-2018 gave rich opportunity for specialists in the field of high-temperature self-propagating synthesis (SHS, also known as combustion synthesis) and powder metallurgy (primarily, spark plasma sintering, hot pressing, etc.) to present their research results and to exchange by new ideas. Such an interdisciplinary character of the meeting produced synergic effect in both fields. We hope that the results of the Conference may turn useful in solving basic problems in materials science and modern R & D.

The conference invited lectures and oral presentations in three sections:

–Mechanisms and prospects of sintering and hot pressing
–Synthesis of powders and nano-powders
–Advanced processes for materials consolidation in electromagnetic fields

For more information about the conference, including photo gallery, go to http://www.ism.ac.ru/events/SCPM-2018/

M. I. Alymov
D. O. Moskovskikh
A. S. Rogachev

IOP Conf. Series: Materials Science and Engineering **558** (2019) 011002 doi:10.1088/1757-899X/558/1/011002

Peer review statement

All papers published in this volume of *IOP Conference Series: Materials Science and Engineering* have been peer reviewed through processes administered by the proceedings Editors. Reviews were conducted by expert referees to the professional and scientific standards expected of a proceedings journal published by IOP Publishing.

IOP Conf. Series: Materials Science and Engineering **558** (2019) 012001 doi:10.1088/1757-899X/558/1/012001

Effect of aluminum additives on the ignition of tungsten–teflon mixtures

M I Alymov, S G Vadchenko and I S Gordopolova

Merzhanov Institute of Structural Macrokinetics and Materials Science, Russian Academy of Sciences, Chernogolovka, 142432 Russia

E-mail: alymov@ism.ac.ru

Abstract. Thermodynamic analysis and studies of combustion and structure formation of products were carried out in the powder mixtures of tungsten and polytetrafluoroethylene (Teflon, Tf) with aluminum additives. The mixture components were selected to fabricate high-density condensed products with high ignition temperature. Aluminum was used as an energetic additive accelerating the ignition process and increasing the mixture combustion temperature. In mixtures, a tungsten/Tf ratio was fixed; the aluminum content was varied according to the formula $(1 - x)(0.8W + 0.2Tf) + xAl$ = const. Mechanically activated mixtures were pressed into the samples and heated in special crucible with controlled heating rate. It was shown that an increase in the heating rate changes insignificantly the ignition temperature of systems, but strongly affects the structure of combustion products. The ignition and combustion of compositions with low aluminum content results a large volume of gaseous products, which fly apart or form a high-porous structure. At high Al concentration, the results of experiments and thermodynamic calculations are sufficiently different that can be explained by the lack of thermodynamic data on tungsten aluminides in the used program and by the fact that real conditions of reaction are far from equilibrium and adiabatic ones. The calculated and experimental data showed that the optimal aluminum content to form melted products with a high density ($\rho(W_2C)$ = 17.2 g/cm^3) is about 10 wt %. At high Al concentration, the main combustion product (tungsten aluminide WAl_4) has the lower density ($\rho(WAl_4)$ = 6.6 g/cm^3) that is insufficiently for practical application.

1. Introduction

Composites based on metal powders and polymers including the structural and functional composite materials are widely used to fabricate products of various applications. Fluoropolymers are commonly used polymeric fillers because they exhibit high thermal stability and enhanced chemical resistance. Fluoropolymers can act as a main or additional oxidizer in the energy-intensive compositions with an active metallic fuel.

One of the most effective method to increase the reaction ability of mixtures is mechanical activation (MA) [1–3]. The MA of metal/fluoropolymer powder mixtures make it possible to prepare highly homogeneous materials owing to the formation of polymeric coatings on the surfaces of metallic particles. The appearance of these films under the action of MA may be associated with elevated concentration of fluoropolymer macroradicals [1, 4]. In addition, the MA of metal/fluoropolymer mixtures can promote to change in the fuel properties of these composites. This is the actual problem for the chemically active metallic powders such as aluminum and magnesium.

In some cases, it is necessary to fabricate the energy-intensive materials with high density. To provide needed density, heavy metals (such as U, W, Th, Ta, Hf, Nb, etc) or heavy metal oxides (WO_2, WO_3, CuO, PbO_2, Pb_3O_4, etc.) are added. One of the most attractive components is tungsten because it has high density and contributes to no increase in the mixture sensibility [5]. There are limited data in literature about the kinetics of reaction between Tf and tungsten, as well as about ignition, combustion, and detonation of these mixtures. Previously, we studied the ignition process of mixtures containing tungsten, Tf, and aluminum [6]. The present work is aimed at the study of the influence of the ratio of components (tungsten, Tf, and aluminum) on the ignition parameters and composition of mechanically activated products.

2. Thermodynamic analysis of reactions for W–Al–Tf system

For thermodynamic calculations, we used the W–Tf-based compositions with aluminum added to initiate the combustion. To fabricate high-density composite, the amount of Tf should be selected in such a way to ensure high compressibility and strength and occurrence of the reaction with a high heat release. The Tf content was limited by 20 wt % in relation to tungsten mass because of its density is 2.1–2.2 g/cm^3. At this Tf/tungsten ratio, the aluminum content was varied according the formula:

$$(1 - x)(0.8\text{W} + 0.2\text{Tf}) + x\text{Al} = \text{const}.$$

Results of thermodynamic calculations carried out using a THERMO program are presented in table 1.

Table 1. Results of thermodynamic calculations.

№	Al, %	W, %	Tf, %	ρ, г/см^3	T_{ad}, K	V, l	S, % (products)	L, % (products)
1	5	76	19	6.93	3050	75	53(C; W_2C)	24(W_2C)
2	10	72	18	6.40	3130	99	2(C)	74(W_2C)
3	20	64	16	5.56	1985	106	71(C; W_2C; Al_4C_3)	0
4	30	56	14	4.91	1660	75	64(Al_4C_3; W_2C)	11(Al)
5	40	48	12	4.34	1520	45	55(Al_4C_3; W_2C)	26(Al)
6	50	40	10	3.98	1460	22	51(Al_4C_3; AlF3; W_2C)	41(Al)
7	60	32	8	3.63	1450	0	45,5(Al_4C_3; AlF_3; W_2C)	54,5(Al)
8	80	16	4	3.10	930	0	100(Al; Al_4C_3; AlF_3; W_2C)	0

In the case of W–Tf mixture, the calculated adiabatic combustion temperature is sufficiently high (T_{ad} =2383.5 K); the content of condensed combustion products is about 56 wt %. However, the initiation process was difficult. The addition of a small amount of aluminum (5–10 wt %) significantly decreases the mixture ignition temperature and increases the combustion temperature. Calculations resulted that the combustion products of the composition *2* may contain up to 74 % melt. The further increase in the aluminum content to 20% leads to a sharp reduction of the combustion temperature and a decrease in the content of melted products. It is worth to note that with increasing aluminum concentration and decreasing combustion temperature, the total mass of condensed products including melt increases. At the same time, the density of the initial mixture and combustion products decreases.

3. Experimental

As the initial components, tungsten powder having a particle size below 10 μm, Teflon (low-molecular weight polytetrafluoroethylene fine powder) with a dispersion below 5 μm, and aluminum powder with a particle size below 20 μm were used. The component ratios are given in table 1. The powders were mixed using an AGO-2 planetary ball mill under hexane atmosphere. The mixture-to-ball mass ratio was equal to as 1:20, a frequency of the drum rotation was 2220 rpm, and an activation time was 5 min. The powder mixtures were pressed under a pressure of 145 MPa into cylindrical samples with a diameter of 3 mm, a height of 0.5–1 mm, and a mass of 0.01–0.02 g. The ignition and combustion processes were studied in argon atmosphere at atmospheric pressure. The ignition temperature of compositions T_c was determined using an experimental setup illustrated in figure 1. The samples were suspended in the boron nitride crucible on the VR5/VR20-type thermocouple welded from 10–15 μm thick wires. The heating of samples were performed by heat irradiation from the crucible heated by an aluminum foil conducting an electric current.

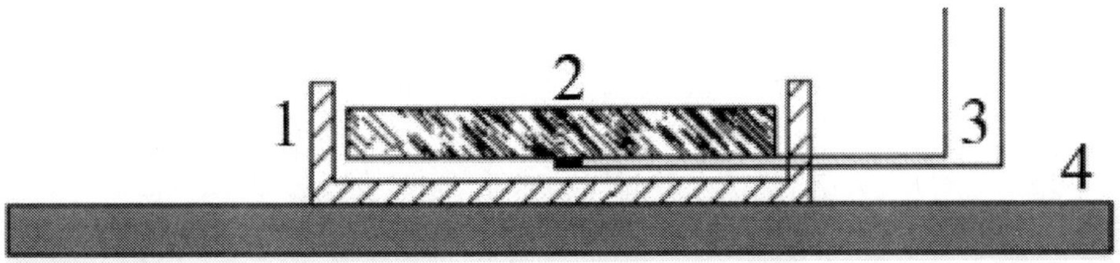

Figure 1. Schematic of experimental setup to determine mixture ignition temperature: *1* – BN crucible; *2* – sample; *3* – thermocouple; *4* – heater; *5* – graphite board.

4. Results and discussion

Figures 2a and 2b show the heating thermograms of samples.

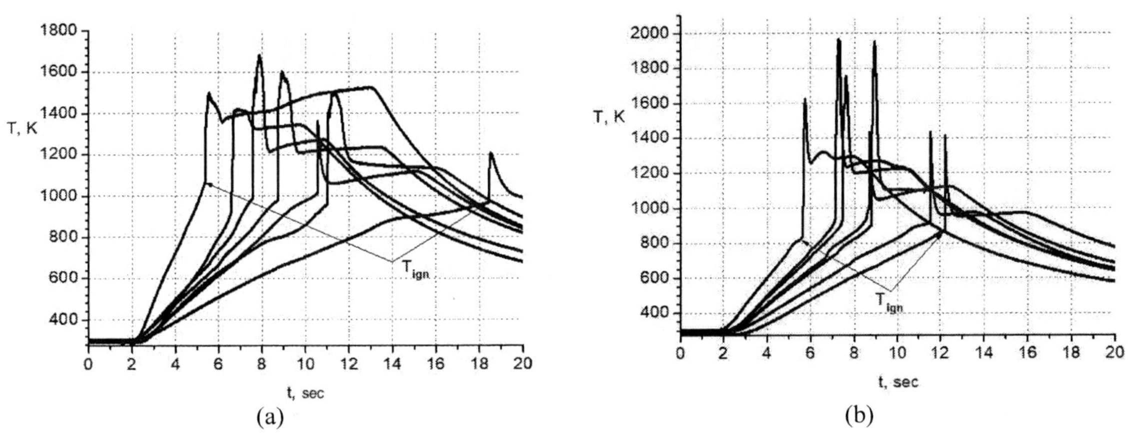

Figure 2. Heating thermograms of (a) 0.95(0.8W + 0.2Tf) + 0.05Al and (b) 0.90(0.8W + 0.2Tf) + 0.1Al mixtures at different heating rates.

At the low heating rate, the sample is heated uniformly and ignited simultaneously. This results a single temperature peak in the thermograms. At the higher heating rate, there are several peaks caused by a nonuniformity of the temperature field in the sample.

Upon heating of mixtures, the fluoropolymer amorphization and melting stages take place in the pre-ignition period. In the video-recording frames, the phase transition process in samples is observed as a concentric wave propagating from the sample side surface to its center. During phase transition, the sample geometric dimensions remain unchanged. The further increase in the heating temperature above PTFE thermal destruction temperature favors a sample foaming (about 873 K) caused by the volatilization of highly volatile polymers and decomposition products (at temperatures higher than the thermal destruction temperature of 678 K). The ignition of Al + Tf mixture begins at temperatures below aluminum melting point. This is caused by the high activity of gaseous products of the PTFE destruction and lead to an initiation of the combustion of tungsten with PTFE. The additional possible reason for the mixture ignition at the temperature of about 873 K can be an exothermic effect observed near this temperature upon DTA study of PTFE, which is explained by an oxidative thermal destruction [7, 8]. In our case, the experiments were carried out in argon and therefore, the oxidation process is not essential one. Figure 3 shows the change in the ignition temperature with increasing heating rate. The curves were plotted by the method of least squares. The analogous dependences were observed for the all studied compositions (figure 4).

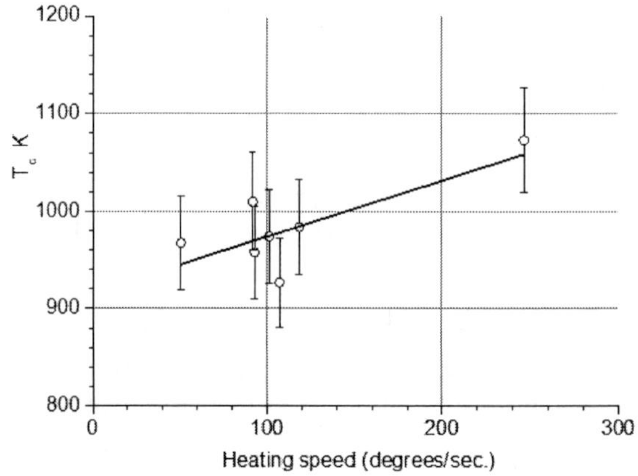

Figure 3. Dependence of the critical ignition temperature on the heating rate.

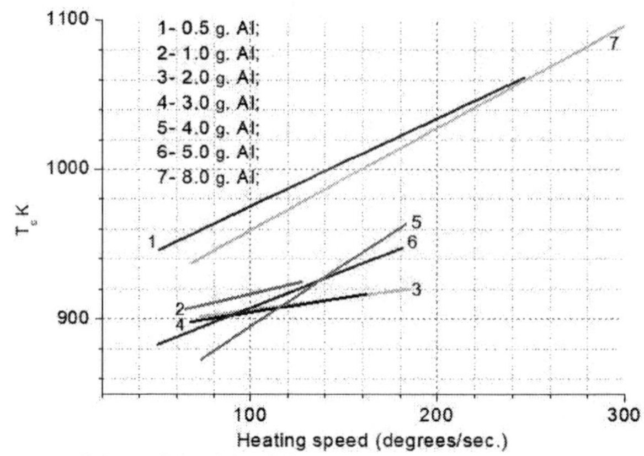

Figure 4. Dependence of the critical ignition temperature on the heating rate (composition numbers *1–7*).

Despite the wide dispersion of experimental data, the tendency to an increase in the ignition temperature for compositions with minimal and maximal aluminum concentrations is observed. This may be due to an increase in the heat conductivity of samples with high aluminum content, the formation of solid products, and a difficulty of the transfer of components in samples with low aluminum content. For all compositions, the ignition temperature of samples increases with increasing heating rate. In this case, conceivably the transition from thermal explosion mode to ignition one might occur.

In [6], we established that that the main combustion product was W_2C. In the compositions with low aluminum content, a large volume of gaseous products is formed as a result of ignition and combustion. This is confirmed by the results of thermodynamic calculations. At 5 wt % Al, these products fly apart. In the range of 10 to 20 wt % Al, they form a high-porous structure. In the case when Al content is higher than 70 wt %, there is no ignition of samples; only small heating take place. No size and shape of samples change because the products remain condensed that is agreement with the results of thermodynamic calculations. This is associated with the fact that the Tf concentration is too low to complete the reaction and therefore, a part of metal powders remains non-reacted. According to the XRD analysis data, the main product is WAl_4 (figure 5).

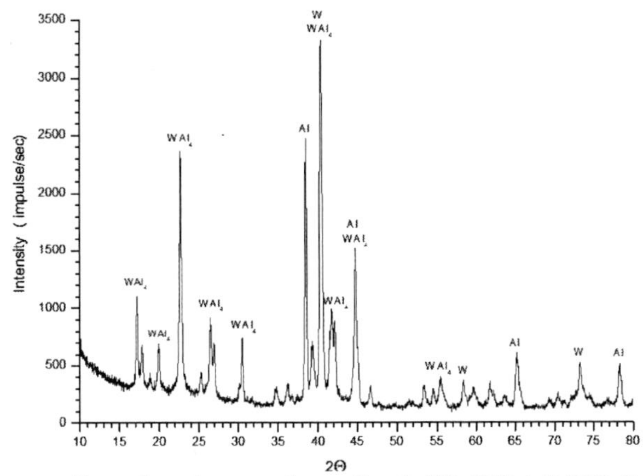

Figure 5. XRD pattern of combustion products for $0.4(0.8W + 0.2(C_2F_4)_n) + 0.6Al$ mixture (composition 7).

At high Al content, the results of our experiments and thermodynamic calculations differ substantially. This is explained by the fact that there are no thermodynamic data on tungsten aluminides in the program. Thermodynamic calculation of this mixture results that one of the combustion products should be tungsten carbide (W_2C), which is not detected in the experiment. The real conditions of the reaction are far from the equilibrium and adiabatic conditions. Nevertheless, the thermodynamic analysis proved to be useful to estimate the combustion temperature and gas release parameters.

5. Conclusions

The calculated and experimental data demonstrate that the optimal aluminum content to form melted products with a high density ($\rho(W_2C) = 17.2$ g/cm^3) is about 10 wt %. At high aluminum concentrations, the density of the main combustion product (tungsten aluminide WAl_4) is sufficiently lower ($\rho(WAl_4) = 6.6$ g/cm^3) that is insufficiently for practical application.

Acknowledgments
The work was supported by the Russian Science Foundation (project no. 16-13-00013).

References

[1] Barambojm N K 1961 *Mechano-chemistry of polymers* (Moscow: Rostehizdat) 252
[2] Boldyrev V V 2009 *Fundamentals of mechanical activation, and mechanosynthesis are considered chemical technologies* ed Avvakumov E G (Novosibirsk: Publishing House of the SB RAS) 342
[3] Malkin A I, Kiselev M R, Kliuev V A, Loznecova N N and Toporov Yu P 2012 Influence mechanical activation on the thermal properties of powder mixtures of aluminium with politetraftorjetilenom *Mater. Sci.* **3** 10
[4] Gorokhovskiy G A 1972 *Surface dispersion dynamically contacting polymers and metals*. (Kiev: Naukova dumka)
[5] Laptenkov V N, Saltanov L S, Svidinskij, A V, Belov V Yu, Baranov G V and Krylov V P 2016 (Patent 2579586 (RF))
[6] Vadchenko S G and Alymov M I 2017 Ignition of W–Teflon–Al Powder Mixtures *Int. J. Self-Propag. High-Temp. Synth.* **26** 137
[7] Buznik V M and Kurjavyj V G 2008 *J. Ros. Chem. Obschestva im. D.I. Mendeleyev* **52** 131
[8] Alikhanyan A S, Archangelskiy I V and Tarasov A V 2009 Interaction of fluoropolymers with transition metals *Inorg. Mater.* **45** 871

IOP Conf. Series: Materials Science and Engineering **558** (2019) 012002 doi:10.1088/1757-899X/558/1/012002

Azide SHS of highly dispersed powder of titanium carbonitride with intermediate partial nitriding or partial carburizing titanium powder

A P Amosov and Yu M Markov

Samara State Technical University, Molodogvardeyskaya 244, Samara, 443100 Russia

E-mail: egundor@yandex.ru

Abstract. It is noted that the powder of titanium carbonitride ($TiC_{0.5}N_{0.5}$) is the basis of the $TiC_{1-x}N_x$/Ni–Mo system cermets, which are considered as the most promising tool materials for the replacement of classical hard alloys based on tungsten carbide of the WC–Co system. The properties of these cermets are the higher the finer is the titanium carbonitride powder from which they are sintered. First, a simple energy-saving process of self-propagating high-temperature synthesis (SHS) was applied for the intermediate partial nitriding of the initial metal powder of titanium to the composition of $TiN_{0.2}$ or partial carburizing to $TiC_{0.3}$ composition in order to ensure fire safety and efficiency of the grinding process in the planetary ball mill to obtain superfine powders of partially nitrided or carburized titanium with ultrafine structure. Further, these powders were used in the azide SHS process for complete nitriding and carburizing to produce titanium carbonitride powder which was composed of superfine agglomerates of ultrafine and nano-particles of $TiC_{0.5}N_{0.5}$.

1. Introduction

The powder of titanium carbonitride ($TiC_{0.5}N_{0.5}$) is the basis of the $TiC_{1-x}N_x$/Ni–Mo system cermets, which are considered as the most promising tool materials for the replacement of classical hard alloys based on tungsten carbide of the WC–Co system [1, 2]. The most complete modern review of cermets based on titanium carbonitride is presented in the article 2018 "Ti(C, N)-Based Cermets: Critical Review of Achievements and Recent Developments" [2]. These cermets are on average twice cheaper and lighter than classical hard alloys, significantly exceeding them in hardness, wear resistance, resistance to oxidation and adhesion to steels, but still noticeably inferior to them in crack resistance and resistance to shock loads. To overcome these shortcomings, various methods are used, mainly, the complexity of the composition of the initial powder mixture, as well as reducing the size of the titanium carbonitride powder, the transition to ultrafine and nanoscale powder. In a number of works it is shown that the properties of these cermets are the higher the finer is the powder of titanium carbonitride from which they are sintered [2–5].

Analysis of existing methods for producing titanium carbonitride powders shows that most of them are energy-intensive, long-term, implemented in complex equipment at high temperatures, which lead to the formation of large particles of the product or strong agglomerates of small particles and the need for subsequent long-term grinding [6]. In this regard, it is of interest to use the method of self-propagating high-temperature synthesis (SHS) or combustion synthesis with low energy consumption, simple equipment, short duration of the synthesis process, high purity of products and great opportunities for controlling the particle size of the synthesized powders [7, 8]. Weakly sintered and subsieve powder of titanium carbonitride is obtained by azide SHS (SHS-Az) method in combustion of (Ti + 0.5C) mixtures with sodium azide NaN_3 and halide salts [9]. In this case, the combustion temperature is relatively low, a lot of gaseous products are formed, separating the powder particles, and therefore the particle size of the titanium carbonitride powder corresponds to the particle size of the original titanium powder (less than 40 μm), and the content of the base material is 96.5–98.5 wt %. Recently, the authors of this article have investigated the possibility of increasing the dispersion of titanium carbonitride powder $TiC_{0.5}N_{0.5}$, obtained by azide SHS technology, by increasing the dispersion of the original titanium powder [6]. However, if the grinding of pure metal titanium of standard grades is carried out in a planetary ball mill, then it becomes fire-hazardous at high dispersion, especially if it is mixed in the charge with oxidants, when its fire and explosion hazard increases many times. Therefore, in order to improve safety, reduce the activity of highly dispersed pure titanium powder, this powder was passivated by partial nitriding to the composition of $TiN_{0.2}$ before grinding in the planetary ball mill [6]. In addition, the partially nitrided titanium, in contrast to pure titanium, is more brittle and better dispersed in the mill.

Partially nitrided titanium of $TiN_{0.2}$ composition was prepared by SHS-Az method under the conditions of the corresponding limited content of sodium azide according to the equation [6]

$$Ti + 0.066NaN_3 + 0.066NH_4F \rightarrow TiN_{0.2} + 0.066NaF + 0.033N_2 + 0.132H_2$$

To increase the efficiency of the SHS-Az process, the halide salt of ammonium fluoride (NH_4F) was added to the charge as it usually done [10]. The process was carried out in the SHS-Az reactor in the gaseous argon medium, not gaseous nitrogen medium, to avoid additional nitriding of titanium with gaseous nitrogen. The synthesis conditions: the diameter of a powdered sample of the original charge of 40 mm; height of 50 mm; bulk density of 0.4, the pressure of argon in the reactor was 4 MPa. The synthesized $TiN_{0.2}$ powder was washed with distilled water to remove the side salt of NaF and then was dried. Grinding with mechanical activation of the synthesized $TiN_{0.2}$ powder was carried out by hard alloy balls for 8 min in the planetary ball mill "Activator-2SL". The scanning electron microscope (SEM) image of the milled particles $TiN_{0.2}$ shown in figure 1a, from which it is seen that the size of the milled $TiN_{0.2}$ particles is in the range from 100 nm to 3 μm, but these particles are combined into agglomerates of a micron size.

(a) (b)

Figure 1. SEM images of powders: (a) $TiN_{0.2}$ and (b) $TiC_{0.5}N_{0.5}$ [6].

Then this $TiN_{0.2}$ powder was used for the synthesis of powder of titanium carbonitride by the SHS-Az process according to the stoichiometric equation:

$$TiN_{0.2} + 0.5C + 0.1(NaN_3 + NH_4F) \rightarrow TiC_{0.5}N_{0.5} + 0.1NaF + 1.08N_2 + 1.3H_2$$

In order to prevent the target product from sintering during the synthesis process and to be obtained as an ultrafine powder, an excess of oxidants NaN_3 and NH_4F was introduced into the initial charge compared to $0.1(NaN_3 + NH_4F)$. They give a large volume of gaseous products and thereby separate the titanium carbonitride particles, preventing them from merging and becoming larger. In addition, the presence of excess of oxidants reduces the combustion temperature of powder mixtures, thereby reducing the size of the synthesized product particles [8]. The excess of oxidants is 100% in the case of $0.2(NaN_3 + NH_4F)$, 200% at 0.3 $(NaN_3 + NH_4F)$, and so on. The SEM image of powder $TiC_{0.5}N_{0.5}$ synthesized under 300% excess of oxidants is shown in figure 1b. This powder was obtained by hand grinding the loose combustion product in a porcelain mortar. Particle sizes of the target powder $TiC_{0.5}N_{0.5}$ were from 40 to 400 nm, but these particles were combined into agglomerates of micron sizes. The X-ray diffraction (XRD) phase analysis of washed titanium carbonitride powder, synthesized in SHS-Az mode using partially nitrided titanium ($TiN_{0.2}$) ground in a planetary mill, confirmed the presence in the combustion product of titanium carbonitride only [6].

This work is a continuation of the work [6]. Its purpose is also to study the possibility of increasing the dispersion of titanium carbonitride powder $TiC_{0.5}N_{0.5}$, obtained by the azide SHS technology, by increasing the dispersion of the starting titanium powder, but using the intermediate partial carburizing of the starting titanium powder instead of its intermediate partial nitriding.

2. Materials and methods of investigations

In the process of experimental study, the following powders were used: the titanium powder of PTM brand (purity ≥ 98.9 wt %, particle size < 45 μm), carbon black of P804T brand (99.5%, < 0.2 μm), sodium azide NaN_3 (98.7%), ammonium fluoride NH_4F (98.5%), ground table salt NaCl of Extra brand (99.7%, < 80 μm), polytetrafluoroethylene $(- C_2F_4 -)_n$ of Ftoroplast-4 brand or Teflon (100–180 μm). The planetary ball mill "Activator-2SL" with centrifugal acceleration of 30 g was used for intensive grinding with mechanical activation of powders. 375 grams of grinding hard alloy balls with a diameter of 7–11 mm and 37.5 g of the powder to be ground were loaded in each of the two drums with a volume of 250 ml. [6].

Powders of initial components were weighed on electronic scales, mixed in a porcelain mortar, filled up (bulk relative density of charge 0.4) in a calc cartridge with a diameter of 30 mm and a height of 45 mm, which was placed in a laboratory reactor SHS-Az of constant pressure with a working volume of 4.5 l. The arrangement of this reactor and the method of work on it with the measurement of linear velocities and the maximum combustion temperature by the thermocouple method using tungsten-rhenium thermocouples TR 5/20 with a diameter of 100 μm are described in detail in [11]. The temperature and velocity of combustion were recorded using an analog-to-digital converter connected to a computer. The reactor was filled with argon in the case of the preparation of partially carburized titanium or with nitrogen in the case of preparation of titanium carbonitride to the required operating pressure. The initial powder mixture (the charge) was ignited by a hot tungsten spiral from an electric DC source. After the end of combustion and short exposure for 10–15 min, the cooled sample was removed from the reactor. In the case of partially carburized titanium, the combustion product was ground in the planetary ball mill "Activator-2SL", and in the case of titanium carbonitride, the combustion product was easily destroyed manually to a loose powder state in a porcelain mortar. The resulting powder was washed from the residue of by-product NaF with distilled water at room temperature for 5–10 min, filtered on a vacuum funnel and dried in a vacuum drying cabinet to a constant weight.

Phase composition of the powdery combustion product was determined on an automated X-ray diffractometer ARL X TRA “ThermoScientific”. Shooting X-ray spectra was performed using Cu-radiation with continuous scanning in the range of angles 2θ = 20°÷80° at a speed of 2 deg/min. The obtained spectra were processed using a special software package WinXRD. The study of the morphology and size of powder particles was performed on the scanning electron microscope JSM-6390A "Jeol".

3. Results and discussion

Our attempts to obtain non-stoichiometric titanium carbide of $TiC_{0.1}$ or $TiC_{0.2}$ compositions in combustion of the charge of (Ti + 0.1C) or (Ti + 0.2C) respectively in the atmosphere of argon at a pressure of 4 MPa, as in the bulk density of the charge, and in an compacted form, have not led to success, as the charge was not ignited by the incandescent tungsten spiral filament. The charge of (Ti + 0.3C) was already ignited, but its combustion was unstable and incomplete. It turned out that the combustion of this charge (Ti + 0.3C) becomes stable with the addition of powdered Teflon. It is known that the addition of Teflon activates the combustion reaction of the charge (Ti + 0.3C) and promotes the synthesis of smaller particles of the combustion product [12]. Our studies have shown that the minimum amount of Teflon, which should be added for sustainable combustion and synthesis of non-stoichiometric titanium carbide $TiC_{0.3}$, is 5 wt % from the mixture of (Ti + 0.3C). The dependence of the combustion parameters (combustion temperature and combustion velocity) of the charge (Ti + xC) at different ratios of titanium and carbon with the addition of 5% Teflon, leading to the synthesis of non-stoichiometric titanium carbide of different composition of TiC_x, was investigated. The results are shown in figure 2a.

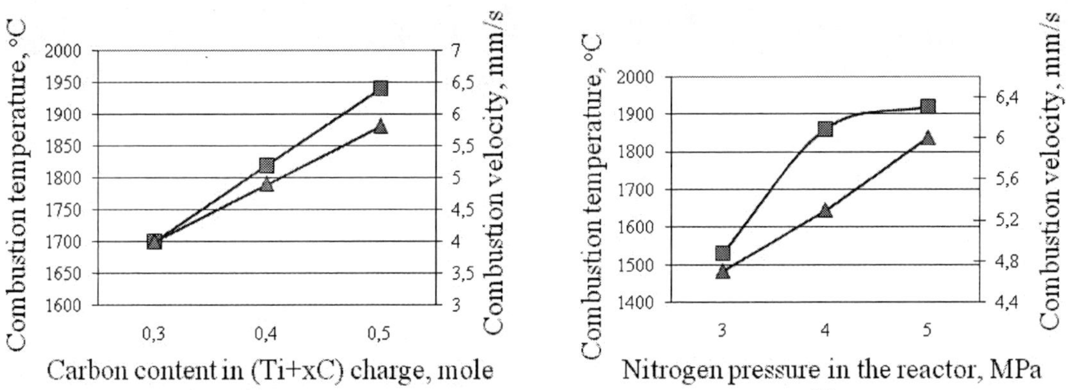

Figure 2. Combustion parameters at SHS of TiC_x from Ti (a) and $TiC_{0.5}N_{0.5}$ from $TiC_{0.3}$ (b).

Figure 3 shows the results of XRD phase analysis of the product of such synthesis that is non-stoichiometric titanium carbide of $TiC_{0.3}$ composition. According to these results, the product contains phase $TiC_{0.3}$ only.

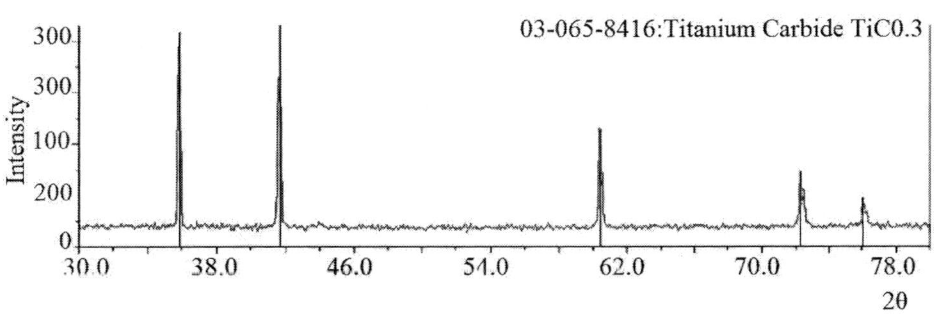

Figure 3. The XRD pattern of the sample of non-stoichiometric titanium carbide.

Synthesis of non-stoichiometric titanium carbide of composition $TiC_{0.3}$ was carried out also with another additive, which was the salt of sodium chloride NaCl, according to the equation $Ti + 0.3C \rightarrow TiC_{0.3}$ with the bulk density of the charge. The NaCl can melt and evaporate during combustion, loosening and reducing the target product $TiC_{0.3}$ in size. The maximum content of NaCl, in which combustion was possible, was 13 wt % from a mixture of (Ti + 0.3C). As can be seen from figure 4a, the particle size of $TiC_{0.3}$, synthesized at 13% NaCl and milled in the planetary ball mill for 15 min, was in the range from 90 to 400 nm.

(a) (b)

Figure 4. SEM images of powders: (a) $TiC_{0.3}$ and (b) $TiC_{0.5}N_{0.5}$.

But the combustion of the charge (Ti + 0.3C) with the addition of 13% NaCl in argon was unstable, so the preparation of non-stoichiometric titanium carbide $TiC_{0.3}$ was carried out at 10% NaCl.

Powder of non-stoichiometric titanium carbide $TiC_{0.3}$, prepared from the charge as with the addition of NaCl, and with the addition of Teflon, has been used to produce the titanium carbonitride of $TiC_{0.5}N_{0.5}$ composition by the SHS-Az method in the atmosphere of nitrogen according to the reaction equation:

$$TiC_{0.3} + 0.2C + 0.16NaN_3 + 0.16NH_4F \rightarrow TiC_{0.5}N_{0.5} + 0.16NaF + 0.08N_2 + 0.32H_2$$

The dependences of temperature and velocity of combustion of this system from the nitrogen pressure in the reactor are shown in figure 2b.

The combustion product was a loose cake, which was easily ground by hand in a porcelain mortar into a very fine powder. The morphology and size of synthesized at nitrogen pressure 4 MPa powder of titanium carbonitride $TiC_{0.5}N_{0.5}$ is presented in figure 4b, which shows that the final product $TiC_{0.5}N_{0.5}$ has the same particle size as the ground non-stoichiometric titanium carbide $TiC_{0.3}$ used for synthesis of $TiC_{0.5}N_{0.5}$. Figure 5 shows the results of the XRD phase analysis of the titanium carbonitride synthesized from $TiC_{0.3}$. It is seen that the final product is $TiC_{0.5}N_{0.5}$ which does not contain impurities.

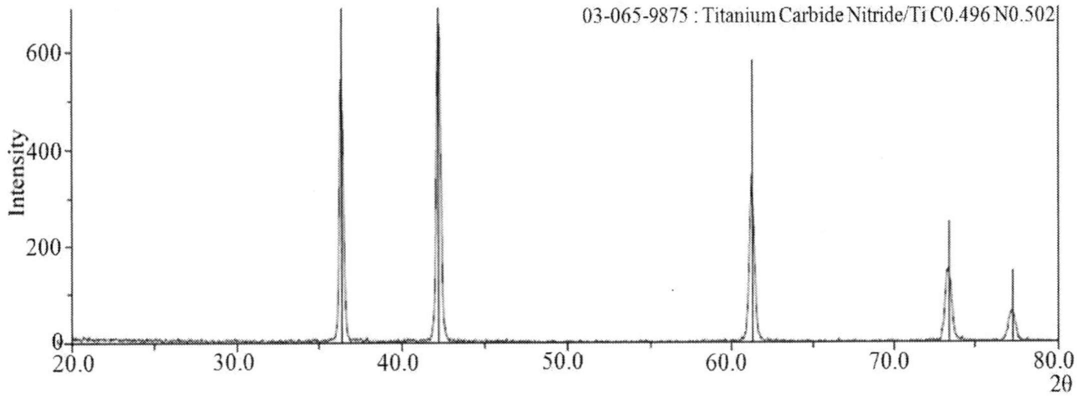

Figure 5. The XRD pattern of the sample of titanium carbonitride synthesized from $TiC_{0.3}$.

4. Conclusion

Thus, the SHS process can be applied for intermediate partial nitriding of the starting metal powder of titanium to the composition of $TiN_{0.2}$ or partial carburizing to the composition of $TiC_{0.3}$ to ensure fire safety and the efficiency of the grinding process in a planetary ball mill for preparation of superfine powders of partially nitrided or carburized titanium with ultrafine structure. It is shown that such powders can be used further in the process of azide SHS to produce the titanium carbonitride powder which was composed of superfine agglomerates of ultrafine- and nano-particles of $TiC_{0.5}N_{0.5}$.

Acknowledgements

The authors thank V A Novikov for assistance in SEM investigations and capturing diffraction patterns for X-ray phase analysis of the products of SHS.

References

[1] Pastor H 1988 *Mater. Sci. Eng.* **A 105/106** 401–409
[2] Lengauer W and Scagnetto F 2018 *Solid State Phen.* **274** 53–100
[3] Liu N, Yin W and Zhu L 2007 *Mater. Sci. Engineer.* **A 445-446** 707–716
[4] Lee D-W, Yu J-H and Jang T 2007 *Solid State Phen.* **124-126** 1225–8
[5] Demoly A, Veitsch C, Lengauer W and Rabitsch K 2011 *Int. J. Refr. Met. Hard Mater.* **29(6)** 716–723
[6] Amosov A P, Markov Yu M, Dobrovolskaya R A and Nikolaeva E N *IOP Conf. Series: Mater. Sci. Eng.* 2017 **177** 012102
[7] Levashov E A, Mukasyan A S, Rogachev A S and Shtansky D V *Int. Mater. Reviews* 2016 **62(4)** 1–37
[8] Amosov A P, Borovinskaya I P, Merzhanov A G and Sytchev A E 2005 *J. Self-Prop. High-Temp. Synth.* 2005 **14(3)** 165–86
[9] Amosov A P, Bichurov G V, Bolshova N F, Erin V M, Makarenko A G and Markov Y M 1992 *J. Self-Prop. High-Temp. Synth.* **1(2)** 239–45
[10] Bichurov G V 2015 *Nitride Ceramics: Combustion Synthesis, Properties, and Applications*, ed A A Gromov and L N Chukhlomina (Weinheim: Wiley-VCH) p 229–63
[11] Shiganova L A, Bichurov G V, Titova Yu V, Ermoshkin A A and Bichurova P G 2011 *Rus. J. Non-Ferr. Met.* **52** 91–5
[12] Kharatyan S L and Nersisyan H H 2002 *Key Eng. Mater.* **217** 83–92

IOP Conf. Series: Materials Science and Engineering **558** (2019) 012003 doi:10.1088/1757-899X/558/1/012003

X-ray powder diffraction analysis of a tungsten carbide-based ceramic

P V Andreev, K E Smetanina and E A Lantsev

Lobachevsky State University of Nizhny Novgorod, Nizhny Novgorod, Russia

e-mail: andreev@phys.unn.ru

Abstract. The results of the qualitative and quantitative X-ray powder diffraction analysis of the W–C system samples are presented. Powder samples were nanopowders of α-WC/W_2C obtained by plasma chemical synthesis. High-density fine-grained structure of ceramics was formed using the method of electro pulse plasma sintering (spark plasma sintering) from the initial powder of the α-WC. The optimal experimental conditions for the X-ray diffraction proceeding of the tungsten carbide based ceramic samples were chosen according to the preliminary experiments. Reference intensity ratio method for quantitative phase analysis was used in this study.

1. Introduction

Tungsten carbide is of interest for various applications (cutting tools, dies, elements of ammunition, etc.) due to its favorable combination of physical and mechanical properties (high melting point, high hardness, low friction coefficient and chemical resistance against corrosion and oxidation). Development of heavy duty constructional ceramics based on pure tungsten carbide is an actual problem [1]. High-density fine-grained structure must be formed to ensure high physical and mechanical properties of tungsten carbide based ceramics.

The method of high-speed electropulse plasma sintering of powder materials (SPS – Spark Plasma Sintering) is one of the most effective ways to obtain high-density fine-grained ceramics based on tungsten carbide. The method represents high-speed heating of powder material by way of transmiting consecutive millisecond impulses of high power current through a conductive graphite compression mold [2–4].

Often initial powders represent the multiphase polycrystalline W–C system [4–6]. The presence in the initial powders of particles of the another crystal phase whose structure is other than monocarbide α-WC has an adverse effect on physical and mechanical properties of sintered ceramics, leading to decrease in crack resistance, bending strength, wear resistances and others [1, 4, 5, 7]. Note also that the area of tungsten monocarbide stability (homogeneity) on the W–C diagram is small [8], therefore, the phase composition of ceramics can change in the process of SPS [4, 5].

In this regard, the relevant task is to develop a method of controlling the phase composition of initial nanopowder compositions of W–C and fine-grained ceramic sample sintered from micron powders.

The study of the qualitative and quantitative phase composition of fine-grained ceramics based on tungsten carbide is usually carried out by the X-ray phase analysis (XRPA). Often researchers make use only of the semi-quantitative corundum number analysis involving well-known data from specialized databanks (for example, PDF–2, 4) [4, 9]. In this case, estimates of the errors and sensitivity of the method are given approximately based on published data.

At the same time, the sensitivity and accuracy of the XRPA depend on the qualitative composition of the sample and the quality of the observed data, which depends on the experimental conditions and the method of sample preparation. This task has particular difficulty and relevance for the XRPA of nanopowder compositions and fine-grained ceramics, the distinctive feature of which is the broadening and decrease in the intensity of X-ray peaks [4, 5, 9].

The purpose of this research was to define the sensitivity of the method of determining the qualitative and quantitative phase composition in fine-grained ceramics based on tungsten carbide.

2. Research objects

Tungsten carbide nanopowders obtained at the Baikov Institute of Metallurgy and Materials Science by plasma chemical synthesis from tungsten oxide and hydrocarbon in a stream of reducing gas and subsequent low-temperature furnace synthesis (reductive annealing in hydrogen) have been taken as objects of research [5, 6, 9].

Also the research deals with ceramic samples with a diameter of 20 mm and a thickness of 5 mm obtained by SPS in vacuum at the temperature of 1625°C (heating rate V = 50°C/min, applied pressure P = 60 MPa) from micron powders. The results of the electron microscopic studies showed that the average grain size in ceramics was 1 μm, just like the average particle size of tungsten carbide in the initial powder. The density of ceramics was 15.608 g/cm^3 (98.97% of the theoretical density of tungsten monocarbide).

3. Description of the used equipment

SPS was conducted using the Dr. Sinter model SPS-625 (SPS Syntex, Japan). Polishing took place on the grinding and polishing machine "Buehler Eco Met 250". X-ray phase analysis was performed with instrument for X-ray powder diffraction Shimadzu XRD-7000 (Japan) (CuKα, λ = 1.54 Å).

All the data of quantitative phase analysis in the present work were calculated by the Reference Intensity Ratio (RIR) method [10] based on the integral intensities of the diffraction peaks.

The powder diffraction patterns of the phases were calculated using the structural parameters of the crystalline phase (unit cell parameters, substance density, mass absorption coefficient, coefficients for approximation of the atomic factor, atomic coordinates, etc.) and the required theoretical values of the intensity ratios were obtained. Structural files were taken from a bank of inorganic structures ICSD: α-WC (ICSD №5212), α-W_2C (ICSD №619097), α-Al_2O_3 (ICSD №31547). The Reference Intensity Ratio used for semi-quantitative phase analysis was $I(W_2C)/I(WC) = 1.407$.

Standard errors of quantitative phase analysis were calculated using the known formulas of mathematical statistics for indirect measurements which were carried out based on the determination error for the relative integrated intensities of X-ray peaks. The measurement errors of the relative integrated intensities of X-ray peaks were calculated using known methods based on data from several (at least three) experiments.

4. Choice of operating modes of the diffractometer

The choice of XRPA experimental conditions of the W – C system samples implies the optimal range of 2θ angle, exposure time and slits system. The selected 2θ angle range (30°–80°) contains peaks (001) and (100) of tungsten monocarbide α-WC and peaks (100), (002) and (101) of tungsten semi-carbide α-W_2C. The following experimental conditions were chosen according to the preliminary experiments: wide-slit mode with a scanning step of 0.02° and exposure time of 3 s. The mass fraction of the W_2C crystal phase in the investigated powder sample under such experimental conditions is 26.0±1.2 wt %.

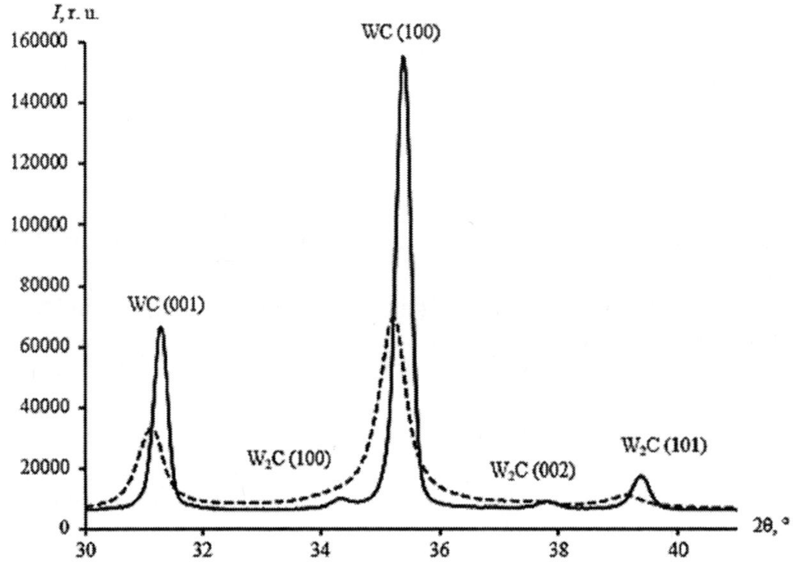

Figure 1. X-ray powder diffraction patterns of ceramics based on tungsten carbide. Dashed line – diffraction pattern after grinding the sample at ~ 250 µm, solid line – after polishing the sample with diamond paste with a grain size 5/3 µm.

5. XRPA of the high density fine-grained ceramics based on tungsten carbide

The investigation of solid samples uniformity was carried out by the layerwise XRPA. First X-ray diffraction experiments was processed at both sides of ceramic sample without any preparation of the surface.

Quantitative calculation showed that the relative intensities of the diffraction peaks α-WC (00*l*) change with sample rotation. It proves the presence of the preferred orientation of crystallites at the surface. Also the tungsten semi-carbide W_2C peaks were not detected.

After that the surface layer of 50 µm was removed by mechanical grinding and XRPA carried out once again. Calculated depth of penetration of CuKα radiation for pure tungsten carbide is less than the selected depth of grinding (~ 7 µm).

Quantitative calculation showed that the considered ceramic sample was homogeneous in depth (starting at 50 µm). The presence of the preferred orientation of crystallites in the sample at such depth was not detect. It is interesting to note that there is a diffraction peak of the W_2C phase on powder diffraction pattern of the grinded sample surface. The mass fraction of the W_2C phase in the surface layer of sintered grinded ceramics is 5.7±0.6 wt %.

At the next stage, the level of surface roughness varied as a result of polishing with a diamond paste of various dispersion level (28/20 µm, 14/10 µm, 5/3 µm, 3/2 µm, 1/0 µm). The diffraction patterns of the ceramic sample mechanically grinded and polished with a diamond paste of 5/3 µm dispersion are shown in figure 1.

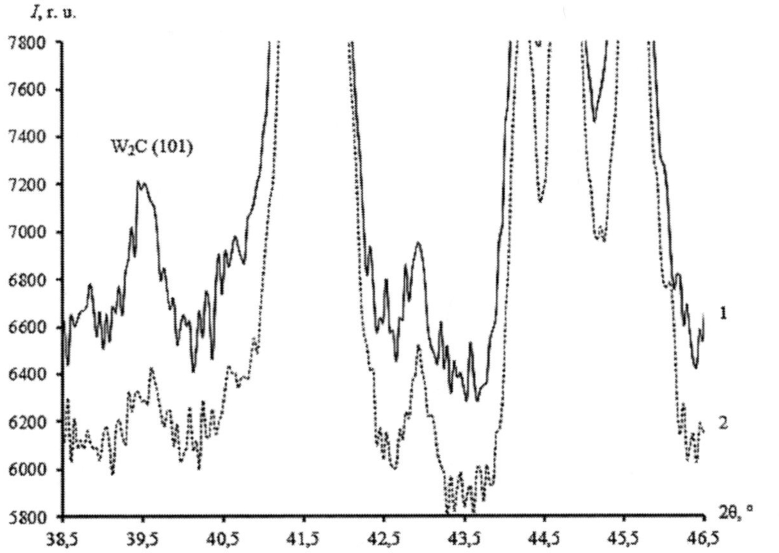

Figure 2. X-ray powder diffraction patterns of mixtures of a powder sample of tungsten carbide with tungsten anhydride WO_3. Diffraction pattern: *1* 80 wt % WO_3, *2* 90 wt % WO_3.

The analysis of the results shows that intensity of the diffraction peaks increases with the decrease in surface roughness level. In our opinion, this is due to the fact that the roughness and strain hardening of a surface lead to "blurring" of the area of X-ray diffraction in the sample, which accounts for the broadening of the diffraction peaks. Ceramic sample surface polishing increases the sensitivity of the used XRPA method and allows specifying the mass fraction of the W_2C phase in the studied sample. The mass fraction of the semi-carbide W_2C in sintered ceramics was refined as 5.25±0.11 wt %. Further polishing with diamond pastes of dispersion level 3/2 μm and 1/0 μm did not lead to any noticeable improvement of the experimental results.

Checking sensitivity on developed method to W_2C phase content in powders of tungsten carbide was conducted at the final stage of work. Tungsten anhydride was added to the studied powder of tungsten carbide to test the sensitivity of the method used. The content of WO_3 in the powder of tungsten carbide was 80 and 90 mass %. The analysis of XRPA results showed that the initial powder sample contains ~ 2 wt % W_2C phase.

Figure 2 shows that the peak of the W_2C phase, corresponding to the reflection (101), could be detected even at 90% WO_3 in the mixture, which corresponds to ~ 0.22±0.04 wt % W_2C phase in the sample.

6. Conclusions

It was found that samples of ceramics based on tungsten carbide WC were characterized by the heterogeneity of the phase composition in depth and the preferred orientation of crystallites only near the surface layer (less than 50 μm). The surface roughness of such samples significantly affects the sensitivity of the used research method. Preliminary polishing of tungsten carbide ceramic samples with diamond paste of 5/3 μm dispersion level is proposed to increase the sensitivity of XRPA.

The developed research showed the sensitivity of the method to the presence of tungsten semi-carbide phase W_2C at the level of 0.2 wt %.

Acknowledgments
The work was financial supported by Russian Science Foundation (RSF), grant no. 18-73-10177.

References

[1] Panov V S and Chuvilin A M 2001 *Technology and properties of cintered carbide alloys and products from them* (Moscow: MISIS)
[2] Somiya S 2013 *Handbook of Advanced Ceramics* (Academic Press)
[3] Olevsky E and Dudina D 2018 *Field-Assisted Sintering* (New York: Springer Int. Publ.)
[4] Chuvil'deev V N, et al 2017 *J. Alloys Comp.* **708** 547
[5] Blagoveshcenskiy Yu, et al 2015 *Inorg. Mater.: Appl. Res.* **6** 415
[6] Blagoveshcenskiy Yu, et al 2018 *Inorg. Mater.: Appl. Res.* **9** 924
[7] Zhao J, Holland T, Unuvar C and Munir Z 2009 *Int. J. Refrac. Met. Hard Mater.* **27** 130
[8] Kurlov A and Gusev A 2014 *Physics and chemistry of tungsten carbides* (Moscow: Fizmatlit)
[9] Isaeva N, et al 2013 *Izvestia Vuzov. Powder metallurgy and functional coatings* **3** 7
[10] Andreev P, Trushin V and Faddeev M 2013 *X-ray phase analysis of polycrystalline materials* (Nizhny Novgorod, Lobachevsy State University)
[11] Anselmi-Tambrini, et al 2005 *Mat. Sci. Eng.* A **394** 139

IOP Conf. Series: Materials Science and Engineering **558** (2019) 012004 doi:10.1088/1757-899X/558/1/012004

Modeling of the distribution of thermal fields during spark plasma sintering of alumina ceramics

N N Berendeev, A A Popov, V V Pyaterikova, M S Boldin, A V Nokhrin, V N Chuvil'deev and E A Lantsev

Lobachevsky State University of Nizhny Novgorod, Research and Development Institute of Physics and Technology, Nizhny Novgorod, 603950 Russia

E-mail: popov@nifti.unn.ru

Abstract. The article shows the use of an experimental-calculation method for the refinement of thermoelectric constants of graphite materials and alumina ceramics by experimentally studying the heating of graphite equipment and refinement thermoelectric constants of materials using the finite element method (ANSYS WORKBENCH). To simulate the thermal fields of graphite equipment based of the data on electric current, voltage and heat exchange conditions, a coupled boundary electric and non-stationary temperature problem was realized. The application of the developed approach made it possible to clarify the thermoelectric constant of materials, which helped to reduce the difference between the experimental and simulated temperatures from 20 to 5%.

1. Introduction

The method of spark plasma sintering (SPS) belongs to a class of methods based on passing low-voltage high-density direct current pulses through a heated volume [1, 2]. Heating of the powder material is provided by the release of Joule heat by passing an electric current through the graphite die with simultaneous application of mechanical pressure (see figure 1a).

The method of electro pulse plasma sintering of powder materials has several advantages compared to traditional sintering methods such as free pressing, hot pressing, isostatic hot pressing [3, 4]. SPS allows to increase the heating rate of the sintered sample tenfold from 5–25°C/min (traditional methods) to 2500°C/min (SPS). High heating rates are especially important when nano- and ultrafine-grained materials are sintered, which allow to limit the rate of grain boundary migration and to obtain high-density metal and/or ceramic materials with high strength, hardness, radiation resistance [5], etc.

One of the features of the SPS method is the control of temperature by a pyrometer focused on an external surface of the graphite die, which makes impossible to estimate the true temperature of the sample during sintering. The presence of radial temperature fields arising under conditions of high-speed heating leads to significant differences in temperature values between the surface of the graphite die and the center of the heated sample [8, 9]. For this reason, a number of researchers mistakenly correlate the controlled sintering temperature with the truth one. To estimate the true difference between the sample temperature and the controlled one, it is necessary to solve the temperature and electrical problems using experimental current and voltage curves for heating the graphite die and the sample. Modeling by the finite element method (FEM) of the SPS process is one of the most effective methods for calculating temperature fields in the graphite die and the sintered sample [1, 6, 7].

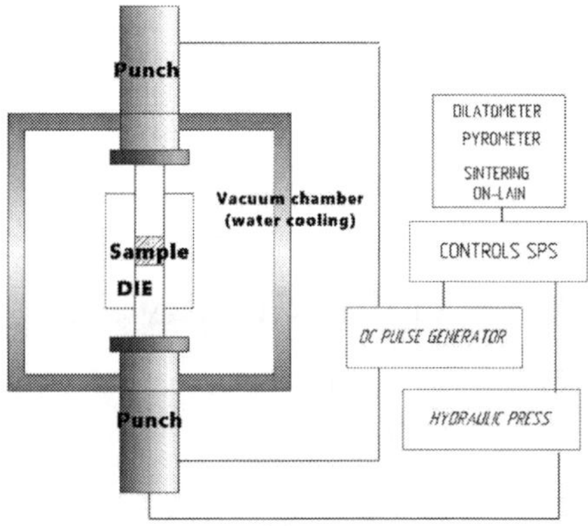

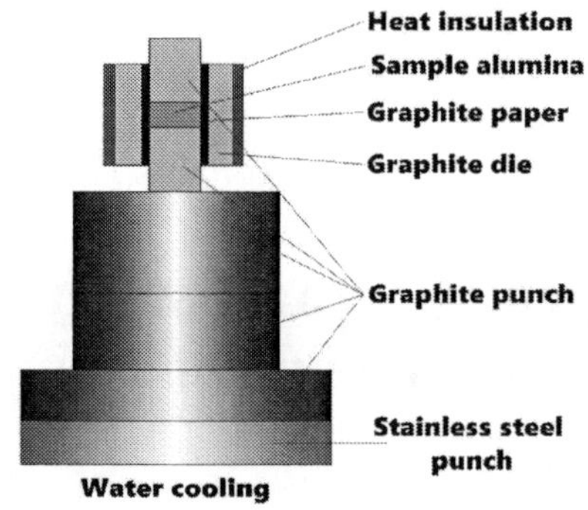

(a) (b)

Figure 1. (a) Schematic of the SPS setup and (b) chamber design.

The results of numerical modeling of temperature fields in [8, 9] show that the temperature difference in the sample area is 1–8% of the maximum temperature. Depending on the thermoelectric properties of the materials being modeled, this difference can either increase or decrease. Often, thermoelectric properties are set the same for all grades of graphite, graphite paper and aluminum oxide, which can lead to the specified errors. Thus, the task of studying the thermoelectric properties of materials is of great interest because it helps to get modeling temperature fields inside the sample and temperature distribution in the graphite die under the heating.

This study focuses on the experimental-calculation method for the refinement of thermoelectric properties of materials (the graphite die and the sample of alumina ceramic) while heating in the SPS.

This work presents an experimentally calculated method for studying the thermoelectric properties of materials heated in SPS. The refinement of the thermoelectric properties of materials is performed by sequential experimental research and modeling of the temperature heating of the graphite die with the sample of aluminum oxide. The choice of aluminum oxide as an object of study is due to the widespread use of this ceramics in the industry, well-known constants of this material, which greatly simplifies the procedure of modeling and verification of the calculation results. Moreover, there is a large number of production data of alumina ceramics using the SPS method and hot pressing [11, 12].

2. Experimental technique

The objects of investigation of thermoelectric properties were graphite MPG-7, graphite paper "Graphlex" and sintered aluminum oxide samples α-Al_2O_3 from powder produced by Alfa Aesar. The properties of graphite materials for modeling are given in table 1. Cylindrical alumina samples had a diameter of 12, 20 and 30 mm and a thickness of 6, 10, and 8 mm, respectively. The density of the samples was 3.99 g/cm^3, the average grain size of ceramics was 1 μm. Samples were obtained using the SPS method of heating α-Al_2O_3 powders (V_H = 100°C/min, T_{sint} = 1150°C, t_{sint} = 2 min, P = 70 MPa). The use of samples previously SPS sintered at higher temperatures allowed us to avoid the uncertainties associated with the effect of porosity on the material properties used in modeling.

The samples are heated in a graphite die according to the standard heating mode: (V_n = 100°C/min, T_{sint} = 800°C, t_{sint} = 0 min, P = 70 MPa). The samples of aluminum oxide were sintered using a DR unit SINTER model SPS-625 Spark Plasma Sintering System (SPS SYNTEX INC. Ltd., Japan). The schematic diagram of the SPS installation is presented in figure 1a.

The calculation method consists of an experimental study of the heating of the graphite die at points of temperature control and the specification of thermoelectric constants of materials using the finite element method.

Thermoelectric constants were refined sequentially for all materials. At the first stage, the properties of graphite on a solid graphite billet were refined exactly in line with the geometric dimensions of the die with a diameter of 12 mm. At the second stage, the properties of graphite paper and sintered aluminum oxide in the die with a sample of 12 mm were refined. At the third stage, the independent experiment verifies the properties of graphite paper and the aluminum oxide sample with a diameter of 20 and 30 mm. Temperature measurement of the graphite die was carried out using a chromel-copel (Cr/Ni–Cu/Ni) thermocouple (TC), connected to the device "METAKON-513". Thermocouple measurement error is ±1°C. The thermocouple was placed in a blind hole 5 mm deep in graphite die. Temperature measurement was carried out with an interval of 15–60 s. The location of the points at which the temperature was measured is shown in figure 2.

Table 1. Tabular properties of materials (with *T* in Kelvin).

Properties \ Materials	Inconel	Graphite MPG-7	Paper "Graphlex"	α-Al_2O_3
Thermal conductivity, $W\,m^{-1}\,K^{-1}$	$10.09+1.57\cdot10^{-2}\cdot T$ [8, 9]	$82.85-0.06\cdot T+2.58\cdot10^{-5}\cdot T^2$ [8, 9]	3-5 [7]	$39.5\cdot T^{-1.26}$ [8, 9]]
Electrical resistivity, Ω m)	$9.82\cdot10^{-7}+1.6\cdot10^{-10}\cdot T$ [8, 9]	$2.14\cdot10^{-5}+1.34\cdot10^{-8}\cdot T+4.42\cdot10^{-12}\cdot T^2$ [8, 9]	1-1.25 [7]	$8.7\cdot10^{19}\cdot T^{-4.82}$ [8, 9]
Heat capacity, $J\,kg^{-1}\,K^{-1}$	$344+0.25\cdot T$ [8, 9]	$34.27+2.72\cdot T-9.6\cdot10^{-5}\cdot T^2$ [8, 9]	-	850 [8, 9]

For the numerical implementation of the thermoelectric problem in a non-stationary formulation, a finite element package ANSYS 18.2 was used. The constants of tables 1 and 2 were chosen as the initial thermoelectric properties of materials. The boundary conditions of this thermoelectric problem are presented in figure 3. On the line B (figure 3a), which bounds the simulated system from above, the current was set as a linear function of time. Along the lower line A (figure 3a), which bounds the simulated system from the bottom, the voltage value was set. The entire volume of the two-dimensional body A (figure 3b) was solved as a part of the electrical problem (the temperatures obtained from the calculation of the electrical problem were solved in the unsteady heat conduction problem). Across the outer contour of the simulated system B (figure 3b), besides the line C, which bounds a part of the graphite die, with the help of the value of specific thermal conductivity the convection conditions were introduced. Along line C (figure 3b), which bounds a part of a graphite die, thermal insulation conditions were imposed, prohibiting heat removal. In processing of modeling, elements of graphite die were considered as isotropic, which made it possible to set an assumption of the uniformity of all physical properties in all directions. The physical properties of the materials used in the model and the alumina constants are listed in table 1.

Figure 2 presents the results of temperature measurement at control points for a single graphite (Figures 2a, 2b) and for die with the alumina sample with diameters of 12, 20, and 30 mm (see figures 2c–2e, respectively).

3. Experimental results and discussion

The results of the research represent the constructed numerical model for a single graphite die. Based on the model, we chose graphite paper constants using the method of passive search in space to minimize the discrepancy of the function of temperature values at the control points in the experiment and numerical calculation. The next part of the calculation was the determination of the constants of the ceramic material - sintered aluminum oxide (sample diameter of 12 mm). Further, the constants obtained earlier were verified on a complete system with ceramic samples of aluminum oxide with diameters of 20 and 30 mm. The refined thermoelectric constants of the materials are given in table 2.

At the control points at the finite period of time corresponding to the experimental end of the sintering process a comparison of experimental data with data obtained as a result of numerical simulation was made. It was established the magnitude of the discrepancy does not exceed 5%, which corresponds to a good numerical result obtained

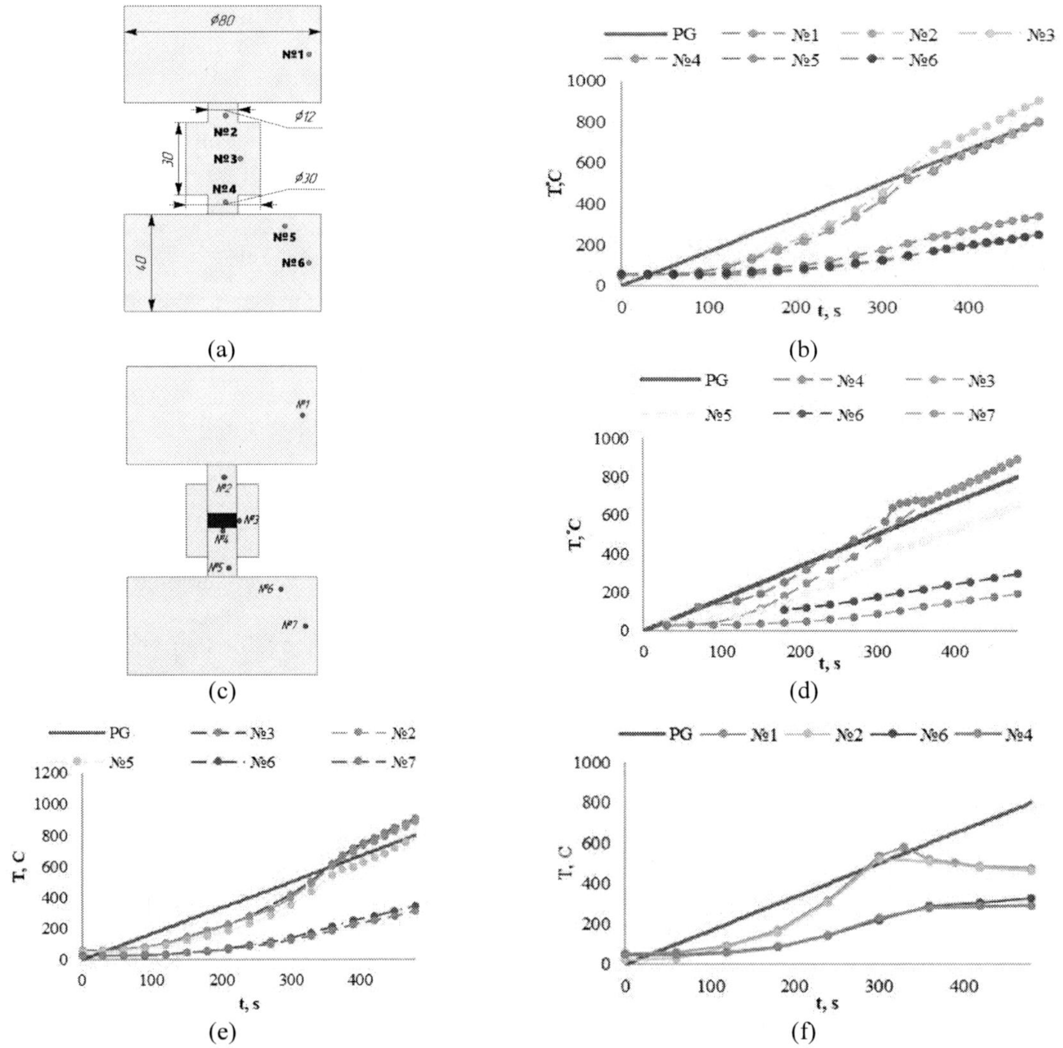

Figure 2. Experimental data: (a) temperature control points for single graphite die, (b) heat curves of single graphite die, (c) temperature control points, (d) heat curves of graphite die and sample 12 mm, (e) heat curves of graphite die and sample 20 mm, (f) heat curves of graphite die and sample 30 mm.

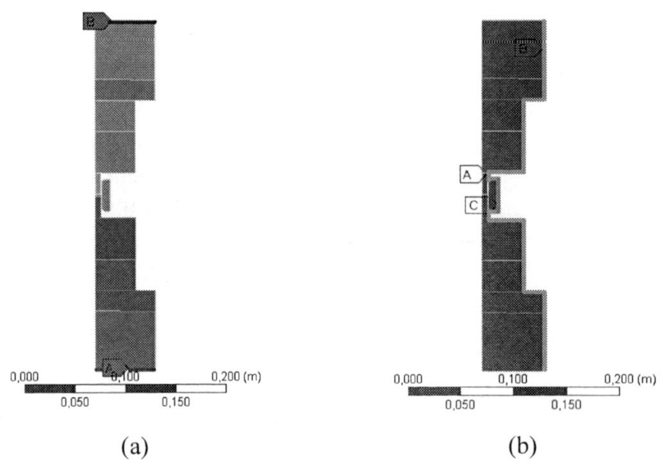

Figure 3. Boundary conditions of model FEM: (a) electrical, (b) temperature.

Table 2. Specify properties of materials (with *T* in Kelvin).

Properties \ Materials	Inconel	Graphite MPG-7	Paper "Graphlex"	α-Al_2O_3
Thermal conductivity, W m^{-1} K^{-1}	$11.4+1.4\cdot10^{-2}\cdot T$	$103.6-0.08\cdot T+3.0\cdot10^{-5}\cdot T^2$	103	$27.6-2.0\cdot10^{-2}\cdot T$
Electrical resistivity, Ω m	$1.18\cdot10^{-7}$	$2.1\cdot10^{-5}$	$3.5\cdot10^{-7}$	$1.0\cdot10^{8}-1.5\cdot10^{-5}\cdot T^{-4.82}$
Heat capacity, J kg^{-1} K^{-1}	$446.5+0.16\cdot T$	$4.2\cdot10^2+1.4\cdot T$	709	$931+0.34\cdot T$

4. Conclusions

The developed experimental-calculation method for the study of thermoelectric constants of materials and the simulation of temperature fields for graphite die made it possible to specify the thermoelectric constants of materials, which helped to reduce the difference between experimental and simulated temperatures from 20 to 5%. The obtained results open up new opportunities for optimizing the SPS regimes to increase the heating uniformity of the sintered alumina samples.

Acknowledgements

The authors wish to thank the Ministry of Education and Science of the Russian Federation (grant no. 11.1114.2017/4.6).

References

[1] Kiyoshi Ichikawa 2001 *Functionally Graded Materials in the 21st Century: A Workshop on Trends and Forecasts* (Kluwer Academic Publishers)
[2] Mamedov V 2002 *Powder Metall.* **45** 1
[3] Olevsky E and L Froyen 2009 *J. Am. Ceram. Soc.* **92** 122
[4] Munir Z A, Anselmi-tamburini U and Ohyanagi M 2006 *J Mater Sci* **41** 763
[5] Rahaman M N 2003 *Ceramic processing and sintering* (New York: Marcel Dekker, Inc.)
[6] Pavia A, et al 2013 *J. Mater. Proc. Tech.* 1327
[7] Maniere C, et al 2016 *J. Eur. Ceram. Soc.* **37** 1593
[8] Giuntini D, et al 2013 *Mater.* **6** 2612
[9] Wei S, et al 2012 *Comput. Mater. Sci.* **60** 168
[10] Molena G, Durand L, Galy J and Couret A 2010 *J. Metall.* 1
[11] Shackelford J F and Doremus R H 2008 *Springer* 1
[12] Demuynck M, et al 2012 *J. Eur. Ceram. Soc.* **32** 1957

IOP Conf. Series: Materials Science and Engineering **558** (2019) 012005 doi:10.1088/1757-899X/558/1/012005

Investigation of the kinetics of spark plasma sintering of alumina ceramics. Part 1. The initial stage of sintering

M S Boldin, A A Popov, E A Lantsev, A V Nokhrin and V N Chuvil'deev

National Research Lobachevsky State University of Nizhny Novgorod
Research and Development Institute of Physics and Technology
Nizhny Novgorod, 603950 Russia

E-mail: boldin@nifti.unn.ru

Abstract. The article presents the results of the investigation of the initial sintering stage of alumina. The authors determined the activation energy of shrinkage at the initial stage of sintering. The study shows that the additives do not change the kinetics of shrinkage at the initial stage of SPS.

1. Introduction

Aluminum oxide is one of the promising ceramic materials for a wide range of structural applications due to the combination of its high hardness, heat and chemical resistance [1, 2]. These properties of alumina can be used only with a fine-grained structure of this compound. The long-term isothermal dwelling applied in the traditional sintering methods (such as vacuum sintering or hot pressing) leads to the production of a coarse-grained structure with low quality.

The quality of alumina-based ceramics can be substantially improved through the reduction of grain growth by using some additives (for example, MgO, CaO, ZrO_2, SiC, TiC, etc.) [3–8].

One of the most effective methods for obtaining a fine-grained structure in alumina ceramics is the method of Spark Plasma Sintering (SPS) [9–13]. This method is based on the controlled heating of the powder material in a graphite mold via a sequence of DC pulses. Sintering is usually carried out in a controlled environment with simultaneous application of uniaxial mechanical stress [13–16].

There are a lot of articles about SPS of the alumina ceramics [11, 17–21], but none of them provides a thorough analysis of the sintering kinetics of each of the stages of the sintering process and the investigation of the role of the additives at each of these stages.

The main purpose of the first part of the article is the investigation of the initial stage of sintering of alumina ceramics during SPS.

2. Experimental

The main material used in this study is α-Al_2O_3 powder ($d = 0.2$ μm, Taimei Chemicals, Japan) and its mixtures with MgO ($d \sim 0.1$ μm, Alfa Aesar, Germany), TiO_2 ($d < 0.5$ μm, UrFU RAS, Russia), ZrO_2 ($d = 0.03$ μm, Pangea Int. Ltd., China).

The additives were selected in accordance with the way they interact with alumina:

- ZrO_2 is not soluble in Al_2O_3 and does not form new phases with it;
- MgO is soluble in Al_2O_3 at low temperature and may form the Al_2MgO_4 phase at $T < 600$°C;

- TiO_2 is not soluble in Al_2O_3 and does not form new phases with it up to $T = 1200°C$. At temperatures higher than 1200°C this compound forms the Al_2TiO_5 phase.

Powder mixtures were obtained by mixing the initial powders in the "FRITSCH-Pulverisette 6" planetary mill. Grinding media and bodies were made of stabilized zirconia. The rotational speed of the grinding cup was 200 rpm. Mixing was done in isopropyl alcohol for 5 h. The removal of alcohol was carried out at a temperature of 70°C for 12 h. The selected mixing modes ensured homogeneity and the absence of impurities in the grinding media.

Sintering of samples with a diameter of 12 mm and a height $h = 3$ mm was carried out by the SPS method in the "DR. SINTER model SPS-625 Spark Plasma Sintering System" (SPS SYNTEX INC. Ltd., Japan). The temperature of the sintering process was measured using Chino IR-AHS2 pyrometer focused on the outer surface of the graphite die with the outer diameter of 30 mm. The actual temperature inside the mold was calculated using the calibration curves of the temperature difference measured by a pyrometer on the outer surface of the graphite mold and temperatures measured by a thermocouple in the central part of the sintered powder compact (for all heating rates used).

The applied uniaxial pressure was 70 MPa. The pressure was applied before the onset of the intense shrinkage stage and maintained at a constant level throughout the entire process. The accuracy of the pressure measurement was ±5 MPa. Sintering was carried out in a vacuum at 6 Pa.

The study of the kinetics of sintering with constant heating rate was carried out at the maximum sintering temperature $T = 1520°C$. The heating rate up to 600°C was 100°C/min; further heating to the sintering temperature was carried out at the heating rate of 10, 50, 100, 250, 350, and 700°C/min. There was no isothermal dwelling at the sintering temperature.

To study the kinetics of sintering under isothermal conditions, we chose the temperature range from 1010 to 1120°C. Samples were heated up to 600°C at the heating rate of 100°C/min, the further heating rate to sintering temperature was carried out at a rate of 25°C/min, after which the samples were dwelled for 60 min.

The shrinkage of the samples during the sintering process was controlled using a precision dilatometer, which is part of the "DR. SINTER model SPS-625 ". The accuracy of the measurement of shrinkage was ±0.05 mm.

The density (ρ) of sintered samples was measured using the Archimedes' Principle in distilled water using Sartorius CPA scales. The accuracy of density calculations was ±0.005 g/cm^3. The theoretical density of alumina (ρ_{th}) was taken as 3.99 g/cm^3.

Vickers hardness (H_V) was measured using the automated microhardness tester "Struers Duramin-5" under a load of 19N. Crack resistance (K_{IC}) was calculated using the Palmqvist method. Measuring accuracy of H_V and K_{IC} was ±1.5 GPa and ±0.5 MPa·m$^{1/2}$, respectfully.

The microstructure of samples was studied using a Jeol JSM-6490 scanning electron microscope. The average grain size in sintered ceramics was calculated after analyzing at least 400 grains.

3. Results and discussion

3.1. Shrinkage behavior at a constant heating rate during SPS

Shrinkage behavior has three stages. This behavior is typical for solid state sintering [22], therefore the processes occurring during SPS sintering can be described as a sequence of initial (I), intermediate (II) and final (III) sintering stages [23].

Let us consider the sintering stages using the samples obtained at a heating rate of 10°C/min. In accordance with [23], the initial stage of sintering — (stage I) is characterized by the formation of contacts between powder particles and fit is observed at temperatures below 1120°C (figure 1). The transition between stages I and II occurs at $\rho/\rho_{th} \sim 0.7$. Stage II ($T \sim$ 1120–1230°C) is characterized by the increase in the contact area between the particles and intensive compaction of the powder. Stage III (T higher than 1230°C) is characterized by the formation of closed porosity ($\rho/\rho_{th} \sim 0.9$), diffusion-controlled dissolution of isolated pores and the growth of grains.

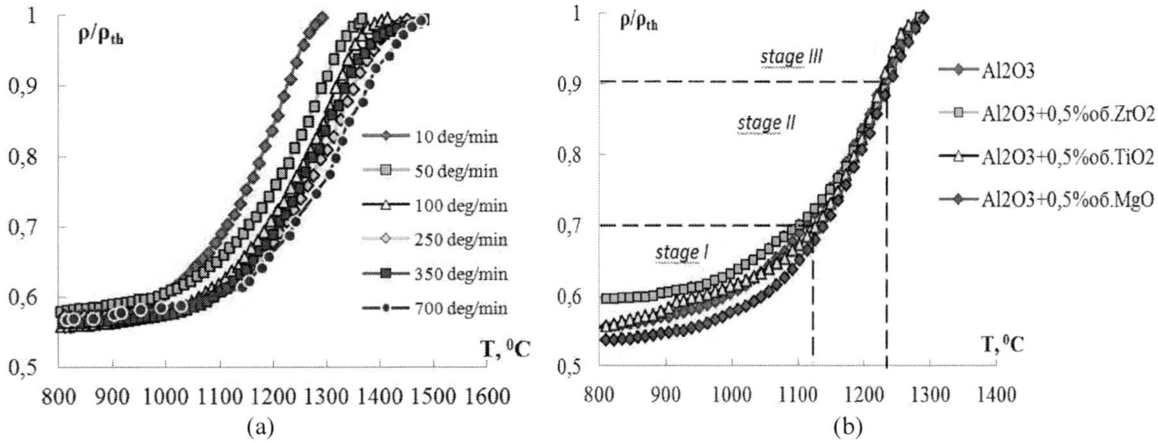

Figure 1. Shrinkage behaviors $\rho/\rho_{th}(T)$: (a) for pure alumina ceramics obtained at heating rates of 10–700°C/min, (b) for alumina-based ceramics obtained at 10°C/min.

It should be noted that the temperature at the beginning of stage II - the stage of intense shrinkage of alumina does not depend on the presence of MgO, ZrO_2 and TiO_2 additives in a concentration of 0.5% (within the experimental error for temperature equal to 50 °C).

3.2. Analysis of shrinkage behavior at the initial stage (stage I). Dependence of shrinkage on isothermal dwelling time during SPS. Sintering of ceramics in isothermal conditions

The analysis of shrinkage behavior under the heating temperature in figure 1 shows that stage I occurs at a temperature lower than 1120°C. In this regard, the study of stage I was carried out at temperatures of 1010, 1070, and 1120°C.

A typical view of shrinkage behavior during the isothermal dwelling, shown in figure 2. The value L_0, equal to the initial height of the sintered powder compact, was 4.5 mm.

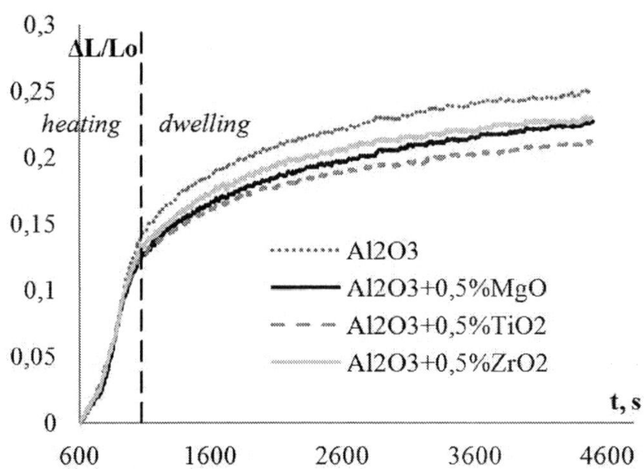

Figure 2. A typical view of shrinkage behavior during the isothermal dwelling, $T = 1010$°C.

Figure 2 shows that at the initial stage of sintering under isothermal conditions (dwelling step), the addition of MgO, TiO_2, and ZrO_2 leads to the decrease in the absolute value of shrinkage compared to pure alumina, while the inclination of the shrinkage curve for the studied compositions is the same within the experimental error.

The dependence of shrinkage on the time of isothermal dwelling can be described using the expression [24]:

$$L_1/L_0 = Bt^q \tag{1}$$

where B and q are constants which depend on the temperature and characteristics of the sintered material.

In the case where the sintering mechanism is determined by diffusion, the constants are related by the following equation [24]:

$$B = \frac{B'}{T^q}D^b = \frac{B''}{T^q}\exp\left(-\frac{qQ}{kT_m}\frac{T_m}{T}\right) \tag{2}$$

where B' and B'' are the constants which are independent of temperature, T_m = 2323 K (melting point for alumina), and Q is the activation energy of a diffusion-controlled process, the intensity of which limits the sintering rate.

From equations (1) and (2), the value of the coefficient q can be found from the slope of the relative shrinkage versus time in logarithmic coordinates ln (L_1/L_0)–ln (t) (figure 3a). The free coefficient in relation ln (L_1/L_0)–ln (t) will determine the value of ln (B). The magnitude of the activation energy of sintering qQ at stage I can be found from the slope of the relation ln (B)–T_m/T (figure 3b).

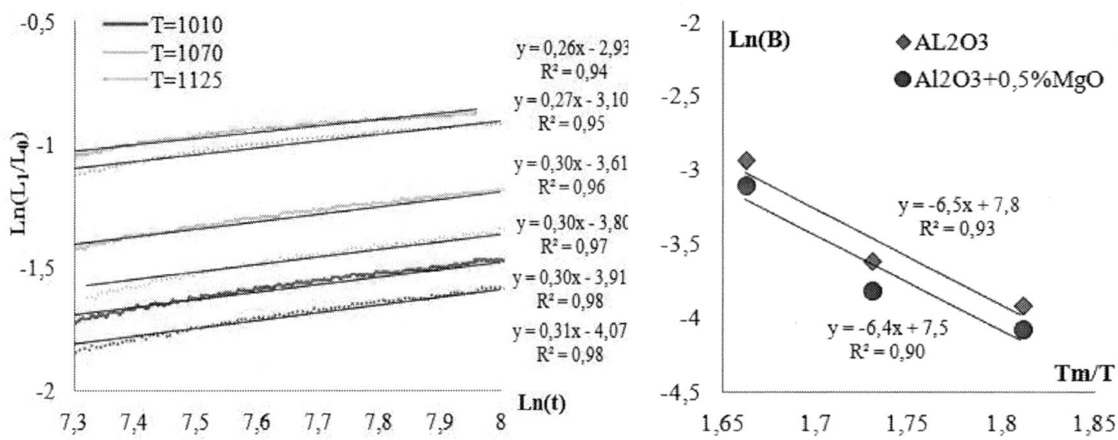

(a) (b)

Figure 3. The shrinkage behavior on initial sintering stage for pure Al_2O_3 (solid lines) and Al_2O_3 + 0.5% vol MgO (dotted lines): a) plot of ln (L_1/L_0))–ln (t); b) plot of ln (B)–T_m/T.

Figure 3 shows that for pure Al_2O_3 and Al_2O_3 + 0.5% vol MgO the value of the coefficient $1/q$ approximately equals to 0.3, and the value qQ = 6.5 kTm (figure 3b). Thus, the activation energy of the initial stage of sintering (stage I) for pure Al_2O_3 and Al_2O_3 + 0.5% MgO Q = 19.5 kTm. The determined value of the energy corresponds to the data of [25] (Q = 19.8 kTm) and agrees with the activation energy of grain boundary diffusion of alumina ($Q = Q_b$ ~ 20 kTm, which is Q = 340 kJ/mol).

Thus, the addition of magnesium oxide in the amount of 0.5% vol does not affect the diffusion properties of alumina at the initial stage of sintering.

Slowing down of the shrinkage process at the initial stage of sintering of alumina with the addition of 0.5% vol ZrO_2, MgO, and TiO_2 may be explained by the fact that the particles of additive powders inhibit the neck growth between the particles of alumina powder due to the action of the Zener force.

4. Conclusions

The authors determined the stages of sintering and their temperature ranges under SPS conditions for alumina-based ceramics and estimated the activation energy of shrinkage processes at the initial stage of sintering. The study shows that the additives do not change the kinetics of shrinkage at the initial stage. The study also determined that the shrinkage process is controlled by grain-boundary diffusion of alumina (Q_b = 340 kJ/mol).

Acknowledgments

The authors wish to thank the Ministry of Education and Science of the Russian Federation (grant no. 11.1114.2017/4.6).

References

[1] Elasser C 2005 *J. Am. Ceram. Soc.* **88** 1
[2] Doremus R H 2006 *J. Appl. Phys.* **100** 1
[3] Baik S and Moon J H 1991 *J. Am. Ceram. Soc.* **74** 819
[4] Niihara K 1991 *J. Ceram. Soc. Japan* **99** 974
[5] Sternitzke M 1998 *J. Eur. Ceram. Soc.* **17** 1061
[6] Sternitzke M, Derby B and Brook R J 1998 *J. Am. Ceram. Soc.* **81** 41
[7] Young M K, Won T K and Young-Wook K 2004 *Ceram. Int.* **30** 2081
[8] Barry J and Byrne G 2001 *Wear* **247** 139
[9] Zhou Y, Hirao K, Yamauchi Y and Kanzaki S 2004 *J. Eur. Ceram. Soc.* **24** 3465
[10] Kim B N, Hiraga K, Morita K and Yoshida H 2007 *Scripta Materialia* **57** 607
[11] Stanciu L A, et al 2001 *Metall. Mater. Trans.* A **32A** 2633
[12] Liu L, et al 2013 *Mater. Sci. Eng.* A **559** 462
[13] Munir Z A, et al 2006 (New York: Wiley–Interscience)
[14] Orru R, et al 2009 *Mater. Sci. Eng.* R **63** 127
[15] Munir Z A 2011 *J. Am. Ceram. Soc.* **94** 1
[16] Chuvil'deev V N, et al 2015 *Acta Astronautica* **109** 172
[17] Galusek D, Sedlacek J, Chovanec J and Michalkova M 2015 *Ceram. Int.* **41** 9692
[18] Kim B N 2012 *J. Ceram. Soc. Japan* **120** 116
[19] Jiang D T 2008 *J. Am. Ceram. Soc.* **91** 151
[20] Fukumoto I, Mekaru S and Shibata S 2006 *JSME Int. J. Series A* **49** 91
[21] Lallemant L, et al 2014 *J. Eur. Ceram. Soc.* **34** 1279
[22] Rahaman M N 2003 *Ceramic processing and sintering* (New York: Marcel Dekker, Inc.)
[23] Fang Z Z 2010 *Sintering of advanced materials* (Cambridge: Woodhead Publishing Limited)
[24] Bacmann J J and Cizeron G J 1967 *Am. Ceram. Soc.* **51** 209
[25] Feng L and Dillon S J 2017 *J. Am. Ceram. Soc.* **00** 1

IOP Conf. Series: Materials Science and Engineering **558** (2019) 012006 doi:10.1088/1757-899X/558/1/012006

Investigation of the kinetics of spark plasma sintering of alumina. Part 2. Intermediate and final stages of sintering

M S Boldin, A A Popov, E A Lantsev, A V Nokhrin and V N Chuvil'deev

National Research Lobachevsky State University of Nizhny Novgorod
Research and Development Institute of Physics and Technology
Nizhny Novgorod, 603950 Russia

E-mail: boldin@nifti.unn.ru

Abstract. The article presents the results of the investigation of the initial sintering stage of alumina. The authors determined the activation energy of shrinkage at the initial. The study shows that the effect of additives is observed only at the final stage of sintering.

1. Introduction

Aluminum oxide is one of the promising ceramic materials for a wide range of structural applications due to the combination of high hardness, heat and chemical resistance [1, 2].

There are a lot of articles about SPS sintering of the alumina ceramics [3–9], but none of them provides a thorough analysis of the sintering kinetics of each of the stages of the sintering process and the investigation of the role of the additives at each of that stages.

The main purpose of the second part of the article is the investigation of the intermediate and final stages of sintering of alumina ceramics during SPS.

2. Experimental

The main material used in this study is α-Al_2O_3 powder (d = 0.2 μm, Taimei Chemicals, Japan) and its mixtures with MgO ($d \sim 0.1$ μm, Alfa Aesar, Germany), TiO_2 ($d < 0.5$ μm, UrFU RAS, Russia), ZrO_2 (d = 0.03 μm, Pangea Int. Ltd., China). A detailed description of the methods which were used in the present work is provided in the first part of the article.

In the study of the kinetics of sintering under a constant heating rate, the heating rate was 100°C/min and sintering temperature T= 1520°C. There was no isothermal dwelling.

To study the grain growth kinetics, the samples were heated to 1320, 1420, and 1520°C with heating rate of 100°C/min and dwelled for 0, 3, 10, and 30 min.

3. Results and discussion

3.1. Analysis of shrinkage behavior at the intermediate stage (stage II). Sintering of ceramics under continuous heating rate mode

As was shown in Part 1 (figure 1b in Part 1), the addition of ZrO_2, MgO, TiO_2 additives does not affect the shrinkage kinetics of the α-Al_2O_3. Thus, the shrinkage curves at stage II can be considered equivalent for all studied composites. The further analysis of the shrinkage curves was made for pure Al_2O_3 oxide and Al_2O_3 + 0.5% vol MgO.

According to [10], the process of shrinkage of a powder material can be described as the process of plastic deformation of a continuous porous material.

In this case, the strain rate can be determined from experimental data by the equation:

$$\dot{\varepsilon} = \frac{S}{L_0} \tag{1}$$

where S is the shrinkage rate and L_0 is the initial height of the powder before sintering.

As experimental data, we used the sintering data of the sample obtained at a heating rate of 100°C/min, which were equal to $L_0 = 6 \times 10^{-3}$ m; $\sigma = 70$ MPa. $S = 14 \times 10^{-6}$ m/s for pure α-Al_2O_3 and $S = 18 \times 10^{-6}$ m/s for Al_2O_3 + 0.5% vol MgO (at $T = 1200$°C).

At this consideration, the dependence of powder shrinkage on time at a fixed temperature and applied pressure can be described using the dependence on the power creep law [11]:

$$\dot{\varepsilon} = A\left(\frac{D_V}{b^2}\right)\left(\frac{G\Omega}{kT}\right)\left(\frac{\sigma}{G}\right)^n \tag{2}$$

where $\dot{\varepsilon}$ is the strain rate, A is the Dorn constant, D_V is the coefficient of bulk diffusion, b is the Burgers vector, G is the shear modulus, Ω is atomic volume, k is the Boltzmann constant, T is the absolute temperature, σ is the stress, and n is the constant which determines the mechanisms of deformation and diffusion (n = 3).

According to [11], the following values can be used for alumina: $A = 3.38$; $b = 4.76 \times 10^{-10}$ m; $\Omega = 4.25 \times 10^{-29}$ m^3; $G = 126$ GPa.

Let us take the expressions (1) and (2) and substitute the above values of constants and experimental values. This way we can obtain $D_V = 3.5 \times 10^{-15}$ m^2/s for pure α-Al_2O_3 and $D_V = 4.5 \times 10^{-15}$ m^2/s for α-Al_2O_3 +0.5% vol MgO.

The expression for the coefficient D_V is the following:

$$D_v = D_{v0}\exp(-\frac{Q_v}{kT}) \tag{3}$$

where D_V is the coefficient of bulk diffusion, D_{V0} is the pre-exponential factor, Q_V is the activation energy of bulk diffusion, k is the Boltzmann constant, and T is the absolute temperature.

By substituting $D_{V0} = 0.19$ m^2/s [11] (for bulk diffusion of oxygen ions), we obtained the value of the activation energy of diffusion at stage II. For pure α-Al_2O_3 ceramic $Q_V = 31.6$ kTm, and for the composite Al_2O_3 + 0.5% vol MgO $Q_V = 31.4$ kTm. Similar values of sintering activation energy ($Q_V \sim 31$ kTm, which corresponds to $Q_V \sim 530$ kJ/mol) at stage II were obtained for all considered systems.

Thus, the studied additives do not affect the sintering kinetics of alumina at the intermediate stage of the SPS process under conditions of constant heating rate.

Analysis of the experimental results shows that the rapid shrinkage of alumina at stage II is caused by plastic deformation carried out under the applied mechanical pressure, controlled by bulk diffusion (power creep). In this case, the activation energy of the shrinkage process at the intermediate stage $Q = \sim 31$ kTm.

To eliminate the possible effect of residual pores on the kinetics of grain growth, further studies were carried out on samples that have reached the density of more than $\rho = 99.5\%$. The study of grain growth kinetics was carried out at the heating rate of 50°C/min and temperatures of 1320, 1420, and 1520°C with subsequent isothermal dwelling for 0, 3, 10, and 30 min The dependences of the grain size (d) on the duration of isothermal dwelling (t) at various temperatures of the SPS are shown in figure 1.

Figure 1 shows that magnesium oxide has a significant impact on the grain growth kinetics at the final stage (stage III). The grain size of ceramic with the addition of MgO is about 10 times smaller than in pure alumina ceramics. Zirconium oxide particles inhibit grain growth less effectively than magnesium oxide. Addition of the titanium oxide leads to a slight increase in the average grain size.

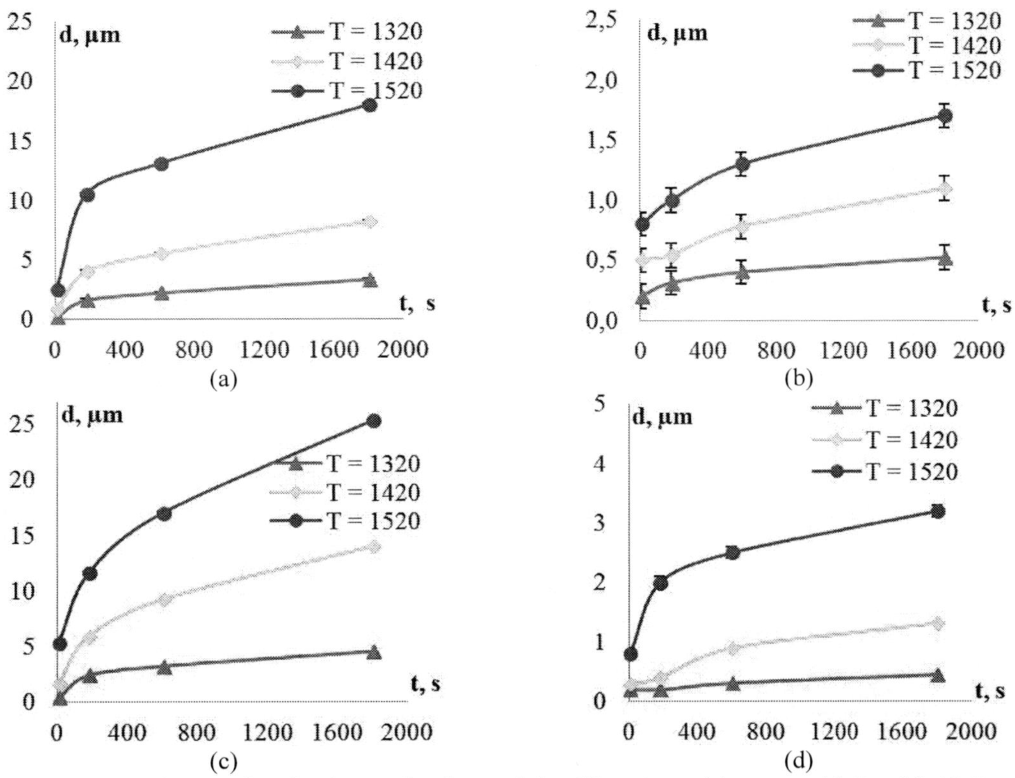

Figure 1. Dependence of grain size on isothermal dwelling time: (a) pure α-Al_2O_3, (b) Al_2O_3 + 0.5% vol MgO, (c) Al_2O_3 + 0.5% vol TiO_2, (d) Al_2O_3 + 0.5% vol ZrO_2.

3.2. Analysis of shrinkage behavior and grain growth at stage III. Sintering of ceramics in isothermal conditions.

3.2.1. Estimation of the grain growth activation energy for pure α-Al_2O_3. The expression describing the grain growth during isothermal dwelling has the following form [12]:

$$d^n - d_0^n = D_b t = D_{b0}\exp(-\frac{Q_b}{kT_m})t \quad (4)$$

where d_0 is the initial grain size, d is the grain size at time t, D_b is the coefficient of grain-boundary diffusion, D_{b0} is the pre-exponential factor, Q_b is the activation energy of grain-boundary diffusion, k is the Boltzmann constant, T_m is the melting temperature, and n is the growth rate constant.

According to [12, 13], the growth rate constant for a homogeneous system, which does not contain second phase particles, $n = 2$.

By substituting the value $n = 2$ into equation (4), which describes the kinetics of grain growth of pure α-Al_2O_3, and by taking a logarithm we can find the value of the diffusion activation energy by the slope of the straight line in the coordinates ln (D)–T_m/T (figure 2).

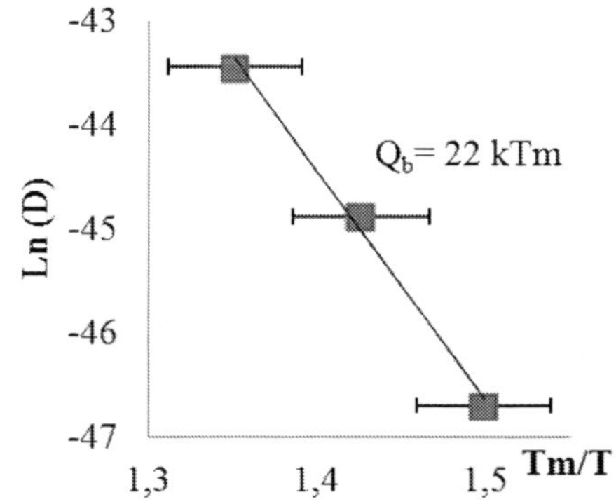

Figure 2. The dependence of ln (D)–T_m/T for pure α-Al_2O_3 ceramic under isothermal dwelling during SPS

As shown in figure 2 for Al_2O_3 the activation energy of grain growth at the final sintering stage $Q = Q_b = 22$ kTm. The found value of the energy corresponds to the data of [14] ($Q = 19.8$ kTm) which is approximately $Q = 340$ kJ/mol).

3.2.2. Estimation of the grain growth activation energy for Al_2O_3 + 0.5% vol ZrO_2. According to the phase diagram, zirconium oxide is not soluble in alumina and does not form any phases with alumina. In ceramics obtained by sintering powder mixtures, zirconium ions are predominantly detected along the grain boundaries of Al_2O_3 [15].

The analysis of the microstructure carried out using SEM showed that the size of ZrO2 particles in sintered samples is approximately 10–100 nm.

Analysis of the sample microstructure parameters at $T = 1520$°C and $t = 30$ min shows that the ratio $d_{\mathrm{ZrO2}} \approx d_{Al2O3+ZrO2} \cdot f_v$ is met. This means that the growth rate of alumina grains is determined by the growth rate of zirconium oxide particles. In this case, the growth rate constant in equation (4) can be 3 when the process of the second phase particles growth is controlled by bulk diffusion (diffusion of Zr ions in the Al_2O_3 lattice), or 4 when the process of the second phase particles growth is controlled by grain boundary diffusion (diffusion of Zr ions along Al_2O_3 grain boundaries). At $n = 3$, the activation energy $Q = 38$ kTm, which is $Q_V \sim 650$ kJ/mol.

Thus, the addition of 0.5 vol % ZrO_2 leads to the increase of the activation energy of grain boundary diffusion from $Q_b = 22$ kTm to $Q_b = 37$ kTm. This increase corresponds to the data [15, 16]. In [16], the values of the activation energy for Al_2O_3 + ZrO_2 systems are equal to $Q = 700$–100 kJ/mol, which are approximately 31–41 kTm.

3.2.3. Estimation of the grain growth activation energy for Al_2O_3 + 0.5% vol MgO. The results of X-ray phase analysis show that Al_2O_3 + 0.5% MgO samples, sintered at a temperature of 1520°C, contain about 2% wt $MgAl_2O_4$ phase. The volume of the unit cell of the $MgAl_2O_4$ phase is 7 times higher than the volume of MgO phase. Therefore, it can be concluded that approximately 0.2% vol MgO was dissolved in the Al_2O_3 lattice and 0.3% vol MgO remained in the form of particles, which formed the $MgAl_2O_4$ phase. The average particle size of the $MgAl_2O_4$ phase is about 0.7 μm.

Analysis of the sample microstructure parameters at $T = 1520$°C and $t = 30$ min shows that the ratio $d_{\mathrm{MgAl2O4}} \geq d_{Al2O3+MgO} \cdot f_{v\_\mathrm{MgAl2O4}}$ is met. This means that particles of $MgAl_2O_4$ may inhibit the growth of alumina grains due to the action of Zener force.

However, the ratio of the volume fraction of particles to their size for all the above temperatures is $(f_{v\_\mathrm{MgAl2O4}}/d_{\mathrm{MgAl2O4}}) \geq 10^{-2} \cdot (1/d_{Al2O3+MgO})$. The Zener effect caused by $MgAl_2O_4$ particles is insignificant and can be neglected. Thus, the growth rate constant in equation (4) for Al_2O_3 + 0.5% MgO can be set at $n = 2$.

The obtained value of the activation energy of grain-boundary diffusion is $Q_b = 26$ kTm (440 kJ/mol). Thus, at a dissolution of 0.2%vol MgO in alumina, the activation energy of grain boundary diffusion increases by 4 kTm, which leads to a decrease of a grain growth rate.

During the interaction between Al_2O_3 and MgO, Al ions are replaced by Mg ions, with the increase in the concentration of Al interstitial ions [13] and the decrease in the diffusion coefficient of oxygen within the boundaries [17–19].

3.2.4. Estimation of the grain growth activation energy for Al_2O_3 + 0.5% vol TiO_2. According to [20], the solubility limit of TiO_2 in Al_2O_3 at temperatures above 1300°C does not exceed 0.25% vol. Thus, in the temperature range used in the experiment, 0.5% vol TiO_2 exists in a form of dissolved Ti ions and particles of Al_2TiO_5.

In the process of isothermal dwelling for 30 min, the particle size of Al_2TiO_5 increases significantly from initial $d \sim 0.1$ µm to $d \sim 0.4$ µm at $T = 1320$°C, $d \sim 1.6$ µm at $T = 1420$°C and $d \sim 10$ µm at $T = 1520$°C.

Analysis of the sample microstructure parameters at $T = 1420$°C and $t = 30$ min shows that the ratio $d_{\mathrm{Al2TiO5}} \geq d_{Al2O3+TiO2} \cdot f_v$ is met. This means that particles of Al_2TiO_5 may also inhibit the growth of alumina grains due to the action of Zener force.

However, the ratio of the volume fraction of particles to their size for all the above temperatures is $(f_{v\_Al2TiO5}/d_{Al2TiO5}) \geq 10^{-3} \cdot (1/d_{Al2O3+TiO2})$. The Zener effect, which is caused by Al_2TiO_5 particles, is insignificant and also can be neglected.

In this case, the grain growth kinetics of alumina grains should be controlled only by the diffusion of Al and O ions in the Al_2O3 lattice; therefore, the growth rate constant should be $n = 2$.

The rapid growth of Al_2TiO_5 particles is due to the fact that the diffusion coefficient of Ti ions in the Al_2O_3 lattice is more than 10 times higher than the diffusion coefficient of Al ions [21].

The obtained value of the activation energy is $Q = 21$ kTm. Thus, after the dissolution of 0.25% vol TiO_2 in alumina lattice, the diffusion activation energy decreases by 1.2 kTm, which accelerates the growth of grains.

During the interaction between Al_2O_3 and TiO_2, Al ions are replaced by Ti ions with the formation of additional Al vacancies. This vacancies increase the bulk diffusion of Al [13] and accelerate grain growth [22–23].

4. Conclusion

The authors estimated the activation energy of shrinkage processes at the intermediate and final stages of sintering.

The study shows that rapid densification of alumina at stage II is caused by plastic deformation carried out under the action of the applied mechanical pressure, controlled by bulk diffusion (power creep law). The activation energy of the shrinkage process at the intermediate stage is ~ 31 kTm (530 kJ/mol).

The effect of additions of MgO, ZrO_2, and TiO_2 (at a concentration of up to 0.5% vol) was observed only at the final stage of sintering of alumina - the stage of grain growth.

Acknowledgments

The authors wish to thank the Ministry of Education and Science of the Russian Federation (grant no. 11.1114.2017/4.6).

References

[1] Elasser C 2005 *J. Am. Ceram. Soc.* **88** 1
[2] Doremus R H 2006 *J. Appl. Phys.* **100** 1
[3] Jeong Y K 2011 *Microstruc. Trans. Non Ferr. Met. Soc. China* **21** s1
[4] Galusek D, Sedlacek J, Chovanec J and Michalkova M 2015 *Ceram. Int.* **41** 9692
[5] Kim B N 2012 *J. Ceram. Soc. Japan* **120** 116

[6] Jiang D T 2008 *J. Am. Ceram. Soc.* **91** 151
[7] Fukumoto I, Mekaru S and Shibata S 2006 *JSME Int. J. Series A* **49** 91
[8] Lallemant L, et al 2014 *J. Eur. Ceram. Soc.* **34** 1279
[9] Chuvil'deev V N, et al 2015 *Inorg. Mater.* **51** 1047
[10] Olevsky E A 1998 *Mater. Sci. Eng.* **R23** 41
[11] Frost H J and Ashby M F 1982 *Deformation-mechanism maps: the plasticity and creep of metals and ceramics* (Pergamon Press)
[12] Gorelik S 1967 *Recrystallization in metals and alloys* (Moscow: Metallurgy)
[13] Rahaman M N 2003 *Ceramic processing and sintering* (New York: Marcel Dekker, Inc.)
[14] Feng L and Dillon S J 2017 *J. Am. Ceram. Soc.* **00** 1.
[15] Wang C M, et al 2000 *Interf. Sci.* **8** 243
[16] Yoshida H, Okada K, Ikuhara Y and Sakuma T 1997 *Philosoph. Magaz. Lett.* **76** 9
[17] Harmer M P and Brook R J 1980 *J. Mater. Sci.* **15** 3017
[18] Greskovich C and Brewer J A 2001 *J. Am. Ceram. Soc.* **84** 420–25
[19] Dillon S J and Harmer M P 2008 *J. Eur. Ceram. Soc.* **28** 1485
[20] Lakshmanan A 2006 *New Emerging Techniques* (Rijeka: InTech)
[21] Weiss M, et al 2000 *J. Chem. Phys.* **113** 5058
[22] Kim Y M, Hong S H and Kim D Y 2000 *J. Am. Ceram. Soc.* **83** 2809
[23] Horn D S and Messing G L 1995 *Mater. Sci. Eng.* A **195** 169

Synthesis and investigation of highly dispersed active phases of intermetallic and supported SHS-catalysts

V N Borshch, E V Pugacheva, S Ya Zhuk, I M Dement'eva, V N Sanin, D E Andreev and V I Yukhvid

Merzhanov Institute of Structural Macrokinetics and Materials Science, Russian Academy of Sciences, Chernogolovka, 142432 Russia

E-mail: borsch@ism.ac.ru

Abstract. In the present study, the active phases (AP) of an Fe–Ni–Co–Mn catalyst produced from SHS-intermetallics (a), a catalyst prepared via formation an intermetallic layer on a mesh surface of chromium-nickel stainless steel (b), and of a Co–Mn catalyst prepared via SHS on silica gel support (c), were isolated and characterized by SEM and XRD. The catalysts studied were highly active in the process of deep oxidation and catalysts (b) and (c) were highly active in the process of CO_2 methanation as well. All AP had oxo-metallic composition and were formed from nanoscale components, but if the structure of these components for catalyst (a) was the same as on the surface of the catalyst, the structures of AP components for catalyst (b) significantly differed from the surface structures. In addition, AP sediment grain nanostructures of catalyst (c) differed from those of catalysts (a) and (b).

1. Active phases of intermetallic catalysts

Previously, a new class of polymetallic catalysts based on complex multicomponent intermetallic compounds, which are the product of self-propagating high-temperature synthesis (SHS), was developed. To obtain the catalysts, intermetallic alloys are crushed, leached in an alkali solution (NaOH, KOH) and stabilized by treatment with a solution of hydrogen peroxide. When leaching and stabilizing, the higher intermetallic phases reduce to disordered nanostructured oxo-metallic formations that form the catalytically active phase (AP) on the support from the remaining unleached lower intermetallic phases. The catalysts have high activity and selectivity in oxidizing and reducing processes, their specific surface reaches 50 m^2/g [1–3].

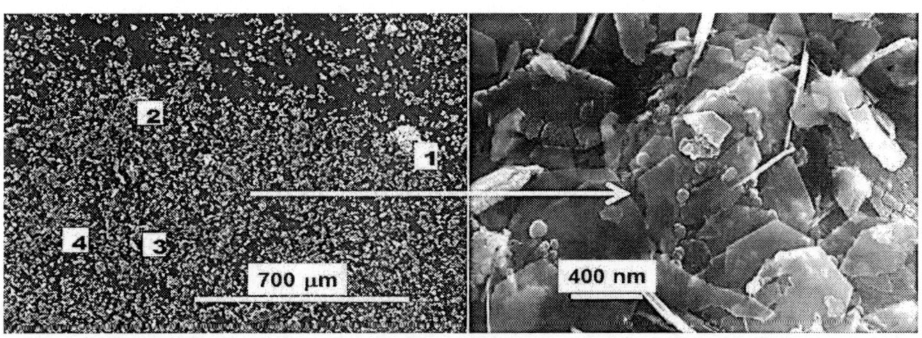

Figure 1. SEM image of AP precipitate of 50%Fe–35%Ni–10%Co–5%Mn catalyst.

In the process of obtaining of the catalysts a small amount of AP forms a highly dispersed suspension, slowly precipitating from the reaction medium. This suspension was isolated and examined by scanning electron microscopy (SEM). Figure 1 presents the morphology of sediment powder and fine structure of an individual grain of AP of the Fe–Ni–Co–Mn catalyst of deep oxidation.

As can be seen in figure 1, the sediment granules are chaotic aggregates of the nanostructures of AP, apparently formed in the process of precipitation from an alkaline medium and subsequent washing and stabilization. It follows from table 1 that the grains of AP are oxo-metallic formations with an admixture of residual alumina (point 4). The nanostructures that make up the grains of precipitated AP are identical to those on the surface of the catalyst granules [3].

Table 1. EDS results of AP particle precipitates in figure 1 (wt %).

Spot	O	Al	Mn	Fe	Co	Ni
1	38.8	4.6	18.7	25.1	5.6	7.2
2	48.2	11.2	2.2	17.8	4.4	16.2
3	16.4	9.5	6.6	44.7	10.3	12.4
4	68.3	20.8	3.4	6.5	-	1.0

Catalysts closely related to the studied class of catalysts were obtained by activating polymetallic support such as stainless steel by forming an intermetallic layer on the surface with its subsequent leaching and stabilization as described above. The intermetallic layer was formed in the process of diffusion saturation of the surface with aluminum at $T \sim 620–640°C$. If intermetallic compounds of other catalytically active metals are added to aluminum, catalysts with AP promoted with these metals can be obtained. The results of testing such catalysts in the processes of deep oxidation of CO and propane, as well as hydrogenation of CO_2, are presented in figures 2 and 3.

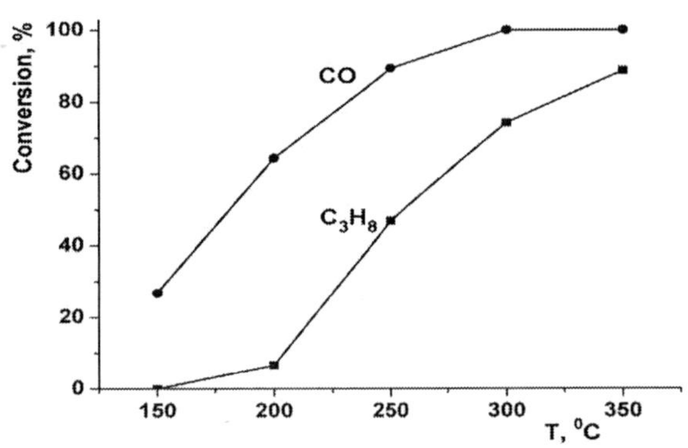

Figure 2. Conversion of CO and propane vs. temperature in the process of deep oxidation on activated chromium-nickel stainless steel wire mesh, promoted with Co and V. Composition of gas mixture (in % vol): propane – 0.2, CO – 0.7, O_2 – 2, N_2 the rest. GHSV = 120000 h^{-1}.

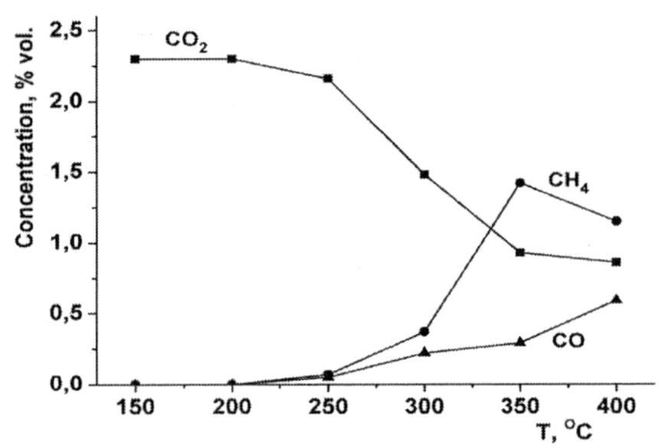

Figure 3. Concentrations of CO_2, CH_4 and CO vs. temperature in the process of CO_2 hydrogenation on activated chromium–nickel stainless steel wire mesh, promoted with Co. Composition of gas mixture (in % vol.): CO_2 – 2.3, H_2 – 9.2, He the rest. GHSV = 3000 h^{-1}.

As can be seen in figure 2, 50% of CO conversion is achieved already at $T < 200°C$, and 50% of propane conversion is achieved at $T < 275°C$. The maximum yield of methane in the hydrogenation of CO_2 is 1.42% at $T = 350°C$ and total conversion of CO_2 equal to 59.6% (figure 3).

Figure 4 shows micrographs (SEM) of the powder and a separate grain of the sediment of the stabilized AP precipitated from the leach solution. As in the previous case, the sediment grains are composed of formations of various shapes, sizes and composition, the elemental composition of the surface of which is shown in table 2.

The XRD results shown in figure 5 also indicate a highly defective, largely amorphized structure of aggregates consisting mainly of oxide phases with an admixture of metallic iron. Note that, in contrast to the previous case, the morphology of individual components of the grains differs markedly from the structures found in the composition of AP on surface of activated support [4].

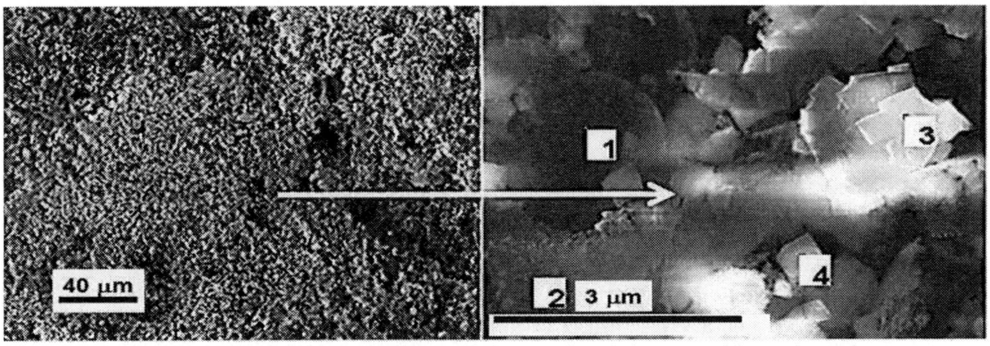

Figure 4. SEM image of AP precipitate of stainless steel based catalyst.

Table 2. EDS results of the AP particle precipitate of stainless steel based catalyst in figure 4 (wt %).

Spot	O	Al	Cr	Mn	Fe	Ni
1	21.6	18.9	7.8	-	48.2	3.4
2	26.1	15.8	6.8	-	48.0	3.2
3	44.9	11.0	3.5	-	35.5	5.2
4	36.0	12.8	3.6	1.2	45.3	1.1

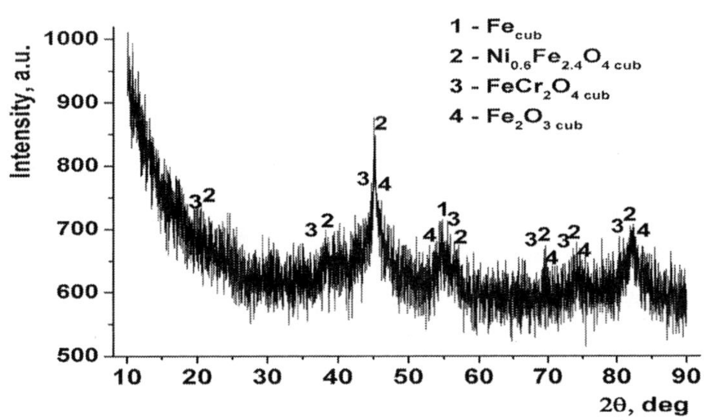

Figure 5. XRD pattern of AP precipitate of stainless steel based catalyst.

2. Active phases of supported catalysts

A number of catalysts on various supports (γ-Al_2O_3, zeolites, silica gel, and haloisite) with AP based on Co, Ni, and Mn, obtained in low-temperature combustion mode, was studied. The supports were impregnated with a calculated amount of a mixture of solutions of metal nitrates (oxidizers) and urea (fuel), dried and heated in an argon atmosphere prior to the initiation of a combustion wave. The temperature in combustion wave reached 300–310°C. As in the previous case, the catalysts were washed and treated with a hydrogen peroxide solution to stabilize the phases of free metal. Their catalytic properties were studied in the processes of deep oxidation and hydrogenation of CO_2. Figures 6 and 7 show the results of testing samples of catalysts with nickel–cobalt and nickel AP on various supports.

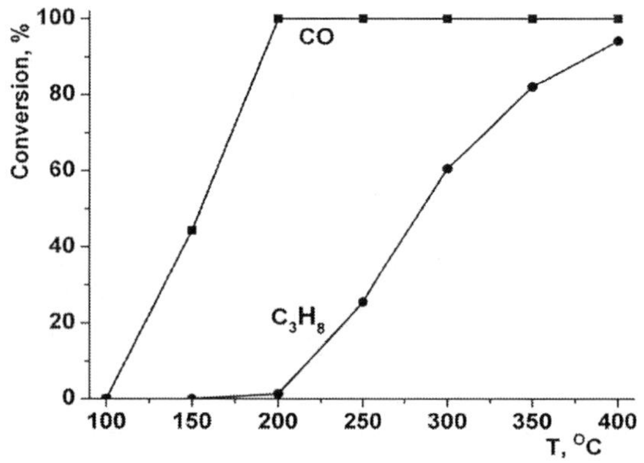

Figure 6. Conversion of CO and propane vs. temperature in the process of deep oxidation on (10%Co+10%Ni)/ZSM-5 catalyst. Composition of gas mixture (% vol): propane – 0.2, CO – 0.7, O_2 – 2, N_2 the rest. GHSV = 120,000 h^{-1}.

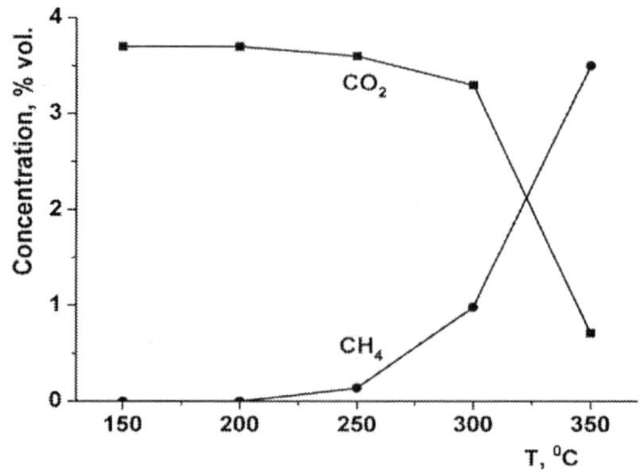

Figure 7. Concentrations of CO_2, CH_4 and CO vs. temperature in the process of CO_2 hydrogenation on Ni10%/haloizite catalyst. Composition of gas mixture (% vol): CO_2 – 3.9, H_2 – 11.8, He the rest. GHSV = 12,000 h^{-1}.

It is possible to note the high activity of the catalyst (10% Co + 10%Ni)/ZSM-5 in the process of deep oxidation, especially CO, for which complete conversion was achieved already at 200°C. The conversion of CO_2 in the process of hydrogenation on the catalyst 10% Ni/haloisite at a temperature of 350°C amounted to 82% with a selectivity to methane of 100%.

The study of the phase composition of AP on the surface of the catalysts is complicated by the fact that the high intensity of the support peaks in the X-ray diffraction spectra masks the weaker and more broadened peaks of AP. For a more detailed analysis of the structure and composition of AP, cobalt-nickel catalysts based on silica gel were synthesized. The catalysts were boiled in a solution of NaOH to dissolve the support and separate AP. The morphology and elemental composition of the washed precipitate of AP of (10%Co + 10%Ni)/SiO_2 catalyst is shown in figure 8 and table 3, respectively.

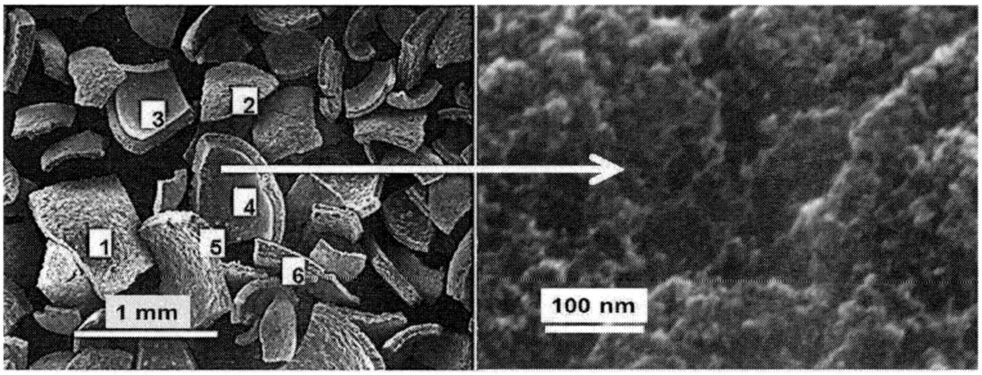

Figure 8. SEM image of AP precipitate of 10%Co + 10%Ni/SiO_2 catalyst.

Table 3. EDS results of AP precipitate particles of (10%Co + 10%Ni)/SiO_2 catalyst in figure 8 (wt %).

Spot	O	Si	Co	Ni
1	7.4	1.0	47.1	44.6
2	21.2	2.7	40.4	35.6
3	18.8	2.2	33.6	45.3
4	4.7	0.7	45.1	49.5
5	25.9	4.5	33.7	35.8
6	36.8	3.7	27.8	31.7

The isolated AP, as can be seen in figure 8, is a rather large (about 1 mm) debris fraction obtained by drying a bulk, poorly precipitating sediment with a large amount of residual water. The formation of such a structure is facilitated by the presence in the composition of the sediment (table 3) of a small amount of, apparently, hydrated silicon oxide, which plays the role of a binder. As follows from the high-resolution image in figure 8, these particles are formed (presumably in the process of precipitation) from granules with dimensions of the order of tens of nm and are highly porous formations. This fact is also confirmed by the nature of XRD spectrum (figure 6). The composition of the active phase, as follows from figure 6, contains both the phases of free metals (Co, Ni) and their oxides.

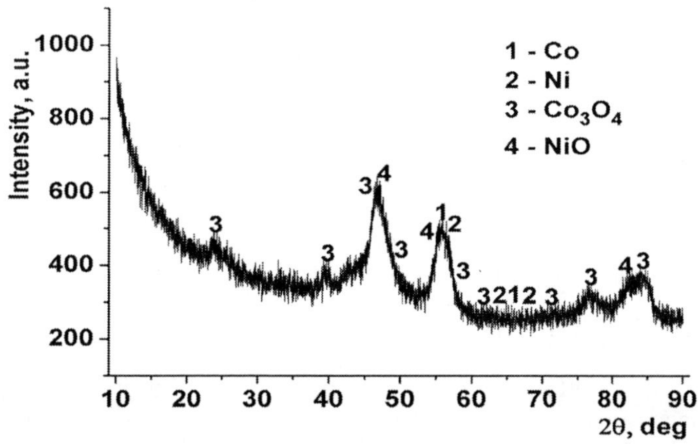

Figure 9. XRD pattern of AP precipitate of (10%Co + 10%Ni)/SiO_2 catalyst.

From the foregoing, it can be concluded that the proposed methods for the synthesis of catalysts, both on the basis of intermetallic compounds and those produced in the combustion process on supports, make it possible to obtain catalysts with highly dispersed, nanostructured AP. All studied AP have oxo-metallic composition. Due in large part to these facts, such catalysts are highly active both in the process of deep oxidation of CO and hydrocarbons, and in the process of hydrogenation of CO_2.

References

[1] Borshch V N, Pugacheva E V, Zhuk S Ya, Andreev D E, Sanin V N and Yukhvid V I 2008 *Dokl. Phys. Chem.* **419** 77
[2] Borshch V N, Pugacheva E V, Zhuk S Ya, Sanin V N, Andreev D E, Yukhvid V I, Eliseev O L, Kazantsev R V, Kolesnikov S I, Kolesnikov I M and Lapidus A L 2015 *Kinetics and Catalysis* **56** 681
[3] Borshch V N, Pugacheva E V, Zhuk S Ya, Sanin V N, Andreev D E and Yukhvid V I 2017 *Int. J. Self-Propag. High-Temp. Synth.* **26** 124
[4] Borshch V N, Zhuk S Ya and Sachkova N V 2018 *Kinetics and Catalysis* **59** 386

IOP Conf. Series: Materials Science and Engineering **558** (2019) 012008 doi:10.1088/1757-899X/558/1/012008

Microstructures and mechanical properties of tungsten-tantalum alloys consolidated by electron beam melting, hot pressing and spark plasma sintering

F Dong[1], Y Chen[1], Z Bai[2], H Wang[1] and L Jiang[1]

[1]School of Materials Science and Engineering, North Minzu University, Yinchuan, 750021 China
[2]CNMC Ningxia orient group Co.L.td, Shizuishan, Ningxia, 753000 China

E-mail: lyhchen@163.com

Abstract. The effect of consolidation process conditions and the tungsten concentration on the microstructures and mechanical properties of tantalum–tungsten alloys has been investigated. Ta–10W (in weight percent) and Ta–20W (in weight percent) alloys have been consolidated by electron beam melting (EBM), hot pressing (HP) and spark plasma sintering (SPS) techniques. The results show that the tantalum–tungsten alloys consolidated by EBM exhibit the highest relative density while by HP or SPS exhibit smaller grain-size compared to that by EBM, which may be attributed from the lower process temperature and the existence of oxide on the grain boundary hindering the growth of grain. The alloys with fine grain size (HP or SPS) have a higher microhardness and compression strength than that with coarse grain size (SPS) under the same tungsten concentration. It has been shown that these two alloy systems exhibit a similar general trend of strength increasing and the ductility decreasing when applying different consolidation process, while an enhanced strength-ductility synergy can be achieved for the Ta–10W alloy.

1. Introduction

Tantalum (Ta) and tantalum alloys are widely used in electronics, chemical industries, industrial furnace, aerospace, atomic energy due to their unique properties such as high density, high melting point, low ductile-to-brittle transition temperature and moderately high elastic modulus [1–6]. Ta–W alloy is a kind of Ta-based solid-solution strengthened refractory alloy with body-centered cubic crystal structure, which is widely used in high-tech applications, such as power, aerospace and nuclear engineering [1, 2].

Ta–W alloys are usually prepared by electron beam melting (EBM) method or powder metallurgy method [7, 8]. The alloys prepared by EBM method have coarse grains and lower content of impurities, compared with those prepared by powder metallurgy method. However, there was very little work reported in the literatures to compare the mechanical properties of Ta–W alloys fabricated by these two methods. Therefore, this study mainly focused on this issue. Ta–10W and Ta–20W alloys have been consolidated by EBM, hot pressing (HP) and spark plasma sintering (SPS) techniques separately. Subsequently, the effect of consolidation process conditions and the tungsten concentration on the microstructures and mechanical properties of tantalum-tungsten alloys were investigated.

2. Experimental

The Ta and W powder had 99.9% nominal purity and an average particle size (*D*) of 15 μm and 10 μm, respectively. Ta–W composites with 10 and 20 wt % of W were consolidated by EBM, HP and SPS.

- EBM: The powder mix was cold isostatically pressed at 100 MPa in order to obtain cylindrical rods, and then the formed billet was sintered in vacuum furnace under 1900°C. The ingot was obtained by melting the sintered billet in the vacuum electron beam melting furnace.
- HP: Powders were mixed in a conventional ball mill for 12 h using ZrO balls with ball to powder ratio (BPR) of 1:1. Then the powders were transferred from mill pots to a graphite die in a glove box which was filled with pure argon gas. HP was performed for 1 h at 2000°C with high purity Ar under 30 MPa. The heating rate was maintained 20°C/min.
- SPS: The samples were sintered by spark plasma sintering (SPS) in vacuum under a pressure of 30 MPa, and the maximum temperature was 1800°C with heating rate of 100°C/min. The holding time was chosen to be 10 min in order to limit grain growth. Afterwards, the surface of all samples was mechanically polished to mirror-like finish. The processing parameters are summarized in table 1.

Table 1. Chemical composition of the Ta–10W and Ta–20W plates (%, mass fraction).

Material	Elements							
	W	Nb	Ti	Mo	Si	C	N	Ta
Ta–10W	10.24	0.003	0.001	0.001	0.001	0.0008	0.0015	Bal
Ta–20W	20.4	0.004	0.002	0.008	0.001	0.0012	0.002	Bal

The consolidation cycles are presented in figure 1.

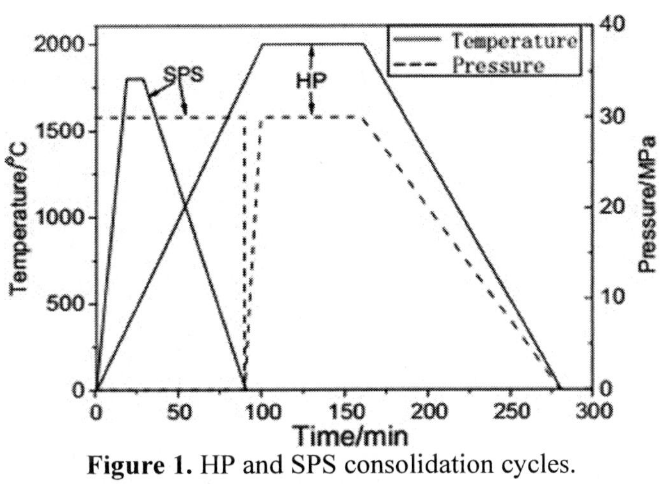

Figure 1. HP and SPS consolidation cycles.

The densification achieved was measured using the Archimedes method.

Microstructure observation was carried out by scanning electron microscope (SEM). The samples for the examination were metallographically polished and etched using hydrofluoric acid solution for 5 s before the microstructural investigation was carried out. Phase compositions were investigated by a X- ray diffractometer (XRD-6000, Shimadzu, Japan) with Cu K_α radiation in a scanning range of 10°–80°. A scanning electron microscope (Sigma 500, Zeiss Gemini, Germany) fixed with energy-dispersive spectroscopy (EDS) was used to investigate the chemistry of the interested phases.

The hardness and compressive properties of the Ta–W alloys were evaluated using Brinell hardness testing, tensile testing, respectively. The specimens for hardness test were ground and polished before carrying out multiple hardness indentations. The hardness values within the range of ±2% were averaged to obtain the hardness value.

A universal compressive testing machine operated at a strain rate of 0.5 mm/min was used for the compressive testing of the composite samples which were machined as a rectangular solid with a dimension of 3 × 3 × 6 mm. The compressive properties evaluated from the engineering stress–strain curves are yield strength, compressive strength and percent elongation. Multiple compressive tests were performed for each test composition in order to guarantee the reliability and reproducibility of the results.

3. Results and discussions

Engineering compressive stress-strain curves are shown in figure 2a. The distinctive behavior of displacement controlled compression was observed in these curves.

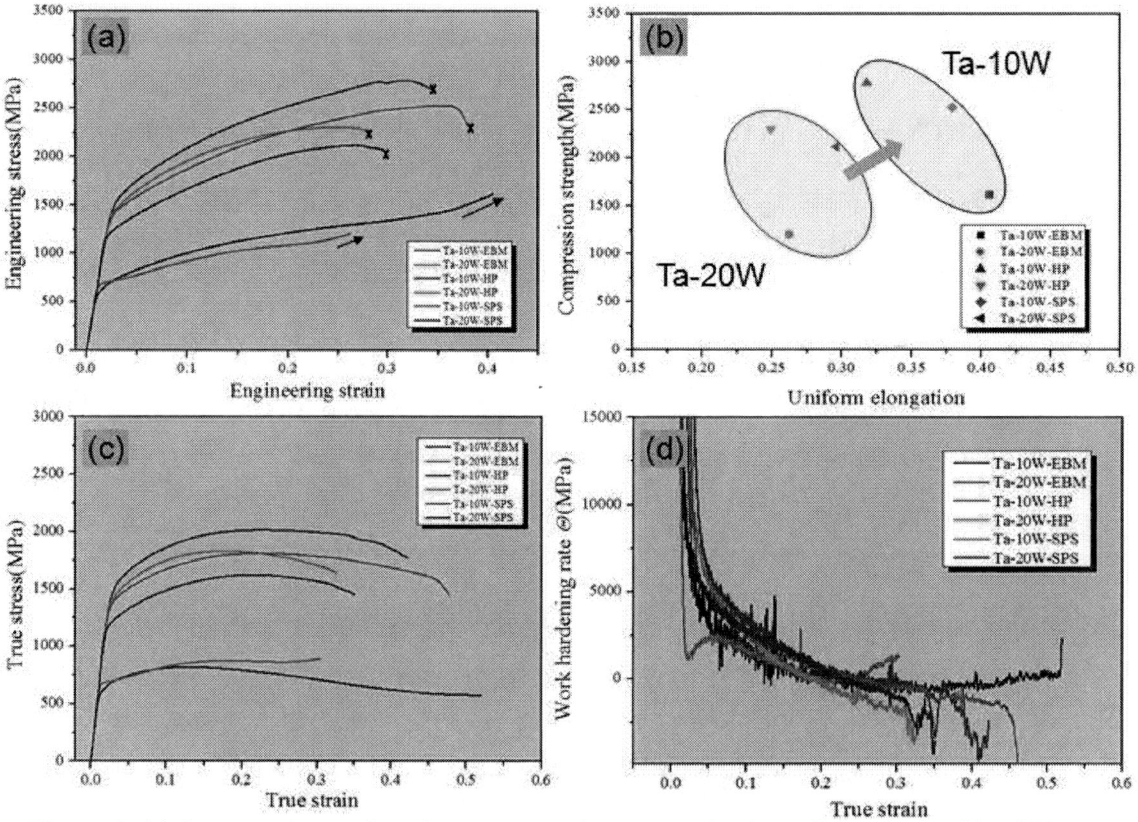

Figure 2. (a) Compressive engineering stress–strain curves of Ta–W alloys obtained by different preparing method; (b) relationship between compression strength and uniform elongation of Ta–W alloys; (c) corresponding true stress-strain curves and (d) work hardening rate (dσ/dε)-true strain curves.

Corresponding compressive properties are listed in table 2. The results show that the range of yield strength (YS) values is from 585 to 1502 MPa, the range of compression strength (CS) is from 1204 to 2785 MPa and the range of elongation varies is from 24.9 to 40.5%.

The Ta–10W and Ta–20W alloys consolidated by HP and SPS exhibited a better strength-ductility combination than consolidated by EBM. More crucially, compared with the Ta–20W, the Ta–10W with lower solution of atoms exhibits simultaneously high strength and ductility consolidated by different method, as illustrated in figure 2b.

Table 2. Mechanical properties of Ta–W alloy samples.

Material	YS, MPa	CS, MPa	UEL	TEL	HV, GPa
Ta–10W-EBM	585	1619	0.405	0.405	2.74
Ta–20W-EBM	669	1204	0.262	0.262	3.41
Ta–10W-HP	1502	2785	0.344	0.318	4.53
Ta–20W-HP	1432	2300	0.278	0.249	4.17
Ta–10W-SPS	1389	2523	0.358	0.379	3.98
Ta–20W-SPS	1198	2115	0.268	0.296	3.68

In order to reveal the work hardening behavior of the present samples, the engineering stress–strain curves in figure 2a were converted into the true stress–strain curves in figure 2c. Work-hardening rate Θ vs. true strain (ε) are shown in figure 2d. It has been widely acknowledged that the ductility of metallic materials is controlled by strain hardening [9]. In macroscale, the Consideré criterion governs the onset of localized deformation [10].

$$\Theta = d\sigma/d\varepsilon \leq \sigma$$

Where σ is true stress and ε is the true strain, it is noted that the mechanical properties of the two Ta–W alloys exhibit a similar trend.

The high strength of the Ta–W alloys consolidated by HP is attributed to the grain size effect in terms of Hall–Petch relation [11]. Figure 3 shows a Hall–Petch plot using measured σ_{ys} of Ta–W alloys consolidated by different methods (as listed in table 2). According to the Hall-Petch relationship, the YS of an alloy can be approximated as

$$\sigma ys = \sigma_0 + kd^{-1/2}$$

Where σ_{ys} is the yield strength, where σ_0 is the friction stress, k is the Hall–Petch parameter, and d is the mean grain size. As for Ta–10W and Ta–20W alloys, the k value was 5.45 MPa·m$^{1/2}$ and 4.24 MPa·m$^{1/2}$; the σ_0 value was 336 MPa and 458 MPa.

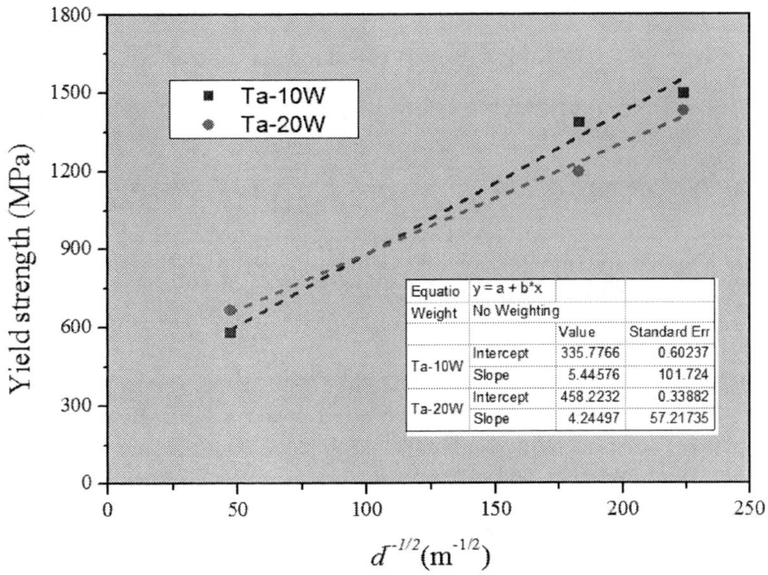

Figure 3. Plots of grain size and yield strength of Ta–W alloys showing the Hall-Petch relation.

Figure 4a shows the X-ray diffraction patterns for the Ta–20W alloy consolidated by EBM, HP and SPS. No peaks of W can be detectable in all of the Ta–20W alloys. The lattice size of the Ta–20W becomes smaller after solid solution W atoms. There are axial offset to high angle for the diffraction peaks, take (200) as an example. And the width of diffraction peaks of alloy after EBM is wider than that after HP and SPS sintering, which can be explained by larger grain size of the Ta–20W after the EBM.

The average grain size, Vickers hardness and density of Ta–W alloy samples obtained by different preparing method are recorded in table 3. The alloys with fine grain size (HP or SPS) have a higher micro-hardness than that with coarse grain size (SPS) under the same tungsten concentration.

Table 3. Characteristics of Ta–W alloy samples.

Material	Average grain size, μm	Density, g·cm^{-3}	Theoretical density, g·cm^{-3}	Relative density, %
Ta–10W-EBM	455	16.91	16.92	99.94
Ta–20W-EBM	487	17.21	17.25	98.23
Ta–10W-HP	23	16.69	16.92	98.64
Ta–20W-HP	21	16.75	17.25	97.1
Ta–10W-SPS	32	16.35	16.92	96.63
Ta–20W-SPS	29	16.52	17.25	95.78

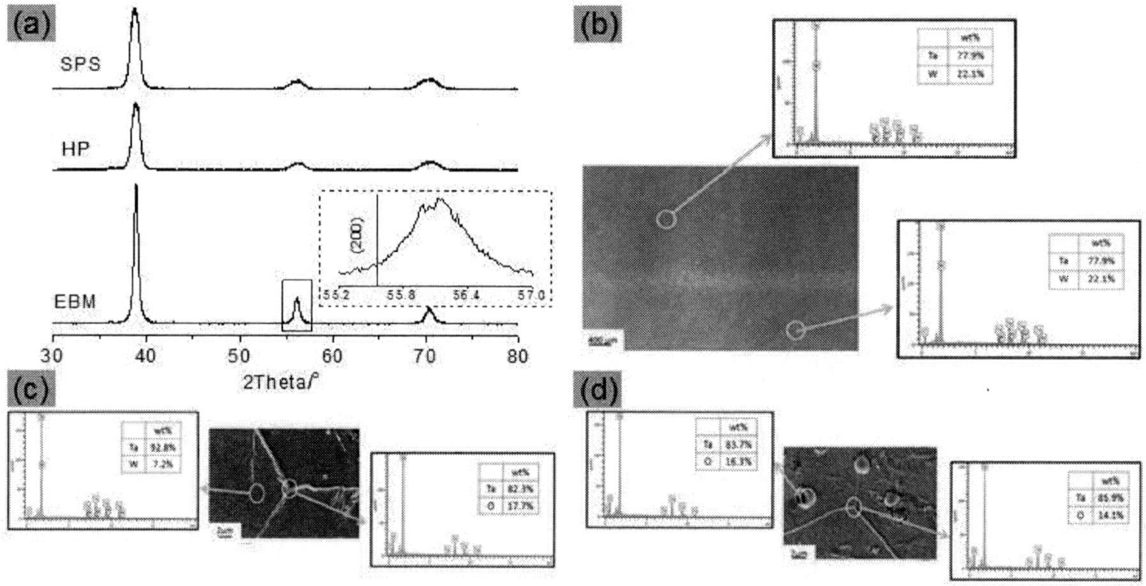

Figure 4. (a) XRD patterns of Ta–20W alloys obtained by different preparing method; (b) SEM and EDS of Ta–20W alloy obtained by EBM; (c) SEM and EDS of Ta–10W alloy obtained by HP; (d) SEM and EDS of Ta–20W alloy obtained by SPS.

Figure 4b showed the SEM images of Ta–20W alloy after EBM. It can be seen that the grain size is about 400–500 μm. The high temperature in the melting process of EBM was the cause of these large particles. The content of W of the alloy under the light color contrast and the dark color contrast was 22.1%, indicating that the chemical homogeneity of the alloy was high. Figures 4c and 4d are SEM images of Ta–10W alloy manufactured by HP or SPS, respectively. The grain size after HP was around 20 μm, the grain size after SPS is about 30μm. The porosity could both be observed after preparation process of HP or SPS, which was mainly formed during the process of solidification. The brittleness of alloy increased due to the existence of these porosities [12]. The surface of the alloy at grain boundary and the porosity is mainly composed of Ta and O two elements, mass ratio was 82.3% and 17.7% respectively. The crystal grain size of the alloy by HP or SPS is small compared to that by EBM. The powder metallurgy method (HP or SPS) of the alloy will form tantalum oxide at the grain boundary. The formation of oxide plays a role of pinning grain boundary, thus inhibiting the grain growth [13].

4. Conclusions

The Ta–W alloys consolidated by SPS exhibits the smallest grain-size and relative density, which may be attributed from the existence of oxide in the grain boundary hindering the growth of grain and the lower process temperature. The alloy with fine grains has a higher micro-hardness and compression strength than that with coarse grains under the same tungsten concentration. It is shown that these two alloy systems exhibit a similar general trend of the strength increasing and the ductility decreasing when applying different consolidation process, while an enhanced strength-ductility synergy can be achieved for the Ta–10W alloy.

Acknowledgments

This work was financially supported by "China National Science Foundation" under Grant no 51464001 and "Natural Science Foundation of Ningxia Province" under Grant no NZ17109.

References

[1] Arsenault R J 1966 *Acta Metall.* **14** 831–838
[2] Gypen L A and Deruyttere A 1982 *Metall. Trans. A* **13** 1015–20
[3] Chen S R and Gray III G T 1996 *Metall. Mater. Trans. A* **27** 2994–3006
[4] Lin Z, Lavernia E J and Mohamed F A 1999 *Acta Mater.* **47** 1181–94
[5] Knezevic M, Beyerlein I J, Lovato M L, Tomé C N, Richards A W and McCabe R J 2014 *Int. J. Plast.* **62** 93–104
[6] Briant C L and Lassila D H 1999 *Transactions of the ASME* **121** 172–7
[7] Sandim H R Z, Padilha A F, Randle V and Blum W 1999 *Int. J Refract Met H* **17** 431–5
[8] Nemat-Nasser S and Kapoor R 2001 *Int. J. Plast.* **17** 1351–66
[9] Jia D, Wang Y M, Ramesh K T, Ma E, Zhu Y T and Valiev R Z 2001 *Appl Phys Lett* **79** 611
[10] Dong F Y, Zhang P, Pang J C, Chen D M, Yang K and Zhang Z F 2013 *Mater. Sci. Eng. A* **587** 185–191
[11] Kashyap B P and Tangri K 1995 *Acta Metall Mater* **43** 3971
[12] Dong F Y, Zhang P, Pang J C, Ren Y B, Yang K and Zhang Z F 2015 *Scr. Mater.* **96** 5–8
[13] Gao J, Thompson R G and Patterson B R 1997 *Acta Mater.* **45** 3653–58

IOP Conf. Series: Materials Science and Engineering **558** (2019) 012009 doi:10.1088/1757-899X/558/1/012009

Chrono-topographical analysis of the laws of the microheterogeneous combustion mode for the production of powders by the SHS method

A V Dolmatov

Ugra State University, Khanty-Mansiysk, 628012 Russia

e-mail: adolmatov@bk.ru

Abstract. The paper presents the capabilities of the chrono-topographical analysis method for studying the discrete nature of the propagation of the SHS process on the scale of the combustion focy and the reaction cell. Using the approach of chrono-topographic analysis, thermal imaging studies of the microheterogeneous mode of combustion in the Ni-Al system were carried out. The regularities of the dynamics of individual foci in the SHS wave are revealed. Using chronographic maps of the SHS process, it is shown that there are two types of foci in the combustion wave, which differ in the nature of the propagation dynamics. On the scale of the reaction cells, the experiment investigated the relationship of the combustion dynamics with the moments of the formation of new reaction sites. According to the results, it was concluded that an increase in the heat deficit in microheterogeneous combustion regimes leads to an increase in the dispersion of the resulting product. The work was supported by the Russian Foundation for Basic Research in research projects no. 18-08-01475, 18-08-01152, as well as grant no. 13-01-20/28 "Leading Scientific School of Ugra State University".

1. Introduction

The propagation of a SHS wave in microheterogeneous modes close to unstable combustion is characterized by the presence of areas with significant depression reactions, which contribute to the separation and even granulation of the products of synthesis [1].The use of this phenomenon for the production of powders with a given dispersion determines the relevance of the study of the laws of the microheterogeneous combustion mode.

The historical development of methods for analyzing experimental data on the SHS process is associated with the study of the macrokinetics of this phenomenon [2]. Studies of the discreteness of the propagation of a combustion wave made it possible to explain the appearance of layered structures in the products of synthesis [1–3]. However, to study the thermodynamics of microheterogeneous processes, registration of high-resolution SHS is required when the inhomogeneity of the combustion wave front becomes comparable to the characteristic size of the field of observation [4]. In this case, there is an uncertainty in the choice of the direction of the measuring cross-section in the images of thermal imaging, which requires the development of new approaches to the analysis of the recorded data. The authors of the article previously proposed an original method of chrono-topographical representation of the results of thermal imaging studies for joint analysis of microheterogeneous processes and macrokinetics of the phenomenon [5].

The purpose of this work was to detect patterns in the thermodynamics of reaction foci in a SHS wave using chrono-topographic analysis (CTA) of high-resolution thermal imaging data.

2. Principles of chrono-topographic analysis

Modern models of the stationary combustion mode link two factors in the equation of state — the depth of chemical transformation of the reaction cells and their temperature [6]. Consequently, a given temperature level can serve as a criterion for determining at the stage of ignition cells with the same depth of response. Moreover, the corresponding elements of the thermal image of the reaction cells, the signal level of which for the first time reached the set value, should be the wave front of the SHS with the appropriate depth of chemical transformation. By determining the coordinates of the front elements in all frames of a thermal imaging survey, you can construct a chronographic map that reflects the space-time distribution of the reaction cells with the same chemical conversion depth in the combustion wave front.

The chronographic map in figure 1 is based on high-resolution thermal imaging: 5.8 μm and 1 ms (figure 2). Her analysis reveals the following properties.

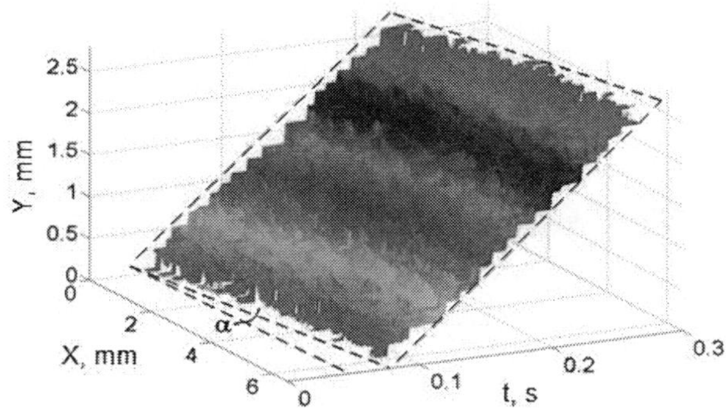

Figure 1. Integral chronographic map of the SHS process.

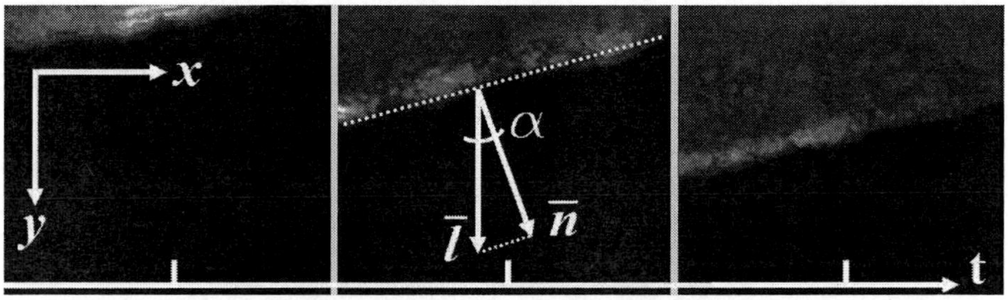

Figure 2. The coordinate system of thermal imaging shooting SHS process.

The experimental data on the position of the SHS wave front $y = f(x,t)$ is very well modeled by the plane. In relation to the CBC process, this means that the estimates of the mathematical expectation of the front speed are equal both at different points in time and at different points in space. From this fact, it follows that the SHS process in the wave mode can be considered not only stationary but also ergodic. Thus, when determining the statistical characteristics of the SHS process, data from different cross-sections of the chronographic map can be combined into a common sample. For modern high-speed cameras, this allows a 3–4 order increase in the sample size and a 1–2 order reduction in the confidence interval of the estimated parameters of the macrokinetics of the process.

Taking into account the ergodicity of the process, it can be argued that the angle between the direction of the columns of the photomatrix $\vec{l}$ and the normal to the front of the combustion wave $\vec{n}$ (figure 2) is equal to the angle α shown in figure 1. If the camera is oriented incorrectly during the

IOP Conf. Series: Materials Science and Engineering **558** (2019) 012009 doi:10.1088/1757-899X/558/1/012009

registration of the SHS process, the burning front will appear in one of the corners of the thermal image earlier than in the other. With a constant front propagation velocity, the value of the time delay will depend linearly on the distance between cross-sections along the axis *X*, and the direction of the normal to the front of the combustion wave can be determined from the average time delay of the occurrence of foci in different cross-sections of thermal imaging images, even in the case of large magnifications of the optical system. Chronographic maps of the SHS process allow us to determine the direction of the normal to the front of the combustion wave, combining two methods: analysis of the spatial data of thermal images; control of the delay time of appearance of the reaction front in different normal cross-sections.

The differential form of the chronographic map (figure 3) makes it possible to identify individual foci in the reaction wave, to study their thermodynamics and interaction. A joint analysis of thermal imaging data and a differential chronographic map allows us to construct a topographic map of the SHS process, which displays the areas where individual combustion centers developed during the whole process of registration (figure 4). The topographic map makes it possible to assess how the combustion depression during the SHS process contributes to the structure of the final product, and to determine the characteristic dimensions of the sample areas with a high depth of response of the initial components. The method of measuring the statistical parameters of the SHS process using chronographic and topographic maps is presented in [7].

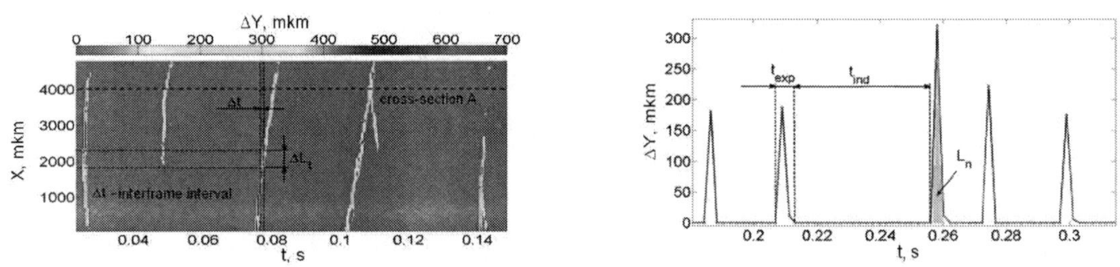

(a) (b)

Figure 3. Differential form (a) and cross section A (b) of the chronographic map of the SHS process (V_t is the velocity of the source propagation along the SHS wave front; t_{exp} is the lifetime of a separate combustion center; t_{ind} is the induction time in the measuring section of a new combustion center; and L_n is the characteristic size of the combustion foci in the direction normal to the reaction front).

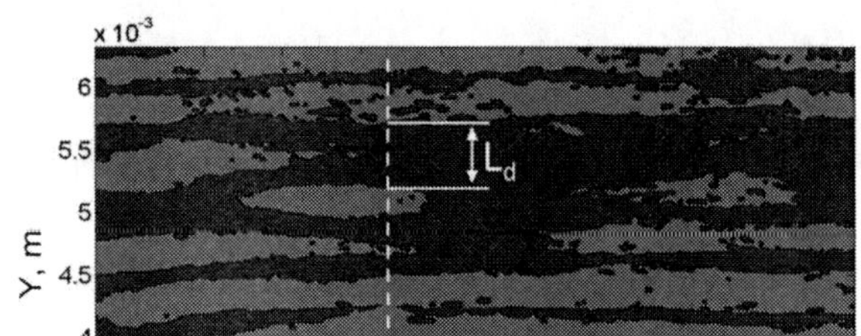

Figure 4. Topographic map of the SHS process (L_d is size of zone of the combustion depression between the foci).

Experimental estimation of the error of measuring the velocity of propagation of a SHS wave by the CTA method using high-resolution thermal imaging data with a sensor size of 1 MPixels was 0.05%. With the use of sensors of higher dimensionality or the approach of tracking thermal imaging control, an error level of $5 \cdot 10^{-4}$% can be achieved.

3. Thermal imaging studies

3.1. Development of foci of combustion in the SHS wave

In the present work, an experimental study of the evolution of individual foci in a SHS wave was carried out on the basis of the Ni–Al system. To control the thermal effect of the reaction in the charge, the mass fraction of Ni was varied from Ni_3Al to NiAl stoichiometry with a step of 0.5%. The density of the mixture at room temperature was chosen so that combustion with the formation of Ni_3Al was critical. The high-temperature synthesis was controlled by the original thermal imaging system with a spatial and temporal resolution of 5.8 μm and 1 ms, respectively [7].

The temperature images of figure 5 show the development of a burning center along the SHS wave front — the X-axis, while the front of the Ni_3Al synthesis reaction itself moved in the Y-axis direction. The size of the observation area in X was 7.3 mm; in Y, 1.87 mm.

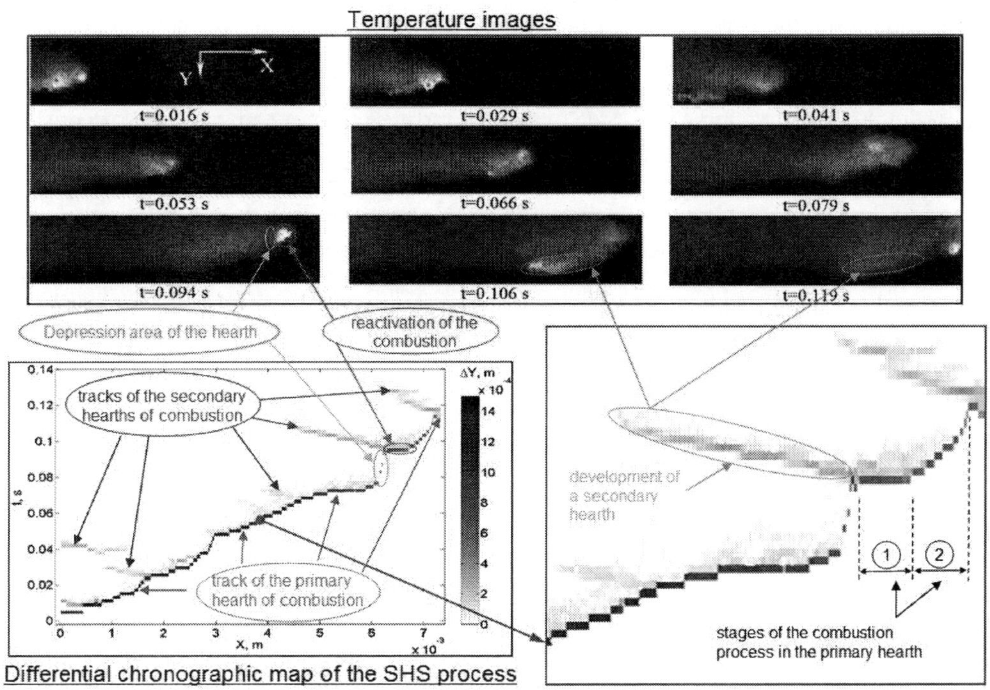

Figure 5. Development of combustion foci along the Ni_3Al SHS wave front.

The differential chronographic map in figure 5 clearly shows that the development of foci along the wave front is similar to the propagation of the reaction front itself - there is a stage of rapid displacement of the boundary of the center in the direction of the X axis with a sharp rise in temperature (flash stage), and a depression stage of the reaction with decreasing temperature. As an example, in figure 5 one of the areas of rapid displacement of the boundary of the source along the SHS wave front and one of the areas of depression of the combustion reaction in the focy are underlined.

A chronographic map of the SHS process also shows that the transition from the depression stage of the reaction to the flash stage is accompanied by the appearance of a secondary focus, which spreads along the wave front of the SHS, but in the opposite direction of the primary focy. An example of the emergence and development of a secondary focus, reflected on the chronographic map, is associated with micro thermal imaging shots that were recorded at time points: 0.094, 0.106 and 0.119 s (figure 5).

In general, based on the analysis of the differential chronographic map of the SHS process, several conclusions can be drawn. First, the velocity of propagation of the boundary of the foci along the wave front of the SHS is finite and can be measured by high-resolution thermal imaging with a small thermal effect of the reaction, which is close to critical. With an increase in the thermal effect, the velocity of propagation of foci along the front increases, but it becomes very difficult to identify the movement of the boundary of the focy. Increasing the temporal resolution of the thermal imaging camera leads to a decrease in the level of the detected signal, up to its loss against the background noise. With a small time resolution, the boundary shifts so quickly that it is not possible to register the phases of this movement in a combustion focy. Reducing the spatial resolution of a thermal imaging system contributes to an increase in the signal level, but leads to a "loss" of the source due to the discreteness of the matrix photo sensor. Secondly, there are at least two types of foci with different propagation dynamics in the front of the SHS wave. In the development of primary foci, two stages can be noted: coherent combustion (flash region) and inhibition of reaction propagation (figure 5). At the stage of inhibition, the dependence of the coordinate of the boundary of the primary burning focy of X_{bp} on time is similar to a logarithmic law. The coordinate of the boundary of the secondary foci of X_{bs}, up to the time of the termination of combustion, linearly depends on time. Thirdly, the flashes observed in the thermal imaging of the SHS wave are most likely not separate foci of combustion, but represent only the development stage of the primary focus, which, like the SHS wave front, also has a discrete propagation pattern. Actually, the developing primary foci of combustion form the wave front of the reaction. Fourth, secondary foci, spreading in the opposite direction in relation to the primary, burn out areas that become an obstacle to the development of the primary focus. This barrier can limit the area of distribution of the primary focus and lead to the fragmentation of the combustion wave front into separate sectors. This hypothesis is confirmed by a topographic map of the SHS process (figure 4). Thus, each product layer can be formed by several primary foci simultaneously.

With a small thermal effect of the reaction in the Ni–Al system, almost all secondary foci go out. With an increase in the magnitude of the thermal effect, there were cases when the secondary focus turned into the primary, i.e. the direction of propagation of the primary and secondary foci was reversed.

3.2. Dynamics of combustion in the reaction cells

The CTA method has been used to study the effect of combustion sources located in the SHS wave front on the thermodynamics of the reaction cells entering the previously formed combustion centers. For this purpose, in the analyzed section X_0, the dynamics of the maximum temperature of individual foci identified in the thermal imaging frames (figure 6) were monitored. Typical results of these studies are presented in figure 7.

It has been established that in modes close to unstable combustion, the thermograms of the reaction cells have several characteristic peaks. Moreover, the occurrence of each peak is very well synchronized in time with the moments of the appearance of new foci of combustion in the X_0 section. Based on these data, it can be assumed that the growth of the product layer between the fuel and oxidizer particles does not occur continuously as a result of reaction diffusion, but is discrete, i.e. the process of diffusion and mixing of the initial components is separated from the combustion process, which is experiencing stages of depression and flash.

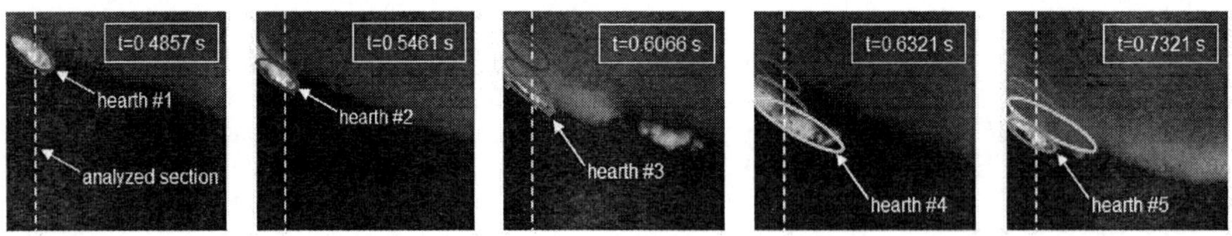

Figure 6. The principle of analysis of the thermodynamics of the reaction cells in temperature images.

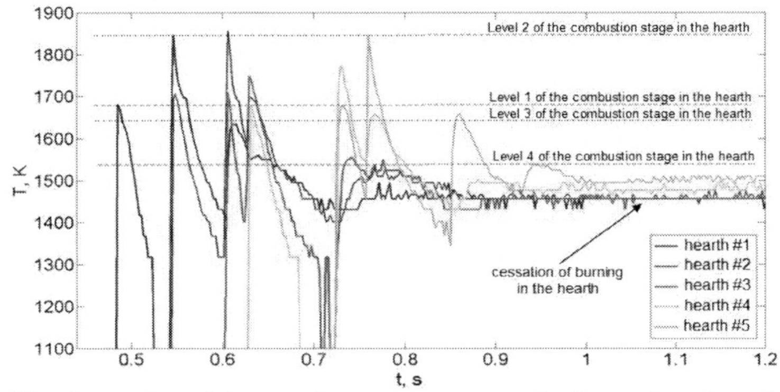

Figure 7. The dynamics of the maximum temperature in the analyzed section of foci.

After the occurrence of a layer of products burning in the cell stops. In accordance with the theory of reaction diffusion, combustion must resume after an interval over which the fuel (and/or oxidizer) overcomes the layer of products. However, the experiment shows that combustion in the cell resumes only with the formation of a new focy in the X_0 section. A similar scenario is likely when the processes in the reaction cell are strongly influenced by both internal and external heat sink. Then the diffusion mixing of the fuel and oxidant in the reaction cell does not lead to burning. The formation of a new source of combustion in the wave front, at a minimum, blocks, from the side of the initial mixture, the external heat sink of the cell entering the region of the previously formed source, and creates in it conditions for the resumption of the combustion reaction, which can occur in the thermal explosion mode.

In the considered case, by changing the perimeter of the sample, it is possible to control the time of diffusion mixing of the fuel and oxidant in the reaction cell. The presence of stages of extinction and explosion in it should lead to product stratification. Therefore, by varying the mixing time, the dispersion of the product can be controlled.

4. Conclusions

The discrete nature of the SHS process is not only manifested when the wave front propagates. In the microheterogeneous mode of combustion, it is detected both with the development of foci forming the front of the synthesis wave, and on the scale of the reaction cells. Based on the knowledge of product stratification in the high-temperature synthesis wave and the results of these experimental studies, it can be concluded that an increase in the heat deficit in microheterogeneous combustion regimes leads to an increase in the dispersion of the resulting product.

Acknowledgements

The work was supported by the Russian Foundation for Basic Research in research projects no. 18-08-01475, 18-08-01152, as well as grant no. 13-01-20 / 28 "Leading Scientific School of Ugra State University".

References

[1] Amosov A P, Borovinskaya I P and Merzhanov A G 2007 *Powder technology of self-propagating high-temperature synthesis of materials* (Moscow, Mashinostroenie-1) 567 p
[2] Merzhanov A G and Mukasyan A S 2007 *Solid flame burning* (Moscow, Torus Press) 336 p
[3] Mukasyan A S and Rogachev A S 2008 *Prog. Energy Comb. Sci.* **34** 377–416
[4] Mukasyan A S, Rogachev A S, Mercedes M and Varma A 2004 *Chem. Eng. Sci.* **59** 5099–5105
[5] Dolmatov A V , Gulyaev P Yu and Milyukova I V 2018 *J. Phys.: Conf. Series* **1115** 042024
[6] Rogachev A S and Mukasyan A S 2013 *Combustion for the synthesis of materials: an introduction to structural macrokinetics* (Moscow, FIZMATLIT) 400 p
[7] Dolmatov A V and Berestok G M 2017 *Bulletin of Ugra State University* **46** 64–73

IOP Conf. Series: Materials Science and Engineering **558** (2019) 012010 doi:10.1088/1757-899X/558/1/012010

Tensile strength of materials obtained by electric pulse consolidation of powders

V Y Goltsev[1,2], E G Grigor'ev[2], N A Gribov[1], A V Osintsev[1] and A S Plotnikov[1,2]

[1]National Research Nuclear University MEPhI, Moscow, 115409 Russia
[2]Merzhanov Institute of Structural Macrokinetics and Materials Science, Russian Academy of Sciences, Chernogolovka, 142432 Russia

E-mail: gvy587@gmail.com

Abstract. The article investigates the possibility of determining the resistance to fracture of materials obtained from powders by electric pulse consolidation method using the test method for bending thin disks on an annular support and testing short cylinders according to the "Brazilian test". Techniques are verified by testing small-sized thin disks and short cylinders made of gray cast iron and graphite. It is revealed a difference in the nature of the destruction of cast iron samples (plastic failure) and samples of graphite (brittle fracture). The good correspondence of the characters of destruction of graphite samples and samples produced from powders by the electric pulse consolidation is shown. The influence of various additives and manufacturing technology of materials on the resistance of the material to brittle fracture is revealed. The optimal modes of high voltage consolidation a heavy alloy based on the results of tests of short cylinders according to the "Brazilian test" scheme were obtained. The effect of the thickness of the cylinders on the strength of aluminum oxide Al_2O_3 is investigated.

1. Introduction

Development of production technology of modern materials by consolidation of powders, welding and sintering regimes [1, 2] is carried out, as a rule, on small samples, the size of which often does not exceed 10–15 mm in diameter, and the thickness can vary from 1 to 10 mm. Standard mechanical test methods are often not applicable to such samples. Alternative test methods have to be used to evaluate the mechanical properties of the material. Examples of such methods can be the bending of thin discs on the annular support and the compression of the disc in the plane of symmetry according to the "Brazilian test" scheme. These methods are commonly used to test brittle materials.

The aim of this work is to substantiate the possibility of their application to assess the tensile strength of materials obtained by methods of electro-pulse consolidation of powders, which are characterized by a low-plastic state.

2. Method of testing of thin disks for bending on the ring support

In these studies, the method of loading the disk on the ring support, presented in [3], was used (figure 1).

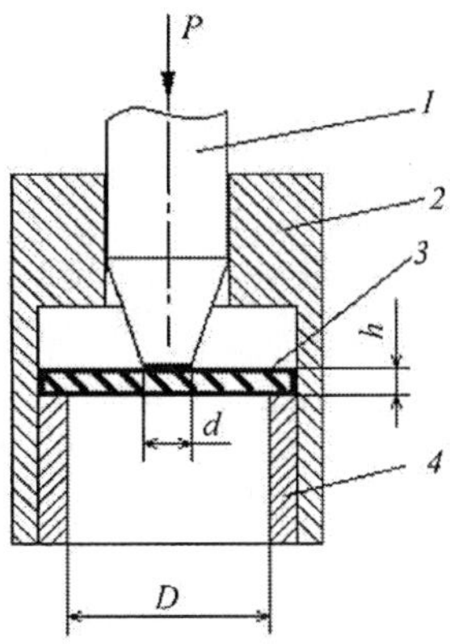

Figure 1. The scheme of testing the disk on the annular support for bending with loading flat indenter: *1* indenter, *2* clip, *3* sample, *4* support ring.

The breaking stress and the resistance of the material to rupture are calculated taking into account the maximum load P_{max} withstood by the sample to failure, with the linear diagram "load P–deflection ω" according to the formula:

$$\sigma_B = \frac{3P_{max}}{8\pi h^2}\left[4-(1-\mu)\left(\frac{d}{D}\right)^2 + 4(1+\mu)\ln\frac{D}{d}\right] \qquad (1)$$

In equation (1) h is the thickness of the disk; d and D are the diameters of the punch and support, respectively; μ is the Poisson's ratio.

Analysis of the stress-strain state of the model, performed using the verified [4, 5] calculation complex ANSYS Mechanical version 16.2, showed the presence of a three-axis stress state in the contact interaction of the indenter and the disk. On the reverse side of the disk, as expected, there is a flat stress state with maximum circumferential tensile stresses, which cause brittle fracture of the sample. Figure 2 shows an almost homogeneous region of maximum tensile stresses under the indenter.

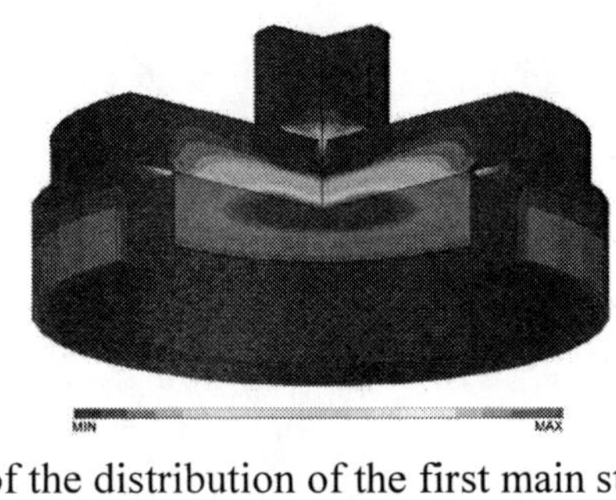

Figure 2. Qualitative picture of the distribution of the first main stresses in the disk sample under the indenter during bending.

The origin of the fracture is possible under the action of maximum shear stresses in the contact loading zone. Verification of methods for small samples was carried out by testing the bending of discs made of gray cast iron and graphite. Mechanical properties of materials are given in table 1.

The typical diagrams of static bending of cast iron and graphite samples MPG-6 on the Instron test machine, which indicate the absence of brittle fracture of cast iron samples and brittle fracture of graphite, and the nature of their destruction were presented in [6]. The results of the test of cast iron disks with a diameter of 10–15 mm and a thickness of 1–1.5 mm showed that the level of stresses calculated by equation (1) is in good agreement with the tensile strength of cast iron [6].

Table 1. The values of ultimate strength in tension/compression and Young's modulus of the materials studied.

Material	σ_u^t, MPa	σ_u^c, MPa	*E*, GPa
Cast iron SCH 10-40	104	404	100
Graphite MPG-6	25	98	10
Graphite ARV-1	13.8	37	11

A completely different picture is observed in the destruction of disks of graphite MPG-6 in the same range of thicknesses. With almost linear machine diagram of static bending up to destruction, there is a brittle dynamic destruction of graphite into numerous parts. The level of destructive stresses calculated by equation (1) is about 20% lower than the tensile strength of the material.

3. Testing of thin disks obtained by spark-plasma sintering

The above method of bending thin disks on the ring support was used to assess the brittle strength of a number of materials obtained by spark-plasma sintering (SPS) of powder materials.

A comparative study of the compacts obtained by the SPS method from nano and ultrafine (UF) aluminum oxide powders with a spherical particle shape [7] showed a higher density, strength, microhardness and homogeneity of the structure of the compacts from the UF powder than the compacts from the nanopowder. In particular, the fracture resistance of UF-powder disks was higher than that of nanopowder, 174 and 141 MPa, respectively. The use of pre-pulsed magnetic pressing of UF-powder increased the strength of the samples to 202 MPa.

For the SiAlON material obtained by the SPS method from powder synthesized in the process of self-propagating high-temperature synthesis (SHS), the influence of various additives and the thickness of the disks on the bending strength were investigated. Discs with a diameter of 10 mm (D = 7.5 mm, d = 3.75 mm) and 15 mm (D = 11.5 mm, d = 3.75 mm) were tested. Figure 3 shows a characteristic diagram of the bending disk of SiAlON. It is almost identical to the graphite bending diagram.

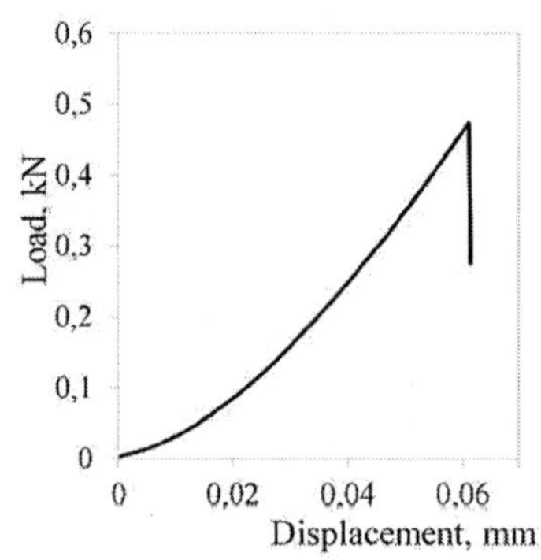

Figure 3. Engine diagram of a bending disk of SiAlON with a diameter of 15 mm.

The destruction of all samples was brittle with the separation of many small fragments. The effect of the sample thickness and various additives on the SiAlON strength is presented in the work [8].

4. Calculation analysis of short cylinder compression according to the "Brazilian test" scheme

The test discs according to the "Brazilian test" scheme [9] is used to determine the tensile strength of rocks when testing samples with a diameter of at least 50 mm with a thickness-to-diameter ratio in the range of 0.2÷0.75. The purpose of this work in relation to this technique was to justify the possibility of its application to assess the tensile strength of materials obtained by the methods of electro-pulsed consolidation of powders.

IOP Conf. Series: Materials Science and Engineering **558** (2019) 012010 doi:10.1088/1757-899X/558/1/012010

Simulation of the disk sample loading process was performed in the verified [4, 5] calculation complex ANSYS Mechanical version 16.2. The full calculation model is shown in figure 4.

ASTM D3967-95a standard offers to determine the strength of the tested material by the formula:

$$\sigma_t = \frac{2P}{\pi t D}. \qquad (2)$$

In equation (2) P is the maximum load sustained by the specimen, t is the thickness of the specimen, and D is its diameter.

The results of the analysis with the study of the influence of the ratio of the cylinder thickness to its diameter on the maximum tensile stresses are presented in [10]. It has been shown that the σ_t parameter determined by the ASTM D3967-95a method is a good estimate of the first principal stress averaged along the x-axis of the disk. Changing the ratio of the disc thickness to its diameter in the range of 0.3–1 has little effect on the maximum tensile stress. It has also been found that the maximum principal stresses at individual points can exceed the average stresses σ_t by 2.5 times, which should affect the applicability of equation (2) for concentration sensitive brittle materials.

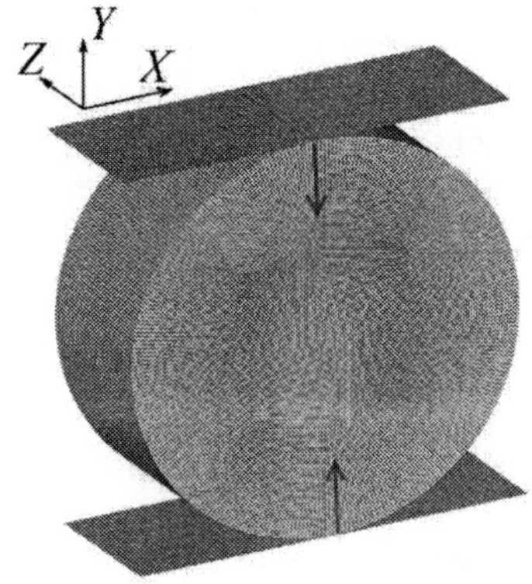

Figure 4. Full geometry of the computational model.

5. Results of experimental study of short cylinders made of model materials

Short cylinders made of gray cast iron and ARV-1 graphite, the characteristics of which are given in table 1, were subjected to experimental study according to the scheme of the "Brazilian test".

5.1. Test cylinders of gray cast iron

The results of testing of small-size discs made of cast iron given in [9] confirmed the possibility of using the calculated equation (2) to determine the tensile strength of the material. Discs made of grey cast iron were about the size of $D \times t$: 10 × 4 and 15 × 4 mm. The diagram of compression of the cast iron disc is presented in figure 5a.

Initial cracks in the cast iron sample are formed in the contact area where the maximum shear stresses are applied. The beginning of crack growth is associated with the maximum load on the sample. Smooth load reduction after the maximum-slow crack propagation is the departure of one half of the sample from the other. No explosive fracture characteristic of a brittle rupture from normal stresses was observed.

5.2. Test cylinders of graphite

Graphite disks were of three sizes $D \times t$: 8 × 4, 8 × 8 and 8 × 12 mm. The characteristic compression diagram of a graphite disk 8 × 8 mm in the plane of symmetry is presented in figure 5b.

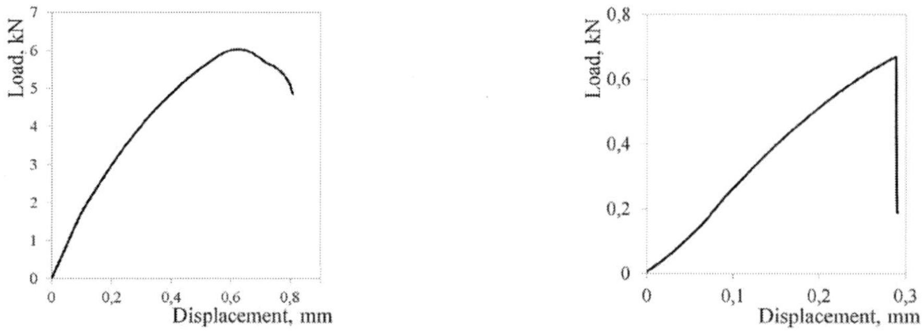

(a) (b)

Figure 5. Compression diagrams of discs of (a) cast iron with a size of 10 × 4 mm and (b) graphite ARV-1 with a size of 8 × 8 mm according to the scheme of "Brazilian test".

The destruction of the graphite disk occurs dynamically on a virtually linear portion of the diagram at maximum load with the separation of the sample sometimes into several fragments. Initial cracks in the contact area were also found in the graphite sample with a size of 8 × 8 mm, which served as a source of destruction of the sample.

Sample tests showed a significant difference in the behavior of these two materials. Graphite turned out to be a more brittle material, the destruction of the samples of which was explosive in the linear deformation diagram. The ratio of strength limits of materials and destructive stresses determined by equation (2) are given in table 2.

Table 2. Ratio of breaking stresses and tensile strengths of cast iron and graphite.

Material	Plastic fracture	Brittle fracture
Cast iron	$\sigma_t \approx \sigma_u^t$,	-
Graphite ARV-1	-	$\sigma_t \approx 0.7\sigma_u^t$,

The results given in table 2 indicate that for graphite samples, the fracture resistance estimated from equation (2), 1.5 times less than the true tensile strength of the material.

6. Testing of samples of materials obtained by methods of electro-pulse consolidation of powders

6.1. Testing of samples of alloy WNiFe obtained by high-voltage consolidation

To obtain samples of heavy alloy WNiFe by high-voltage consolidation used industrial powder composition 90W–7Ni–3Fe, obtained by mechanical mixing of the component. Data on the composition and density of the powder are given in table 3.

Table 3. Composition of industrial WNiFe powder.

Element	ρ, g/cm^3	Weight, %	ρ_{theor}, g/cm^3
W	19.25	residue	
Ni	8.902	6.93	17.13
Fe	7.874	3.12	

Various modes of consolidation in the manufacture of samples of alloy WNiFe were investigated. The supplied voltage varied from 4.5 to 5.8 kV and the pressure from 100 to 250 kPa. Testing of samples allowed us to determine the levels of pressure and stress pulses during consolidation, providing optimal strength characteristics of the alloy WNiFe. Figure 6 shows the dependences of the material resistance to rupture on the voltage pulse level for samples sintered at the same pressures of 200 and 250 kPa. In all cases, the material fracture resistance was calculated using equation (2).

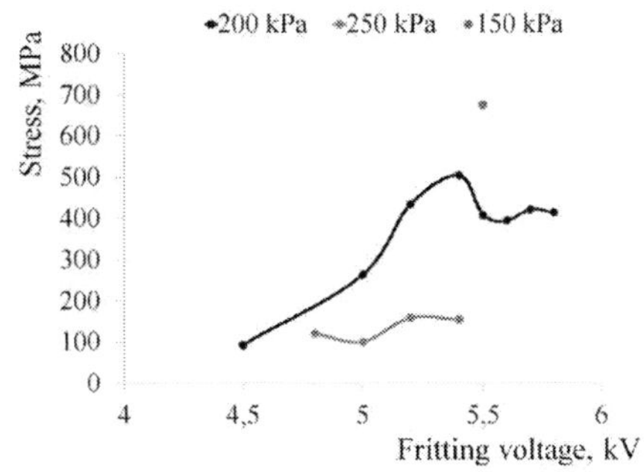

Figure 6. The dependence of the fracture resistance of the applied voltage at sintering.

At a pressure of 200 kPa, there is an increase in the strength of the samples with an increase in the voltage pulse to a value of 5.4 kV, then a decline. The growth and decrease in strength occurred against the background of the development of plastic defo rmations that preceded the brittle destruction of the samples. At a pressure of 250 kPa, we obtained an absolutely brittle fracture without noticeable traces of plastic deformation.

6.2. Testing of samples of aluminum oxide obtained by SPS

The results of the "Brazilian test" of cylindrical samples with a diameter $D = 10$ mm made of aluminum oxide powder by the method of electro pulse consolidation and calculated by the formula (2) are presented in figure7.

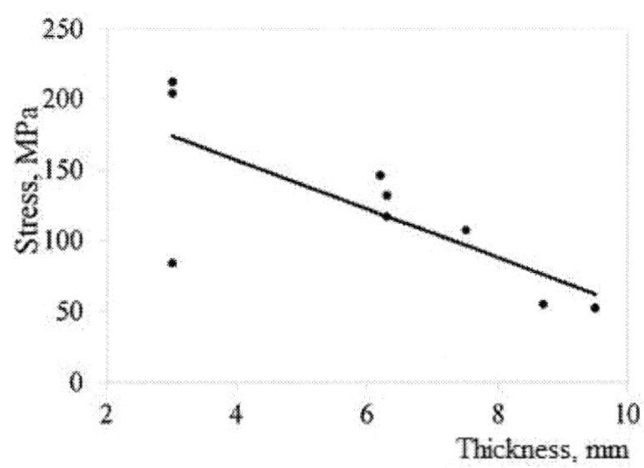

Figure 7. The test results of samples of aluminum oxide.

We note a good correspondence between the results obtained by the test of short cylinders according to the "Brazilian test" scheme and the results of the test of thin aluminum oxide disks with a diameter of 15 mm and a thickness of 1.5 mm for bending on the annular support mentioned above.

7. Conclusions

Thus, the method, which combines calculations using FEM and experimental test results accor ding to the following schemes: "bending on the ring support" and "Brazilian test", makes it possible to determine the fracture resistance of materials obtained by consolidation from powders usingthe method of electric pulse action. Small samples in the form of thin discs are tested for bending on the ring support, and short cylinders (thick discs) - according to the scheme of the “Brazilian test”. The influence of a number of technological factors on the strength of consolidated materials is studied using the proposed methods.

Acknowledgement

The study was supported by a grant from the Russian science Foundation (project no. 16-19-10213).

References

[1] Grigoryev E G 2011 *Int. J. Mach. Tech. Mater.* **8** 8–10

[2] Bashlykov S S, Demenyuk D V, Grigoryev E G, Olevsky E A and Yurlova M S 2013 *Fizika i himiya obrabotki materialov [Physics and chemistry of materials processing]* **5** 77–83

[3] Tumanov A T 1974 *Methods of testing control and research of engineering materials. Volume II. Methods of investigation of mechanical properties of metals* (Moscow, Mashinostroenie) 320 p

[4] Release 16.2 Documentation for ANSYS [electronic document] / ANSYS Inc. Electronic data and software (104019 files: 10660130531 bytes)

[5] NAFEMS search engineering analysis and simulation FEA, Finite Element Analysis, CFD, Computational Fluid Dynamics and Simulation. NAFEMS Ltd., Hamilton, United Kingdom

[6] Goltsev V Y and Gribov N A 2017 *15th School-Conf. "New materials of innovative energy: development, characterization methods and application*; *KnE Life Science* 148–154 DOI: 1018502/kms.v4h.2139

[7] Zholnin A G, Kovaleva I V, Pahilo–Darial I O, Litvinova I C, Grigoryev E G, Olevsky E A, Goltsev V Yu and Rytenko V R 2016 *Fizikai khimiya obrabotki materialov [Physics and chemistry of materials processing]* **2** 5–10

[8] Goltsev V Y and Gribov N A 2018 *IOP: Conference Series (Current issue)*

[9] ASTM D3967-95a. Standard Test Method for Splitting Tensile Strength of Intact Rock Core Specimens

[10] Osintsev A V, Goltsev V Y and Plotnikov A S 2017 *Pis'ma o materialah [Letters on materials]* **7** 21–25 DOI: 10.22226/2410-3535-2017-1-21-25

IOP Conf. Series: Materials Science and Engineering **558** (2019) 012011 doi:10.1088/1757-899X/558/1/012011

Determination of tensile strength of brittle materials by bending thin discs on the annular support

V Y Goltsev[1,2] and N A Gribov[1]

[1]National Research Nuclear University MEPhI, Moscow, 115409 Russia
[2]Merzhanov Institute of Structural Macrokinetics and Materials Science, Russian Academy of Sciences, Chernogolovka, 142432 Russia

E-mail: griboet@ya.ru

Abstract. It is proposed to use the test of bending small thin disk specimens simply supported along the contour, to determine the resistance of material to tensile strain. The results of computer analysis of stress-strain state and test a thin disk specimens made of brittle materials are cast iron and graphite as a possible model, and directly samples made by electric pulse methods are presented. It is shown the effect of size of specimens on the resistance to their destruction and different character of deformation and destruction of cast iron and graphite samples. The possibility of application of thin disc samples for the determination the resistance to tensile strain of the composite ceramics based on SiAlON with various additives Y_2O_3, SiC, and TiN is confirmed.

1. Introduction.

Powder metallurgy plays the important role in the development of technologies of creation of materials with desired properties. Along with traditional methods of sintering, methods of electric pulse impact (EPI) are actively developing in the production of materials. The method of obtains materials with desired properties by EPI is based on researching small samples with thickness from 1 mm and a diameter from 10 to 15 mm. Standard methods for determination of materials resistance to rupture under tension to samples of such small size is not applicable. We used the method of bending a thin disk on an annular support to determine the brittle fracture of consolidated materials [1, 2].

2. Computational analysis of bending small thin discs on the ring support

The bending scheme of the thin disk on the annular support is shown in figure 1.

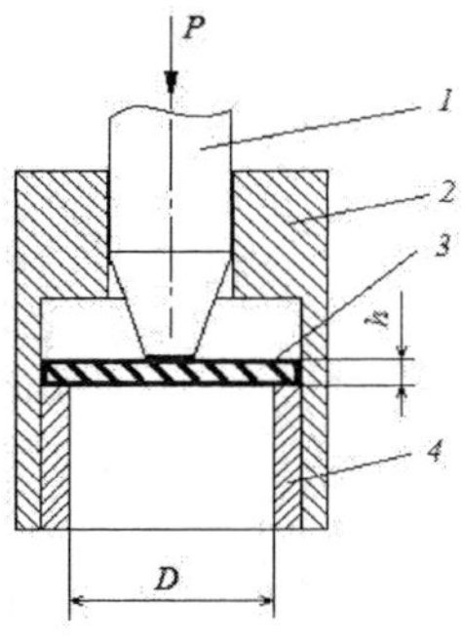

Figure 1. Diagram of bend test drive, simply supported along the contour: *1* punch, *2* holder, *3* sample, *4* foot ring.

IOP Conf. Series: Materials Science and Engineering **558** (2019) 012011 doi:10.1088/1757-899X/558/1/012011

The calculation of the breaking stress and the resistance of the material to rupture is carried out taking into account the maximum load P_{max} withstood by the sample to failure, with a linear load diagram P–deflection ω according to the formula [1]:

$$\sigma_B = \frac{3P_{\max}}{8\pi h^2}\left[4-(1-\mu)\left(\frac{d}{D}\right)^2+4(1+\mu)\ln\frac{D}{d}\right] \quad (1)$$

in which h is the thickness of the disk; d and D are the diameters of the punch and support, respectively; μ is the Poisson's ratio.

Modeling of the process of loading a disk sample was produced in the verified computational complex ANSYS Mechanical [3]. The results of this study are presented in the paper [4]. It was shown that three-axis compression occurs in the contact zone of the indenter and the disk, while on the opposite side there is a plane stress with maximum tensile stresses, which cause brittle fracture of the sample. The origin of the destruction is possible under the action of the maximum tangential stresses in the contact zone. Figure 2 shows the stress distribution through the thickness of the sample. The beginning of the abscissa axis corresponds to the contact zone of the punch and disc.

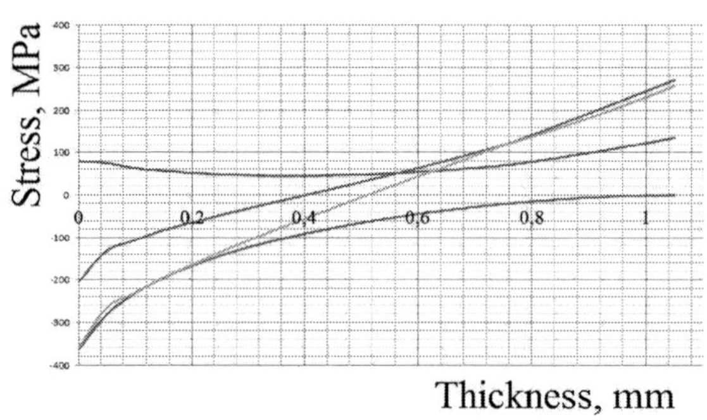

Figure 2. The stress distribution through the thickness of the sample
● – σ_1, ● – σ_2, ● – σ_3, ● – τ_{max}

In the contact area, the maximum shear stresses prevail, and on the reverse side of the disk, the main tensile stresses are maximal. The main compressive stresses on the reverse side of the disk tend to zero, which confirms the presence of a biaxial stress state.

3. Testing of model materials for circuit bending on the ring support

Cast iron SCH 10–40 and graphite MPG-6 were selected as model materials, whose physical and mechanical properties are well studied,. The mechanical properties of the tested materials are given in table 1.

Table 1. Mechanical and physical properties of tested materials

Material	σ_u^t, MPa	σ_u^c, MPa	E, GPa
Cast iron SCH 10-40	104	404	100
Graphite MPG-6	25	98	10

3.1. Tests of cast iron samples

In developing methods of testing, disks from cast iron had diameters of 10 and 15 mm, the thickness was changed within 1–2 mm. Comparison of numerical analysis of the stress-strain state and experimental results allows us to conclude that the cast iron in this scheme of loading is destroyed viscous under the action of tensile stresses. At first, the maximum shear stresses lead to the destruction of cut, trying to push the "plug" but when the first cracks appear they are disclosed under the action of tensile stresses. These statements were confirmed by the emergence of pushed through "crater" in the contact zone and the embossed shape of the stamp from the back side of the disc (figure 3a).

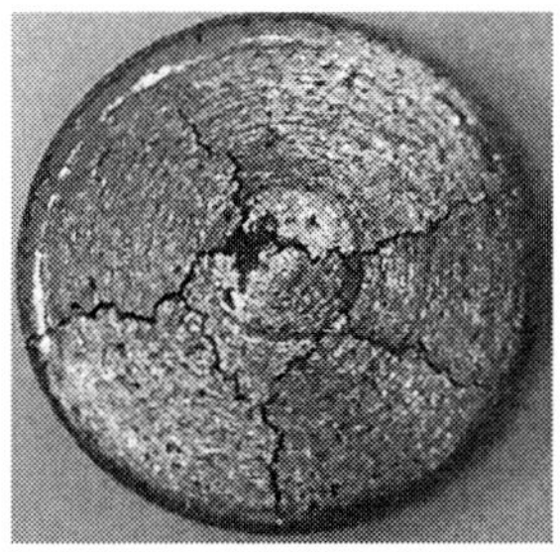

(a)

(b)

Figure 3. Character of destruction of (a) cast iron and (b) graphite discs 15 mm in diameter.

The presence on the diagrams (figure 4a) of cast iron sections a gradual decrease of load after the maximum indicates the gradual divergence of the destroyed parts of the sample. Crack propagation in the radial direction suggests that they are disclosed under the action of tensile stresses. There is slow crack propagation.

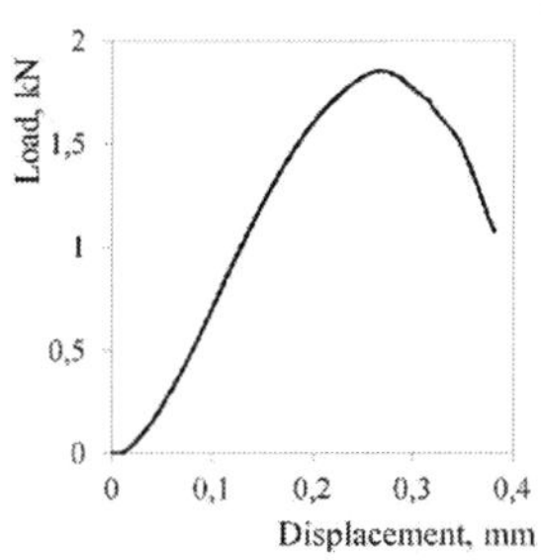

(a)

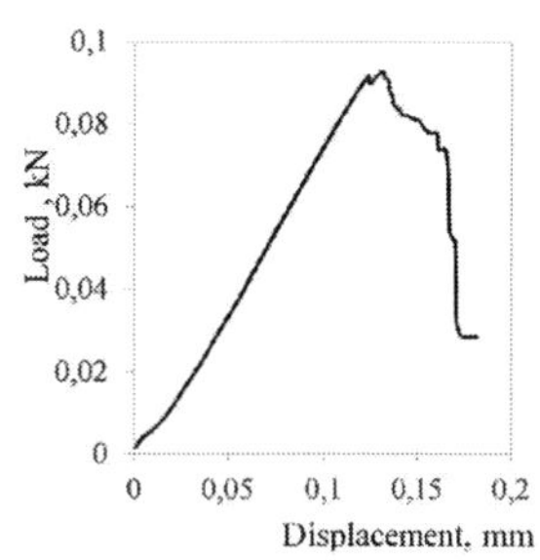

(b)

Figure 4. Machine deformation diagram of bending of (a) cast iron and (b) graphite disks.

The results of the test of cast iron disks with a diameter of 10–15 mm and a thickness of 1–1.5 mm showed that the level of stresses calculated by equation (1) is in good agreement with the tensile strength of cast iron [4].

3.2. Test graphite samples

Discs of graphite were two sizes Ø × *t*: 15 × 1.8 mm and 15 × 1.7 mm. Machine characteristic deformation diagram of a graphite disk is shown in figure 4b. We can see that the diagrams substantially differ from the diagrams of samples from cast iron. The destruction of the graphite disc is at maximum load. Stages in the diagram indicate stepwise crack growth. We observe the explosive nature of the destruction and the final dynamic crack propagation in the sample (figure 3b). The destructive stresses of graphite samples were 20% less than the tensile strength of the material.

4. Tests of thin disks obtained by spark-plasma sintering

he method of bending samples of small size is used in the study of the tear resistance of the materials SiAlON obtained by spark-plasma sintering.

Thin disks of SiAlON destroyed completely fragile with linear load-displacement diagram (figure 5) before break, separating into many pieces.

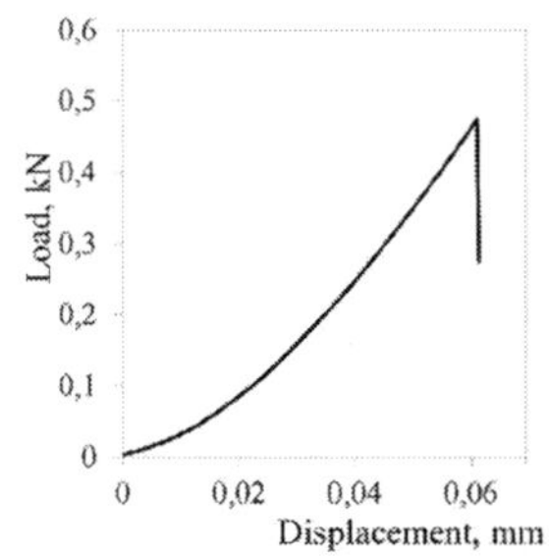

Figure 5. Engine diagram of a bending disk of SiAlON with a diameter of 15 mm.

The destruction of all samples was brittle with the separation of many small fragments. For samples with the same composition and the same sintering regime, the analysis of the dependence of the breaking stress on the thickness of the sample was carried out (figure 6).

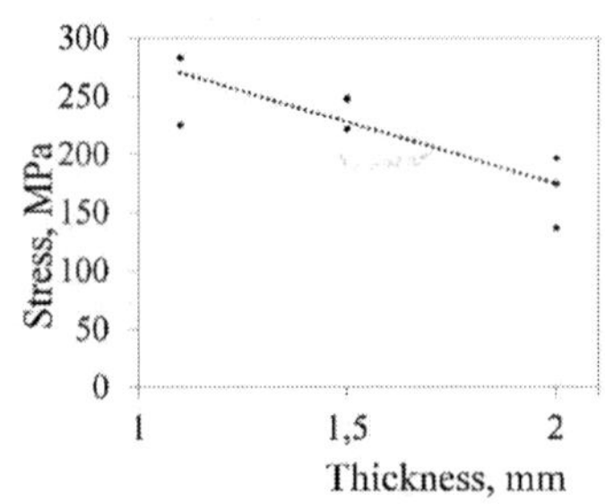

Figure 6. The dependence of the breaking stress on the thickness of the samples β-SiAlON + BN 30% +Y_2O_3 8%.

There was a slight decrease in the fracture resistance characteristics with a change in the thickness of the samples in the range of 1–2 mm.

A weak dependence of strength on the percentage of yttrium oxide Y_2O_3 in β-SiAlON in the studied range was observed (figure 7).

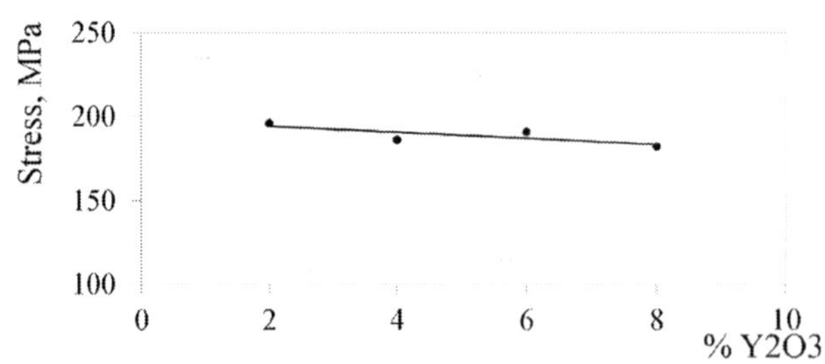

Figure 7. The dependence of the breaking stress on the percentage of yttrium oxide Y_2O_3 in β- SiAlON.

The dependence of the strength on the 2–8% of inclusions of yttrium oxide Y_2O_3 in a β-SiAlON structure was weak. It was also investigated the influence of addition of 20–40% SiC and TiN on the strength characteristics at various temperatures of sintering. The increase of sintering temperature in the investigated range for all cases gives a more solid structure (figure 8).

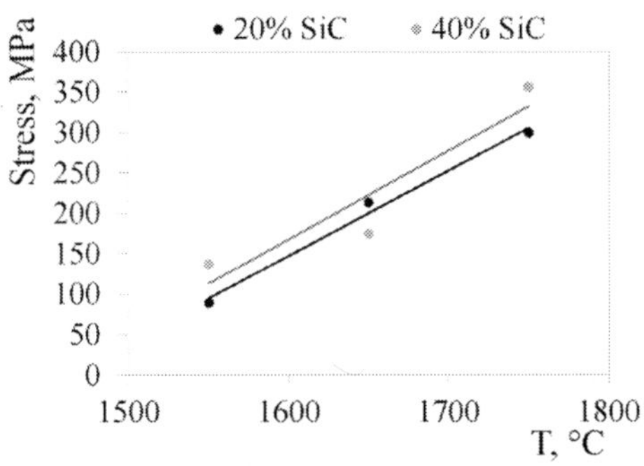

(a)

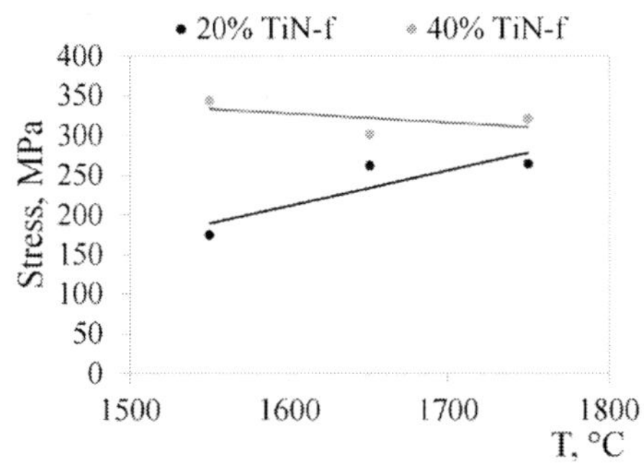

(b)

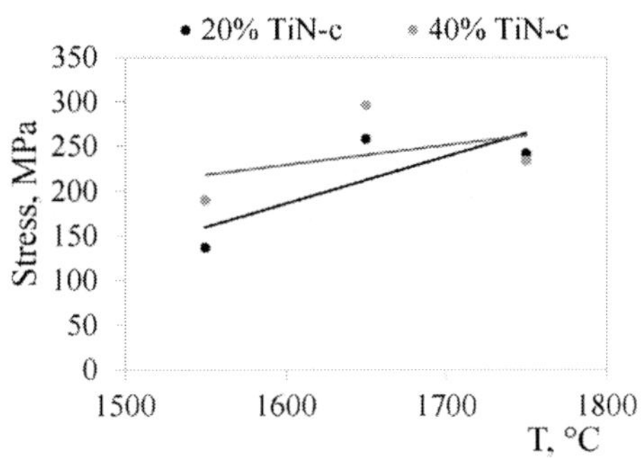

(c)

Figure 8. Influence of the sintering temperature of β-Sialon with different content of additives on the fracture resistance: (a) 20 and 40% silicon carbide SiC, (b) 20 and 40% fine-grained titanium nitride (TiN-f), (c) 20 and 40% coarse-grained titanium nitride (TiN-c).

5. Conclusions

The possibility of determining the brittle strength of the material of thin disks tested for bending on an annular support is demonstrated numerically and experimentally. The results of the study of a range of materials, resulting electric pulse impact method are shown.

Acknowledgments

The presented research was supported by Grant RSF no. 16-19-10213 from May 6, 2016.

References

[1] Tumanov A T 1974 *Methods of testing control and research of engineering materials, Volume II, Methods of investigation of mechanical properties of metals* (M., Mashinostroenie) 320 p

[2] Zholnin A G, Kovaleva I V, Pahilo–Darial I O, Litvinova I C, Grigoryev E G, Olevsky E A, Goltsev V Yu and Rytenko V R 2016 *Fizika i himiya obrabotki materialov [Physics and chemistry of materials processing* **2** 5–10

[3] Release 16.2 Documentation for ANSYS [electronic document] / ANSYS Inc. Electronic data and software (104019 files: 10660130531 bytes).

[4] Goltsev V Y and Gribov N A 2017 *15th School-Conf. "New materials of innovative energy: development, characterization methods and application; KnE Life Science* 148–154 DOI: 10.18502/kms.v4i1.2139

IOP Conf. Series: Materials Science and Engineering **558** (2019) 012012 doi:10.1088/1757-899X/558/1/012012

Application of the "Brazilian test" to determine the strength of materials obtained by electric pulse consolidation of powders

V Y Goltsev[1,2], E G Grigor'ev[2], N A Gribov[1], A V Osintsev[1] and A S Plotnikov[1,2]

[1]National Research Nuclear University MEPhI, Moscow, 115409 Russia
[2]Merzhanov Institute of Structural Macrokinetics and Materials Science, Russian Academy of Sciences, Chernogolovka, 142432 Russia

E-mail: gvy587@gmail.com

Abstract. Testing of short cylinders according to the "Brazilian test" scheme is considered as a method of assessing the tensile strength of brittle materials obtained by electric pulse consolidation of powders. The paper presents a calculated using FEM and an experimental analysis of the behavior of short cylinders (thick disks) when loaded according to the "Brazilian test" scheme. The calculation shows the weak influence of the disk thickness on the average and maximum tensile stresses in the plane of symmetry of the disk. Results of testing of small samples made of cast iron and graphite are presented. It is shown a good correspondence between the results of the test samples from cast iron when using formula recommended by the ASTM Standard D3967-95a, and the strength of the material when tested in tension. The results of testing samples of graphite were 1.5–2 times less than its tensile strength. The results of testing of small-size samples of heavy alloy WNiFe and aluminum oxide obtained by electric pulse consolidation of powders are presented. The influence of the pressure and current pulse on the fracture resistance of the alloy WNiFe and the sample thickness on the strength of Al_2O_3 is investigated. Thus, the possibility of testing small samples according to the "Brazilian test" scheme to determine the tensile strength of materials obtained by electric pulse consolidation of powders is demonstrated.

1. Introduction

Development of production technology of modern materials by consolidation of powders, welding and sintering regimes [1, 2] is carried out, as a rule, on small samples, the size of which often does not exceed 10–15 mm in diameter, and the thickness can vary from 1 to 10 mm. Standard mechanical test methods are often not applicable to such samples. Alternative test methods have to be used to evaluate the mechanical properties of the material. An example of such methods is compression of short cylinders in the plane of symmetry according to the "Brazilian test" scheme [3]. This method is used to test brittle materials. The method is standardized in the USA [4] to determine the tensile strength of rocks when testing samples with a diameter of at least 50 mm with a thickness-to-diameter ratio in the range of0.2÷0.75.

The aim of this work is to study the possibility of using the test of small samples according to the "Brazilian test" scheme to assess the tensile strength of brittle materials obtained by the method of electro-pulse consolidation of powders.

2. Calculation analysis of short cylinder compression according to the "Brazilian test" scheme

Simulation of the disk sample loading process was performed in the verified [5, 6] calculation complex ANSYS Mechanical version 16.2. The full calculation model is shown in figure 1.

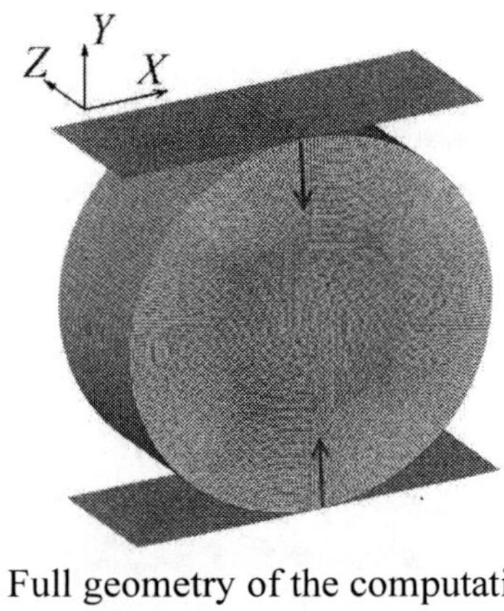

Figure 1. Full geometry of the computational model.

Figure 2 presents the results of FEM modeling of the disk compression process: the distribution of ε_x strains on the front surface of the sample and the distribution of the first main stresses on the front surface and the cross section of the sample. As a result of simulation, it was found that the maximum principal stresses at individual points can exceed the average stresses σ_t by 2.5 times, which should affect the applicability of the formula recommended by ASTM D3967–95a for concentration sensitive brittle materials.

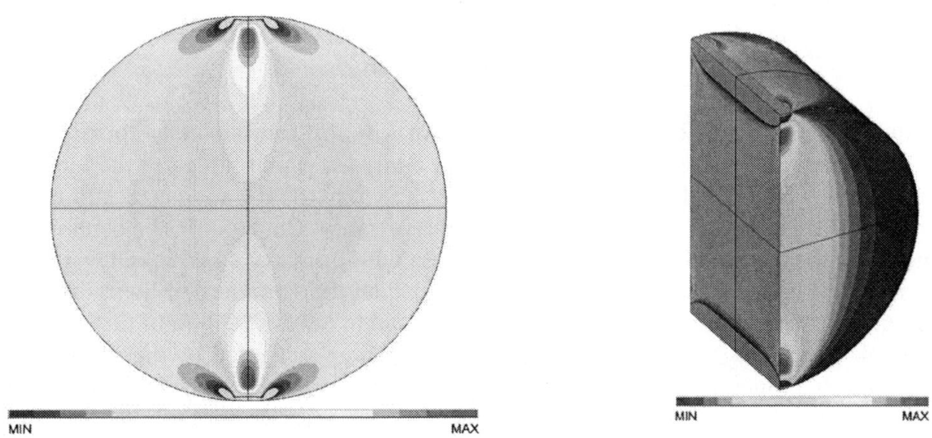

(a) (b)

Figure 2. Qualitative patterns of strain distribution ε_x (a) and the first principal stresses (gray areas where $\sigma_1 < 0$) (b) in the disk sample under compression.

The results of the analysis with the study of the influence of the ratio of the cylinder thickness to its diameter on the maximum tensile stresses are presented in [7]. It is shown that in the plane of symmetry of the disk YZ coinciding with the plane of its loading, there are tensile stresses σ_x that lead to brittle fracture of the sample. Changing the ratio of the disc thickness to its diameter in the range of 0.3–1 has little effect on the maximum tensile stress.

3. Results of experimental study of short cylinders of model materials

Table 1 presents the mechanical properties and modulus of elasticity of the tested of model materials.

Table 1. Tensile/compressive strength and modulus of elasticity of the materials studied

Material	σ_u^t, MPa	σ_u^c, MPa	*E*, GPa
Cast iron SCH 10-40	104	404	100
Graphite ARV-1	13.8	37	11

3.1. Test cylinders of gray cast iron

The cylinders are made of grey cast iron with three sizes $D \times t$: 10 × 4, 15 × 4, and 10 × 5 mm.

Loading of the samples was carried out using an INSTRON test machine in a special device providing strict parallelism of the reference planes. Simultaneously with the loading of samples using the method of digital image correlation [8, 9] and the use of equipment Vic-3D RT recorded the movement of surface points of each sample. Distribution patterns of deformation ε_x surface points of the cast iron sample at the initial stage of loading, the stage preceding the destruction and the destruction stage are shown in figure 3.

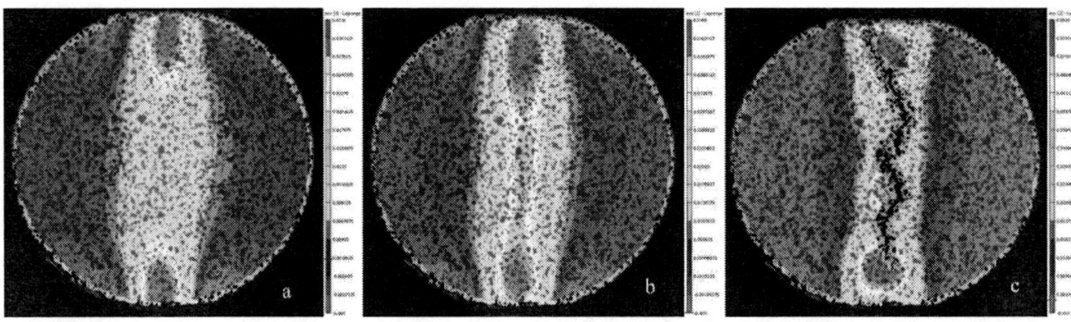

Figure 3. The color pattern of the strain distribution ε_x in the surface of iron sample points: (a) initial stage of loading, (b) stage preceding the destruction; (c) stage destruction.

The distribution of tensile strains ε_x (red zone - maximum strain) exactly corresponds to the pattern of distribution of tensile stresses in the calculation analysis, which indicates the identity of the results obtained by calculation and experimental methods. Symmetrical patterns of strain distribution with respect to the X and Y axes are observed at all stages of sample loading. In the contact area of the stamp and wheel surface points is implemented plane stress with principal stresses σ_x tensile and compressive stresses σ_y, and as a result, significant shear stresses equal the semi-sum of their absolute values. Shear stresses cause shear cracks, which, when connected, lead to the destruction of the sample in the plane of symmetrical, where the maximum tensile stresses, which is confirmed by the nature of the destruction of the sample. Step fracture of the sample indicates that the rupture of the sample occurs under the action of shear stresses. No explosive fracture characteristic of a brittle rupture from normal stresses was observed. We observed a slow departure of one half of the sample from the other, which is usually inherent in plastic destruction. The beginning of destruction of samples is associated with the maximum loads recorded by the test machine.

Table 2 summarizes the test results of 9 cast iron cylinders: the maximum load corresponding to the beginning of the destruction of all tested samples, and the breaking stresses calculated by the formulas:

$$\sigma_t = \frac{2P}{\pi t D} \quad (1)$$

$$S_0 = \frac{P}{tD} \quad (2)$$

where *P* is the maximum load sustained by the sample, *t* is the thickness of the sample, and *D* is its diameter. The average stress values calculated by formula (1) were close to the tensile strength values of the material. Thus, the formula (1) recommended by ASTM D3967–95a can be used to calculate the destructive stresses when testing metal brittle discs according to the "Brazilian test" scheme.

3.2 Graphite cylinder tests

Cylinders made of graphite were three sizes $D \times t$: 8 × 4, 8 × 8, and 8 × 12 mm. The destruction of the graphite sample occurs at maximum load in the linear nature of the curves until fracture of the specimen. Table 3 shows the test results of the graphite samples and the resistance of the graphite fracture.

Table 2. The fracture resistance of cast iron, determined by the formulas (1) and (2).

Sample	Size, mm	P, kN	σ_t, MPa	S_0, MPa
1	10 × 4	5.3	85	134
2	10 × 4	6	96	151
3	10 × 4	7	111	174
4	10 × 4	6.8	109	171
5	10 × 4	7	112	176
6	10 × 5	9	115	180
7	10 × 5	8.8	112	176
8	15 × 4	8.3	88	138
9	15 × 4	7.5	80	126
The average stress values			101	158.4

Table 3. Graphite fracture resistance determined by formulas (1) and (2).

Sample	Size, mm	P, kN	σ_t, MPa	S_0, MPa
1	8 × 4	307	6.1	9.6
2	8 × 8	660	6.6	10.3
3	8 × 8	695	6.9	10.9
4	8 × 8	595	5.9	9.3
5	8 × 12	1100	7.4	11.5
6	8 × 12	1108	7.4	11.5

The strength of graphite, calculated by the formula (1), was about 1.5–2 times lower than the true (see table 1), which can be explained by the more fragile state of graphite. This is confirmed by the difference between the diagrams of deformation of samples of cast iron and graphite: the diagram of plastic deformation and fracture of cast iron and brittle fracture of graphite (figure 4).

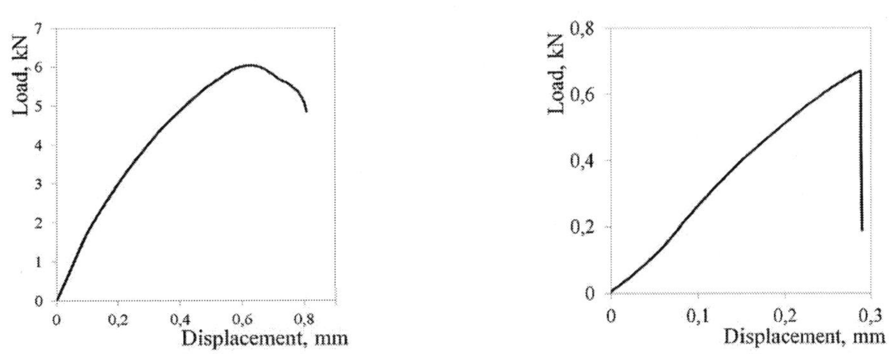

(a) (b)

Figure 4. Test charts for cylinders of (a) cast iron (10 × 4 mm) and (b) graphite (8 × 8 mm).

4. The test of cylinders obtained via the high-voltage consolidation of powders

4.1. Study of samples of alloy WNiFe

To obtain samples of heavy alloy WNiFe by high-voltage consolidation used industrial powder composition 90W–7Ni–3Fe, obtained by mechanical mixing of the components. Data on the composition and density of the powder are given in table 4.

Table 4. The composition of industrial powder WNiFe.

Element	ρ, g/cm^3	Weight, %	ρ_{theor}, g/cm^3
W	19.25	residue	
Ni	8.902	6.93	17.13
Fe	7.874	3.12	

Various modes of sintering in the manufacture of cylindrical samples from the alloy WNiFe were investigated. The supplied voltage varied from 4.5 to 5.8 kV and the pressure from 150 to 250 kPa. The beginning of destruction of samples is associated with the maximum loads recorded by the test machine. Tests of samples allowed determine the levels of pressure and voltage pulses during their compaction, providing optimal strength characteristics of the alloy WNiFe. Figure 5 shows the dependence of the material resistance to rupture of the voltage pulse level for samples sintered under the same pressure of 200 and 250kPa.

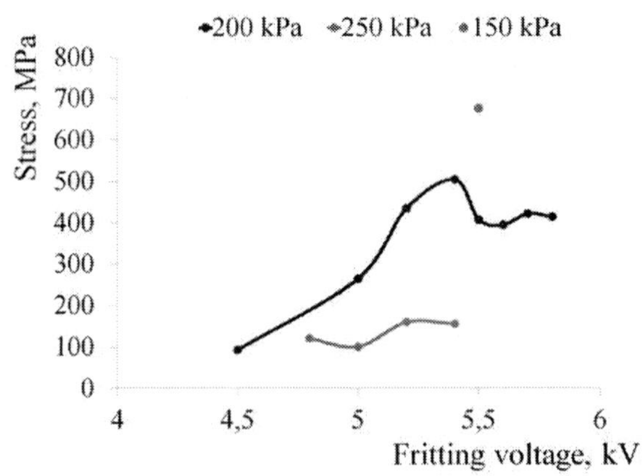

Figure 5. The dependence of the fracture resistance of the applied voltage at sintering at the same pressure.

At a pressure of 200 kPa, there is an increase in the strength of the samples with an increase in the voltage pulse to a value of 5.4 kV, then a decline. The growth and decrease in strength occurred against the background of the development of plastic deformations that preceded the brittle destruction of the samples.

At a pressure of 250 kPa, we obtained an absolutely brittle fracture without noticeable traces of plastic deformation.

4.2. Testing of samples of aluminum oxide obtained by SPS

Al_2O_3 powder was placed in a graphite matrix and pressed down by punches on both sides with a constant force, and was heated by passing a pulsed current through the matrix.

The results of the "Brazilian test" of cylindrical samples with a diameter $D = 10$ mm made of aluminum oxide powder by the method of electro pulse consolidation and calculated by the formula (1) are presented in figure 6.

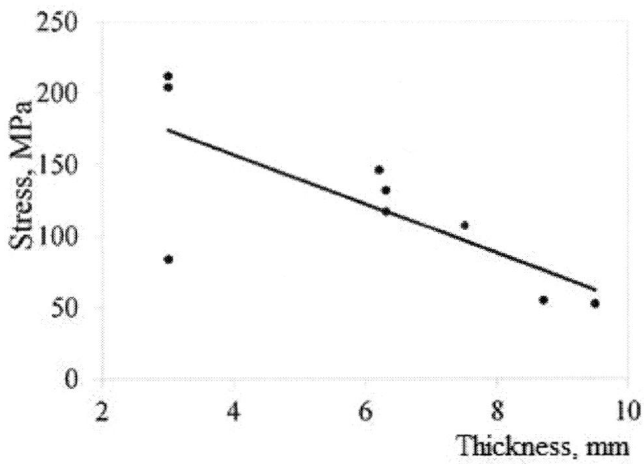

Figure 6. The test results of samples of aluminum oxide.

We note a good correspondence between the results obtained by the test of short cylinders according to the "Brazilian test" scheme and the results of the test of thin aluminum oxide disks [10] with a diameter of 15 mm and a thickness of 1.5 mm for bending on the annular support. Thin discs made by the SPS of ultradispersed (UD) powder spherical, collapsed under the stress of 171 MPa. The use of pre-pulsed magnetic pressing UD powder increased its strength to 202 MPa.

5. Conclusions

Thus, the possibility of testing small cylindrical samples according to the scheme of the "Brazilian test" and indirect determination of tensile strength of materials obtained from powders by the method of electric impulse action is demonstrated by calculation using FEM and experimentally.

Acknowledgements

The study was supported by a grant from the Russian science Foundation (project no. 16-19-10213).

References

[1] Grigoryev E G 2011 *Int. J. Machin. Tech. Mater.* **8** 8–10
[2] Bashlykov S S, Demenyuk D V, Grigoryev E G, Olevsky E A and Yurlova M S 2013 *Fizika i himiya obrabotki materialov [Physics and chemistry of materials processing]* **5** 77–83
[3] Wang Q Z, Jiaa X M, Koub S Q, Zhangh Z X and Lindgvistg P A 2004 *Int. J. Rock Mechan. Mining Sci.* **41** 245–253
[4] ASTM D3967-95a Standard Test Method for Splitting Tensile Strength of Intact Rock Core Specimens
[5] Release 16.2 Documentation for ANSYS [electronic document] / ANSYS Inc. Electronic data and software (104019 files: 10660130531 bytes)
[6] NAFEMS search engineering analysis and simulation FEA, Finite Element Analysis, CFD, Computational Fluid Dynamics and Simulation. NAFEMS Ltd., Hamilton, United Kingdom
[7] Osintsev A V, Goltsev V Y and Plotnikov A S 2017 *Pis'ma o materialah [Letters on materials]* **7** 21–25 DOI: 10.22226/2410-3535-2017-1-21-25
[8] Sutton M A, Orteu J J and Schreier H W 2009 *Image correlation for shape, motion and deformation measurements: basic concepts, theory and applications* (Springer) 321 p

[9] Chu T C, Ranson W F, Sutton M A and Peters W H 1985 *Exper. Mechan.* **25** 232–245
[10] Zholnin A G, Kovaleva I V, Pahilo–Darial I O, Litvinova I C, Grigoryev E G, Olevsky E A, Goltsev V Yu and Rytenko V R 2016 *Fizika i himiya obrabotki materialov [Physics and chemistry of materials processing]* **2** 5–10

IOP Conf. Series: Materials Science and Engineering **558** (2019) 012013 doi:10.1088/1757-899X/558/1/012013

Evaluation of tensile strength of nuclear fuel samples

V Y Goltsev[1,2] and A V Osintsev[1]

[1]National Research Nuclear University MEPhI, Moscow, 115409 Russia
[2]Merzhanov Institute of Structural Macrokinetics and Materials Science, Russian Academy of Sciences, Chernogolovka, 142432 Russia

E-mail: gvy587@gmail.com

Abstract. For estimation of the strength of the nuclear fuel material as well and for further refinement of nuclear fuel test technology we propose to use small discoid samples, similar in shape to the elements of nuclear fuel in the context of "Brazilian test". We present here the results of testing small discoid specimens with and without center hole, made of brittle materials, such as cast iron and graphite (both being considered as possible model materials for the nuclear fuel). We compared these materials to nuclear fuel itself (as represented by uranium dioxide). The type of deformation and fracture found in samples made from cast iron suggests that this material cannot be used as a model for the nuclear fuel. In addition, the effect of the specimen size on resistance to destruction was investigated. The results obtained in tests on samples of graphite ARV-1 were in good agreement with the results of tests on uranium dioxide. Using the data obtained in this study, a calculation formula for determining the strength of the nuclear fuel material based on the "Brazilian test" results is proposed.

1. Introduction.

Determination of the tensile strength of the nuclear fuel material involves mechanical testing of small samples with a thickness of up to 10 mm and a diameter of 7–10 mm. In the present work to assess the strength of this material, we used the method of compressing the sample disc (fuel tablet) in the symmetric plane following the "Brazilian test" approach. The method is standardized in the United States to determine the brittle strength of rocks [1, 2]. Recommended sample size: diameter is not less than 50 mm, thickness – 0.2–0.75 of the sample diameter. We have shown that the "Brazilian test" of small samples with a diameter of 10–15 mm and a thickness of up to 10 mm can be successfully used to determine the brittle strength of both metal and non-metal materials, such as cast iron, graphite [3].

2. Testing of brittle discs under the scheme of the "Brazilian test"

A detailed computational analysis of loading small solid disk under the scheme of the "Brazilian test" is presented in our prior study [3]. It was shown that the maximum tensile stresses developing upon loading of the sample disk in its symmetrical plane lead to brittle fracture of low-plasticity materials. The calculation of destructive stress can be done according to the formula recommended by the ASTM D3967-95a:

$$\sigma_t = \frac{2P}{\pi t D} \quad (1)$$

In equation (1) P is the maximum load on the stamp, compressing the disk, t the thickness of the disk, and D its external diameter. The stress determined by this formula is the average tensile stress in the symmetrical plane of the disk.

Brittle materials as grey cast iron and graphite were selected for testing according to the "Brazilian test". Mechanical properties of model materials are shown in table 1.

Table 1. Mechanical and physical properties of the materials studied.

Material	σ_u^t, MPa	σ_u^c, MPa	E, GPa
Cast iron SCH 10-40	104	404	100
Graphite ARV-1	13.8	37	11

2.1. Testing solid disks made of grey cast iron

Prior results by our group [3] of testing small-size disk made of cast iron confirmed the possibility of applying the eq. (1) determine the strength of the material in tension. Discs made of grey cast iron were of the following $D \times t$ dimensions: 10 × 4 and 15 × 4 mm. Diagram of compression of cast iron disc is shown in figure 1a.

The initial cracks are formed in the contact region, where maximum shear stresses are observed. The beginning of the crack growth is associated with the maximum load on the sample. Gradual decline in load after the maximum is associated with slow crack propagation and dissociation of one half of sample from each other. We did not find any explosive fracture characteristic of brittle fracture from normal stresses. There is an assumption about the destruction of cast iron samples under the action of maximum shear stresses, as evidenced by their step fracture [3].

2.2. Testing solid disks of graphite

Discs of graphite were of the following $D \times t$ dimensions: 8 × 4, 8 × 8, and 8 × 12 mm. Characteristic diagram of the compression of the graphite disc with a size of 8 × 8 mm in the symmetrical plane is shown in figure 1b. The destruction of the graphite disc happens dynamically at a maximum load. Table 2 shows the breaking strength of the graphite samples calculated using formula (1) and formula

$$S_0 = \frac{P}{tD} \tag{2}$$

The average value of the resistance of graphite to fracture was approximately 2 times lower than the “true” material resistance (see table 1) and slightly increases (by 10%) with increasing sample thickness twice. This difference between the specific sample resistance and the resistance as defined by the nature of the material is likely to be associated with more fragile condition of graphite as compared to cast iron evident from the charts of compression observed in samples built from these materials (figure 1).

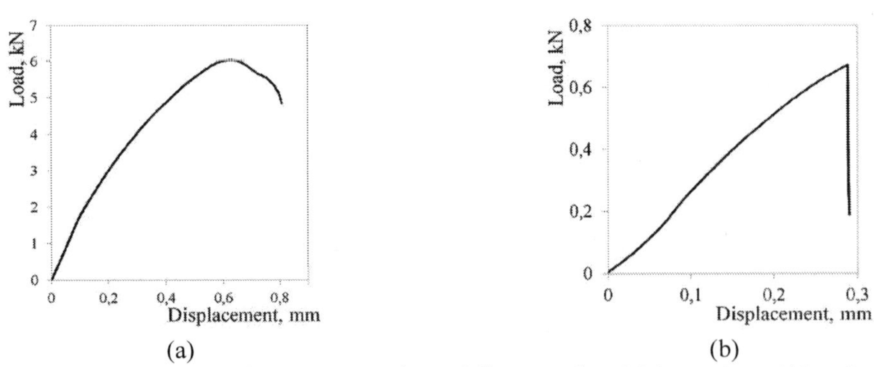

Figure 1. Diagrams of the compression of discs made of (a) cast iron (10 × 4 mm) and (b) graphite (8 × 8 mm).

Table 2. The resistance of graphite to destruction.

Sample	Size, mm	P, N	σ_t, MPa	S_0, MPa
1	8 × 4	307	6.1	9.6
2	8 × 8	660	6.6	10.3
3	8 × 8	695	6.9	10.9
4	8 × 8	595	5.9	9.3
5	8 × 12	1100	7.4	11.5
6	8 × 12	1108	7.4	11.5

3. Testing analog fuel tablets

There is a central hole in fuel tablets, designed to drain the reactor gases. Accordingly tests were performed on disks with a central hole. The presence of a central hole in the sample leads to a stress concentration in the hole area.

3.1. Test disks made of grey cast iron and having a central hole

Cast iron disks had the following dimensions: diameter 7.2 mm, thickness 4 mm, central hole diameter $d = 1.2$ mm. The destruction of a disk with a hole occurs along the vertical diameter similar to the disk without the hole. The crack originates on the contour of the central hole at the point of intersection with the vertical diameter and extends along the diameter of the complete separation of the specimen. Fracture character is slow controlled crack propagation similar to the process of destruction of solid samples. Table 3 presents the results of tests on cast iron disc specimens with a central hole. The resistance of the material to destruction σ_{th} and S_1 were calculated according to the formula

$$S_0 = \frac{2P}{\pi t(D - d)} \tag{3}$$

and

$$S_1 = \frac{P}{t(D - d)} \tag{4}$$

The fracture resistance of cast iron specimens was 1.5 times less than the true fracture resistance pertinent to the material (see table 1). This effect was associated with the concentration of stresses at the central hole.

Table 3. The resistance of cast iron to destruction.

Sample	Size, mm	P, N	σ_{th}, MPa	S_1, MPa
1	7.2 × 4	3103	83	130
2	7.2 × 4	2555	68	107
3	7.2 × 4	3040	80	126
Average values			77	121

3.2. Test disks made of graphite with central hole

Tested graphite disks were of the following three sizes: the diameter of all disks was 8 mm, thickness was 4, 8 and 12 mm. Diameter of central hole was 1 mm for all disks. Table 4 shows the results of the tests. The compression diagram for graphite samples with a central hole is linear up until the point of brittle fracture, similar to observed in samples without hole. The samples collapsed in a completely brittle way in the diametrical plane of the disk. The destruction was found to originate from the central hole. In some samples, we observed several additional cracks, along with the central one. The average tear resistance of samples with a central hole as derived using the formula (3) was almost 2.5, and with formula (4) – 1.5 times less than true tear resistance (see table 1). Breaking stress for brittle materials such as graphite, as defined by the formula (4), is 1.5 times less than the true strength of the material as defined by tensile tests according to GOST 1497. Figure 2 shows the dependence of the average values of S_0 and S_1 from the thickness for the graphite specimens with and without a central hole.

Table 4. The resistance of graphite to destruction for discs with a central hole.

Sample	Size, mm	*P*, N	σ_{th}, MPa	S_1, MPa
1	8 × 4	321	7.3	11.5
2	8 × 4	283	6.5	10.2
3	8 × 4	233	5.3	8.3
4	8 × 8	479	5.5	8.6
5	8 × 8	514	5.8	9.1
6	8 × 12	611	4.7	7.4
7	8 × 12	820	6.2	9.7
8	8 × 12	710	5.4	8.5

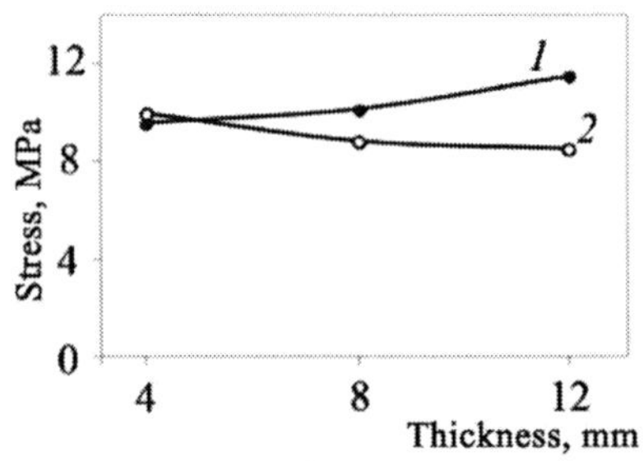

Figure 2. Dependence of the average values of S_0 and S_1 from the thickness of the graphite specimen without central hole (1) and with a central hole (2).

The analysis of figure 2 it can be concluded that at a thickness of graphite samples equal to 5 mm, the fracture resistance estimated by the formulas (2) and (4) is 1.5 times less than the true resistance of the material to rupture and is equal for solid samples and samples with a central hole. Increasing the thickness of the sample twice can lead to the error of the result is not more than 10%. Thus, the results of the test samples according to the "Brazilian test" scheme, having a linear compression machine diagram, similar to the diagram of graphite samples testing, can be used to form a calculation formula for estimating the tensile strength of the material. The true tensile strength of the material could be evaluated according to the formula

$$\sigma_B = \frac{1{,}3P}{tD} \tag{5}$$

for testing solid samples of nuclear fuel without central hole or

$$\sigma_B = \frac{1{,}5P}{t(D - d)} \tag{6}$$

for testing specimens with central hole.

4. Testing of samples of nuclear fuel.

Fuel tablets made of uranium dioxide, the size of which *Dxt* is given in table 5, have been tested according to the "Brazilian test" scheme. Table 5 shows the results of the tests performed on fuel pellets.

Figure 3 shows the diagram of the test of the sample № 7.

The destruction originates from the central hole. We observed absolutely brittle fracture of samples in the symmetrical plane of the disk in the area of tensile stresses. Note the similarity of the destruction for both the graphite specimen and the real fuel pellets. The levels of destructive stress for the fuel pellets are within the order of magnitude agreement with the values of the breaking stress of the graphite samples. And accordingly the estimated level of strength agrees with the true strength of graphite ARV-1. This supports the possibility of using graphite ARV-1 as a material for modeling the behavior of uranium dioxide. This is also confirmed by the coincidence of the nature of the destruction of graphite samples and fuel tablets, shown in figure 4. Samples with a central hole just before their destruction were registered by digital image correlation [5].

IOP Conf. Series: Materials Science and Engineering **558** (2019) 012013 doi:10.1088/1757-899X/558/1/012013

Table 5. The results of the test fuel pellets.

№ sample	Size, mm	*P*, N	σ_{th}, MPa	S_1, MPa
1	6.0 × 10	406	5.17	8.12
2	6.0 × 10	531	6.75	10.6
3	7.2 × 13	796	6.31	9.90
4	7.2 × 13	1488	11.8	18.5
5	8.3 × 10	351	3.06	4.80
6	8.3 × 10	382	3.33	5.23
7	8.3 × 10	612	8.50	12.7

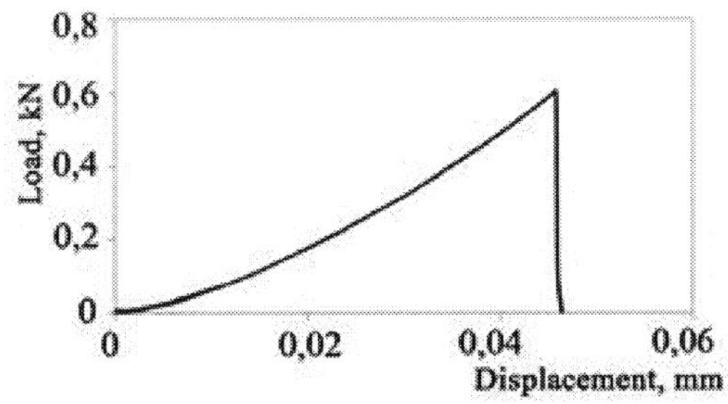

Figure 3. The diagram of the compression sample №7 of uranium dioxide.

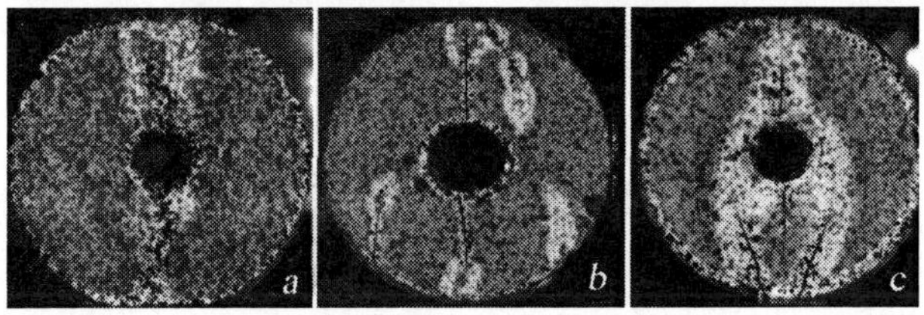

Figure 4. These are fields of transverse relative deformations recorded immediately before the destruction of samples of cast iron – *a*, graphite – *b*, uranium dioxide – *c*.

Thus, we showed the possibility of testing the nuclear fuel (fuel pellets) according to the "Brazilian test" with indirect assessment of the strength of brittle material.

5. Conclusions

The test results of the samples according to the “Brazilian test " scheme show that the fracture resistance determined by the formula recommended by ASTM d3967-95a for solid cast iron samples is practically the same, and from graphite differs by 1.5 times from the true strength of materials. When testing samples with a central hole, the tensile strength differs by 1.5 and almost 2.5 times, respectively, from the true strength of the materials. The possibility of testing nuclear fuel (fuel pellets) according to the “Brazilian test” with an indirect estimate of the material brittle fracture resistance is demonstrated. The formula for indirect evaluation of tensile strength of brittle components of the nuclear fuel based on results of the tests done according to the “Brazilian test” approach is proposed.

Acknowledgement.
The presented research was supported by Grant RSF no. 16-19-10213

References

[1] ASTM D3967-95a Standard Test Method for Splitting Tensile Strength of Intact Rock Core Specimens

[2] Wang Q Z, Jiaa X M, Koub S Q, Zhangh Z X and Lindgvistg P A 2004 *Int. J. Rock Mechan. Mining Sci.* **41** 245–253
[3] Goltsev Y V, Osintsev A V and Plotnikov A S 2017 *Pis'ma o materialakh [Letters on materials]* **7** 21–25 DOI: 10.22226/2410-3535-2017-1-21-25
[4] Golstev V I, Grigoriev E G, Osintsev A V, Ermakova N S, Plotnikov A S and Gribov N A 2017 *IOP Conf. Ser.: Mater. Sci. Eng.* **218** 012007 https://doi.org/10.1088/1757-899X/218/1/012007
[5] Sutton M A, Orteu J J and Schreier H W 2009 *Image correlation for shape, motion and deformation measurements: basic concepts, theory and applications* (Springer) 321 p

IOP Conf. Series: Materials Science and Engineering **558** (2019) 012014 doi:10.1088/1757-899X/558/1/012014

Advanced methods for consolidation of powder materials by impulse electromagnetic fields

E G Grigoryev

Merzhanov Institute of Structural Macrokinetics and Materials Science,Russian Academy of Sciences, Chernogolovka, 142432 Russia

E-mail: eugengrig@mail.ru

Abstract. Advanced technologies for the production of new materials and, in particular, nanostructured materials, using powder technologies require fundamentally new approaches for the formation and preservation of a given structural-phase state. Precision control of the state of materials in the process of consolidating powders of nanomaterials can be carried out using pulsed electromagnetic fields. The aim of the project is to study the effect of high-voltage and low-voltage pulsed electromagnetic fields in the technologies of powder consolidation. The experimental devices of spark-plasma sintering, flash-sintering, high-voltage consolidation and magnetic-pulse compaction are used in our laboratory for the production of advanced materials from metal powders, ceramic and composite powders. We can produce boron carbide, silicon carbide, uranium nitride, tungsten carbide - cobalt - diamond composites, tungsten heavy alloys, and others by electromagnetic methods of powder consolidation. Experimental results to consolidation metal powders, ceramic and composites powders by electromagnetic methods presage fruitful results.

1. Introduction

Advanced technologies of consolidation of powder materials, based on various techniques using of electric current pulses and mechanical pressure, are widely studied in many research laboratories [1]. The interest in these methods is motivated by their ability to consolidate a large variety of nanostructure powder materials to high densities within short periods of time, without having to increase initial grain sizes. The unique potential of pulse electric consolidation methods of powders is reflected in the ever-growing number of scientific publications studying these technological approaches. The wide range of possible electrical and mechanical treatment modes of powder has resulted in a large number of these methods, including high-voltage electric discharge consolidation (HVEDC, also sometimes referred to as capacitor discharge sintering (CDS)), spark-plasma sintering (SPS), magnetic-pulse compaction (MPC), etc. [1–3].

The advantages of these techniques can be exploited only through the optimization of the consolidation parameters since excessive energy dissipation during this type of processing can lead to the instability of the compaction process, to the formation of an undesirable heterogeneous material structure, and even to the destruction of the sintered specimens and of the equipment used. The time dependence of the associated thermal processes at the interparticle contacts plays a key role in electric pulse powder consolidation [4].

To obtain materials with required properties, one has to know the macroscopic processes occurring in the volume of a consolidated sample. Indeed, the kinetics of the consolidation of powder materials in various electric field-assisted methods is significantly different, and their duration changes from several tens of minutes for electric-discharge sintering and spark plasma sintering [1–3] to several milliseconds for high voltage electric discharge consolidation [4] and magnetic-pulse compaction [5].

In this paper, we report the results of studying the multiscale physical processes during the powder consolidation by pulse electric current methods. These physical processes are taking into account both as a whole specimen and in the particle contact zones. Also, we describe the experimental equipment for implementation spark-plasma sintering, flash sintering, high-voltage electric discharge consolidation and magnetic-pulse compaction of powder materials.

2. Research method

A mathematical model describing the multiscale physical processes occurring under the powder consolidation by pulse electric current methods is based on the conservation laws of mass (1), momentum (2), energy (3) and the electrodynamics equations (4, 5) for consolidated powder.

$$\frac{\partial \rho}{\partial t} + div\left(\rho \vec{v}\right) = 0 \qquad (1)$$

$$\rho\left(\frac{\partial \vec{v}}{\partial t} + (\vec{v}, \nabla)\vec{v}\right)_i = \left(\frac{\partial \sigma_{ik}}{\partial x_k}\right) + F_i \qquad (2)$$

$$\frac{\partial}{\partial t}\rho\left(\varepsilon + \frac{\vec{v}^2}{2}\right) = -div\left(\rho\vec{v}\left(w + \frac{v^2}{2}\right) - (\vec{v}, \hat{\sigma}') - \kappa \nabla T\right) + \frac{\vec{j}^2}{\sigma} \qquad (3)$$

$$rot\ \vec{E} = -\frac{\partial \vec{B}}{\partial t},\quad rot\ \vec{H} = \vec{j},\quad div\ \vec{B} = 0, \qquad (4)$$

$$\vec{F} = \left[\vec{j}, \vec{B}\right],\quad \vec{j} = \sigma\left(\vec{E} + \left[\vec{v}, \vec{B}\right]\right) \qquad (5)$$

where ρ is the density, $\vec{v}$ is the velocity, $\hat{\sigma}$ is the internal stress tensor, ε is the internal energy, *w* is the enthalpy, $\hat{\sigma}'$ is the viscoplasticity tensor, *T* is the temperature, $\vec{j}$ is the electrical current density, $\vec{E}$, $\vec{H}$ are the tensors of the electrical and magnetic fields, respectively, $\vec{B}$ is the magnetic field induction, $\vec{F}$ is the Ampere force; *k* is the thermal conductivity; σ is the conductivity of the powder material. The system of equations (1–5) should be supplemented by the state equations of the powder material and electrodes-punches material. It is assumed that the electrode-punch material obeys Hooke's law. We used a visco-plastic material model [13] for the description of the powder compaction process.

$$P = \frac{2}{3}\sigma_T \ln\frac{\alpha}{(\alpha-1)} - \frac{4}{3}\eta\frac{\dot{\alpha}}{\alpha(\alpha-1)} - \frac{\rho_m b^2}{3(\alpha_0-1)^{2/3}}\frac{d}{d\alpha}\left\{\frac{\dot{\alpha}^2}{2}\left[(\alpha-1)^{-1/3} - \alpha^{-1/3}\right]\right\} \qquad (6)$$

where: *P* is the compaction pressure, σ_T is the yield stress of powder material, η is the viscosity of a powder material, *b* is the initial size of pores, ρ_0, ρ_m are the initial and theoretical density of powder material, respectively, $\alpha = \rho_m/\rho$ ($\dot{\alpha} = d\alpha/dt$, $\alpha_0 = \rho_m/\rho_0$).

Set of equations (1–6) with the appropriate initial and boundary conditions allows the establishment of the laws governing the process of powder consolidation by pulse electric current of conductive powder materials. Preliminary numerical estimates can simplify the solution of system (1–6) while maintaining an acceptable accuracy of the results. Joule energy input into the powder sample was determined by the parameters of the pulse current: amplitude $J_0 \leq 500$ kA, pulse frequency $\omega \sim 10^4 \div 10^5$ Hz, pulse duration $\tau_0 < 10^{-3}$ s. Cooling time in the consolidated sample is determined by the heat sink in the electrodes-punches and the die, and depends on the thermal conductivity of the material of punches and dies, as well as on the geometrical dimensions of the sample. The duration of the compaction process of the powder material depends on the parameters of the pressing system, which creates pressure applied to the consolidated sample.

The simulation of the thermal processes in the interparticle contacts has identified the critical amplitude of the pulse current density, at which there is an electric thermal explosion of contact [4]:

$$j_0 = \sqrt{\frac{2\varsigma\sigma}{\rho h} T_b^2} \tag{7}$$

here: j_0 is the pulse electric current amplitude, $\zeta \leq 1$, σ is the Stefan–Boltzmann constant, T_b is the boiling point of the material, ρ is the electrical resistivity of the contact spot, and h is the thickness of the contact area between two powder particles.

Equation (7) is obtained from the analysis of the heat balance at the initial period of time when the pulse electric current density rapidly increases during the pulse (at its leading edge). The heat balance assumes the equivalence of the Joule heat generation rate and the heat dissipation by heat transfer through radiation. Criterion (7) is obtained based on equations (1–5) for the temperature of the interparticle contact zone. If under the maximum value of the electric current density j0, the contact temperature reaches boiling point (or exceeds it), then this results in a contact thermal explosion. Figure 1 shows the experimental points for SPS of heat-resistant steel (EP-741) powder (under a load of 10^{-2} N between two particles) and the respective calculated data based on equation (7) for the current amplitude at the explosion contact as a function of particle size [4].

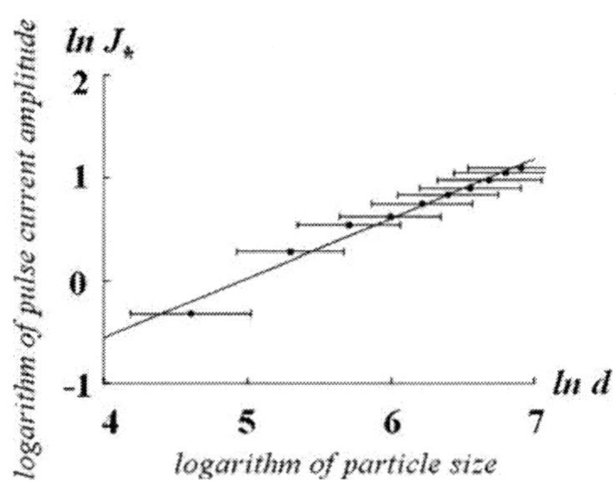

Figure 1. Experimental points and the theoretical prediction for the pulse current amplitude J_* of the contact explosion as a function of particle size d.

The simulation results indicate the possibility of the localization of heat in the interparticle contacts for certain parameter values of the pulse electric current. An upper critical level has been determined for the pulse current amplitude beyond which the interparticle contacts in powder material disintegrate via an electrothermal explosion.

For the implementation spark plasma sintering, microwave sintering, flash sintering, high voltage consolidation and magnetic pulsed compaction of powder nanostructured materials we used experimental equipment available in our lab.

3. Results of the study

We have done (by SPS) the processes of consolidation of high-strength high-refractory compounds, maintaining the grain size of the powders (figure 2):

We have conducted successful experiments on consolidation using spark plasma sintering, and high voltage compaction of iron-titanium composite powders, of vanadium carbide powders, tantalum powders, zirconium nitride powders, alumina powders, ferritic-martensitic steels (ODS) powders with unique radiation-protective properties, tungsten carbide tools (figure 3).

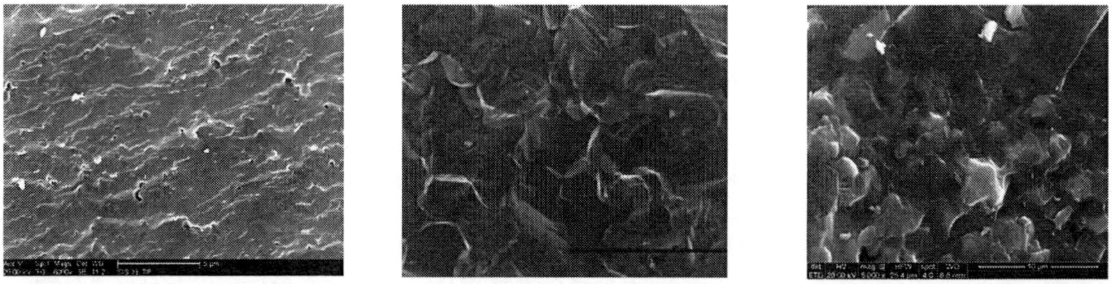

Boron carbide (15 min) Zirconium carbide (5 min) Silicon carbide (less then 1s)

Figure 2. Most significant experimental results of the study.

Tungsten carbide (WC) tools Tungsten carbide + diamond tools Uranium mononitride

Figure 3. Most important application results.

We have shown the possibility of manufacturing samples with relative density of 85–95% for titanium nitride and uranium mononitride in less than 1 s. The resulting samples had a fine-grained microstructure with a grain size of about 1–2 μm.

4. Conclusions

For both techniques (SPS and HVEDC) there is an upper level for the local Joule heating of the inter-particle contacts beyond which the processing instability may occur. While for HVEDC the duration of a single pulse and the amplitude are process controlling parameters, SPS apparently can be controlled by multiple pulse on and off frequency. It is shown that an optimum current amplitude and pulse time are necessary to generate sufficient heat for producing strong inter-particle joining and to avoid local overheating phenomena. An important expression for the critical amplitude of the pulsed electric current is derived. This upper critical level has been assessed for the high voltage pulse current amplitude beyond which the inter-particle contacts in powder material disintegrate by an electro-thermal explosion.

Experimental results to consolidation metal powders, ceramic and composites powders by electromagnetic methods presage fruitful results.

Acknowledgments

The research was supported by Russian Science Foundation (project no. 16-19-10213).

References

[1] Orrù R, Licheri R, Locci A M, Cincotti A and Cao G 2009 *Mater. Sci. Eng.* **R63** 127
[2] Manière C 2017 *Sci. Rep.* **7** 15071
[3] Olevsky E A, Aleksandrova E V, Ilyina A M, Dudina D V, Novoselov A N, Pelve K Y and Grigoryev E G 2013 *Mater.* **6** 4375
[4] Grigoryev E G and Olevsky E A 2012 *Scripta Mater.* **66** 662
[5] Kovaleva I V 2015 *Physics Procedia* **72** 386

IOP Conf. Series: Materials Science and Engineering **558** (2019) 012015 doi:10.1088/1757-899X/558/1/012015

Synthesis of Ti_3SiC_2 MAX phase ceramic materials using macrosized non-powder forms of titanium metal

P V Istomin, A V Nadutkin, E I Istomina and V E Grass

Federal Research Centre "Komi Science Centre of the Ural Branch of the Russian Academy of Sciences", Kommunisticheskaya 24, Syktyvkar, Komi Republic 167982, Russia

E-mail: istomin-pv@yandex.ru

Abstract. A novel approach to the synthesis of Ti_3SiC_2 MAX phase ceramic materials using macrosized non-powder forms of titanium metal, such as foils, sheets, rods, tubes, etc., as a reactant was developed. According to this method, dispersions of SiC particles in slurries or pastes as well as tape casted polymer films filled with SiC particles can be used as a second component of the reaction system. Heat treatment under oxygen-free conditions at 1350–1500°C enables the SHS and RMI processes to occur in the reactive assemblies, allowing them to be sintered into a dense ceramic composites with a damage tolerant Ti_3SiC_2-based matrix reinforced with hard SiC particles. It was demonstrated that the proposed approach can be applied for fabricating Ti_3SiC_2/SiC ceramic matrix composites with specified internal structures including cellular and cannular ones. Since there is no need for a high pressure assistance during the sintering process, the method is expected to provide the Ti_3SiC_2 MAX phase ceramic materials with important competitive advantages through an increase in manufacturing productivity.

Various methods for high-temperature synthesis are presently used to obtain bulk Ti_3SiC_2 MAX phase ceramics and ceramic composites. They include furnace synthesis, hot pressing, spark plasma sintering, self-propagating high-temperature synthesis (SHS), microwave sintering, etc. [1–3]. Most of them are often combined with mechanical activation of starting materials. It is important to note that all these methods, regardless of particular technical and technological details, are commonly associated with reaction sintering of powder mixtures of elements as well as binary compounds of the Ti–Si–C system. This ensures both a good homogeneity of the powder reactant mixtures and high rates of solid-state reactions due to the small size of the reactant particles. However, an external pressure assistance is usually required in order to obtain a fully dense material. Commonly, a rather high loading pressure of several tens of MPa is needed to be used for this purpose. Such a requirement implies a significant decrease in productivity and makes the final products more expensive than competing materials. This, in large part, explains why Ti_3SiC_2-based materials, in spite of their promising properties, have not yet found wide application beyond laboratory scale.

An alternative way for the synthesis of Ti_3SiC_2-based ceramic materials is through the use of macrosized non-powder forms of titanium metal, such as foils, sheets, rods, tubes, etc., whose characteristic sizes are larger than 50–100 μm. In this case, titanium items can be manipulated individually, allowing them to be positioned at a desired orientation and distance with respect to each other, and thereby giving opportunities to build reaction assemblies with a specific spatial arrangement of the reactants. This makes it possible to design materials with complex internal structures (cellular, cannular, tubular, annular, lamellar, granular, etc.) as well as gradient and spatially modulated materials. Another advantage of using non-powder titanium metal items is that due to their high strength and good stiffness, they can act as a kind of load-bearing component, which significantly reduces the risk of damage to the green body under mechanical stresses. This point becomes of crucial importance when Ti_3SiC_2-based ceramic parts of large size and complex shape have to be produced. Besides, the use of the macrosized titanium metal items, which originally contain no pores, minimizes total pore volume in the reaction assemblies, thereby making their sintering to full density more facile.

However, there is a serious obstacle for solid non-powder reactants with a characteristic size larger than 50–100 μm to be used in the synthesis of ceramic materials. In this case, the product layer can grow up to several tens of micrometers in thickness, making diffusion path of the reactants too long. As a result, rates of the solid state reactions decrease dramatically. So, in order to overcome this limitation an alternative path has to be provided for mass transfer. To address this problem, a kind of reactive melt infiltration (RMI) process comprising rapid in situ melting of the titanium metal non-powder reactants followed by capillary spreading of the melt over other components of the reaction system can be employed. In addition, some exothermic SHS reaction is desirable to assist the RMI process, since a large amount of heat energy is needed to melt titanium metal items.

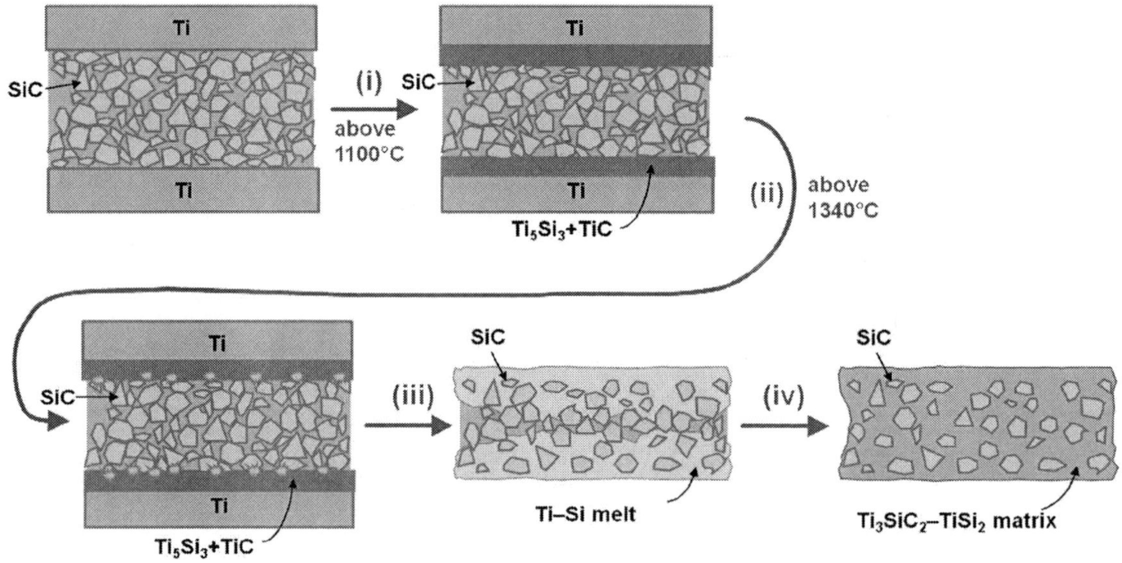

Figure 1. Mechanism for the formation of the Ti_3SiC_2/SiC ceramic matrix composites.

In line with this concept, we have recently developed a novel approach to fabricate Ti_3SiC_2/SiC ceramic matrix composites through the use of macrosized non-powder forms of titanium metal as a reactant [4–9]. Dispersions of SiC particles in slurries or pastes as well as tape casted polymer films filled with SiC particles can be used as a second component of the reaction system. These dispersions and films can optionally contain TiC and carbon additions acting as chemical modifiers. For the synthesis, the reactant components are assembled into a green body comprising of regularly arranged macrosized titanium metal items with a space between which is filled with SiC particles as well as TiC and carbon additions (if required). Heat treatment under oxygen-free conditions at 1350–1500°C enables the SHS and RMI processes to occur in the reactive assembly thus prepared, allowing it to be sintered into a dense ceramic composite with a damage tolerant Ti_3SiC_2-based matrix reinforced with hard SiC particles. Beside these major constituents, a minor amount of $TiSi_2$ inclusions are also present in the matrix. It is worthy to note that the presence of $TiSi_2$ in the composite is not desirable because of its poor mechanical properties. In order to decrease $TiSi_2$ content, carbon-containing compounds (e.g. carbon black or TiC particles) should be added to the reaction system. Thus, the phase composition of the obtained materials corresponds to the overall reaction:

$$(3 + y - z)Ti + (1 + x + 2y)SiC + (1 - 2y - z)C + zTiC = [Ti_3SiC_2 + yTiSi_2]_{matrix} + xSiC_p$$

The mechanism for the formation of the Ti_3SiC_2/SiC composites, as shown in figure 1, can be described in the following stages: i) a primary diffusion-controlled solid-state exothermic reaction between the initial reactants above 1100°C giving rise to the formation of intermediate product layer consisting of TiC_x and $Ti_5Si_3C_x$; ii) triggering the SHS reaction above 1340°C, corresponding to the Ti–Ti_5Si_3 eutectic temperature in the binary Ti–Si system, resulting in rapid melting of the titanium metal items; iii) rapid infiltration of the entire area occupied by SiC particles with the titanium melt due to the capillary effect; iv) crystallization of Ti_3SiC_2 MAX phase together with a minor amount of $TiSi_2$ by a peritectic reaction of the melt with SiC particles resulting in the formation of a dense ceramic matrix enclosing the SiC particles that remained after the previous stages. SEM image and XRD pattern of a typical sample of the Ti_3SiC_2/SiC composite are illustrated in figures 2 and 3, respectively.

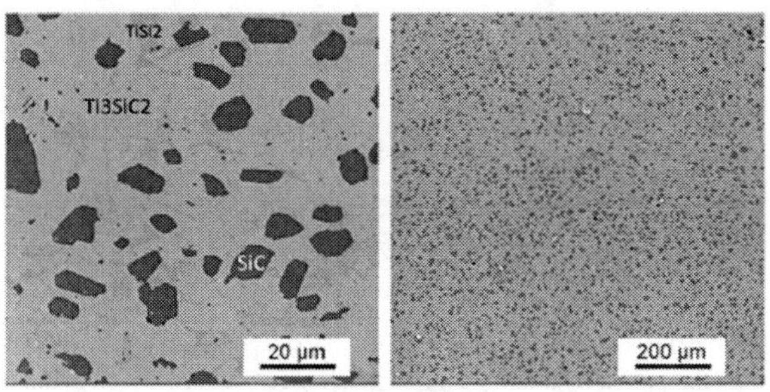

Figure 2. Cross-sectional SEM images of the Ti_3SiC_2/SiC ceramic matrix composite synthesized using non-powder titanium metal items.

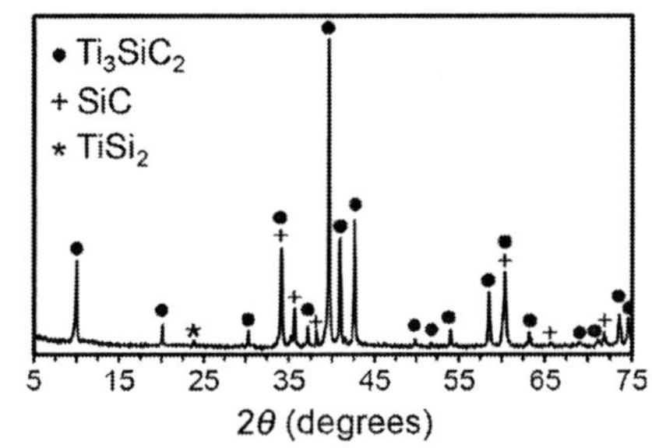

Figure 3. XRD pattern of the Ti_3SiC_2/SiC ceramic matrix composite synthesized using non-powder titanium metal items.

Based on the proposed approach, particular techniques for fabricating Ti_3SiC_2/SiC ceramic matrix composites with specified internal structures, some of which are given in figure 4, have been developed.

Solid plates and sheets. The Ti_3SiC_2/SiC ceramic composites in a shape of solid flat plates, sheets, slabs and other similar ones can be fabricated by low pressure assisted reactive sintering of multilayer sandwich-like assemblies consisting of titanium foils and tape casted polymer films filled with SiC particles, as shown in figure 4a.

Hollow-core plates. The Ti_3SiC_2/SiC ceramic composites in a shape of hollow-core flat plates, slabs and other similar ones can be fabricated by low pressure assisted reactive sintering of sandwich assemblies consisting of titanium foils, array of parallel spaced titanium rods and tape casted polymer films filled with SiC particles, as shown in figure 4b.

Blocks with arrayed hollow channels. The Ti_3SiC_2/SiC ceramic composites in a shape of blocks, slabs and other similar ones containing arrayed hollow channels can be fabricated by pressureless reactive sintering of assemblies consisting of an array of parallel aligned titanium rods with a space between which is filled with a slurry of SiC particles, as shown in figure 4c.

Cellular blocks. The Ti_3SiC_2/SiC ceramic composites in a shape of cellular blocks, slabs and other similar ones can be fabricated by pressureless reactive sintering of assemblies consisting of thin-walled titanium tubes of different cross-sections including rectangular ones, bound together with a paste of SiC particles, as shown in figure 4d.

It is especially important that all the proposed techniques do not require high loading pressure to be applied during the sintering process. In fact, there is needed only a small uniaxial load commonly less than 1–2 MPa that suppresses the warping deformation of thin titanium metal items in the first stage of heat treatment. Such a decrease in the requirement for pressure assistance during the synthesis is expected to enable the Ti_3SiC_2 MAX phase ceramic materials to gain competitive advantages through an increase in productivity of manufacturing processes.

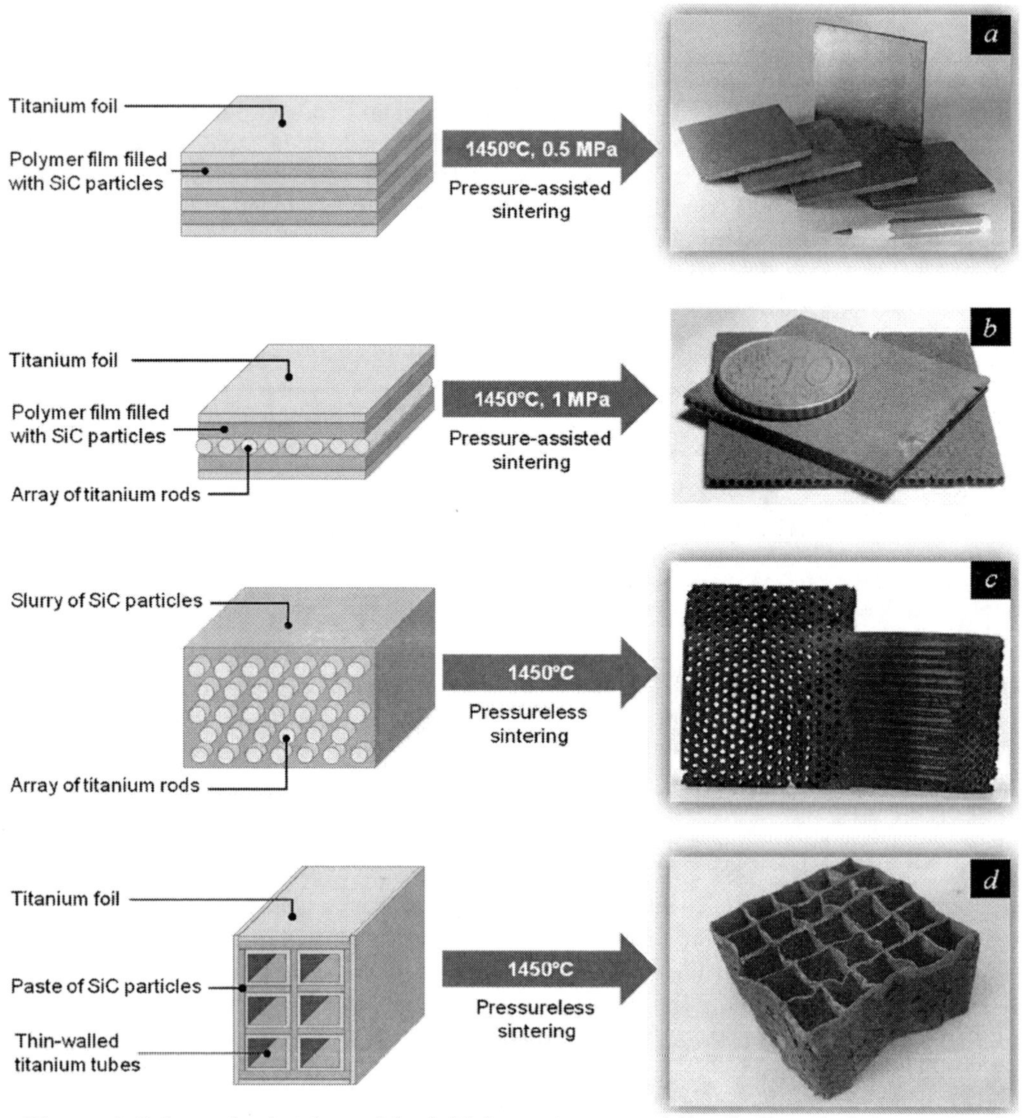

Figure 4. Schematic sketches of the initial reactive assemblies (left) and photographs of the prepared samples of the Ti_3SiC_2/SiC ceramic matrix composites (right): (a) solid flat plates, (b) hollow-core flat plates, (c) blocks containing arrayed hollow channels, and (d) cellular block.

Acknowledgments
The authors are grateful to the Russian Foundation for Basic Research for grant no. 18-08-01460. The study was performed using the equipment of the Center for Shared Use of Scientific Equipment "Khimiya" of the Institute of Chemistry, Komi Science Center, Ural Branch, Russian Academy of Sciences.

References

[1] Barsoum M W 2013 *MAX Phases: Properties of Machinable Ternary Carbides and Nitrides* (Wiley-VCH Verlag GmbH & Co. KGaA, Weinheim)
[2] Sun Z M 2011 Progress in research and development on MAX phases: a family of layered ternary compounds *Int. Mater. Rev.* **56** 143
[3] Zhang H B, Bao Y W and Zhou Y C 2009 Current status in layered ternary carbide Ti_3SiC_2, a review *J. Mater. Sci. Technol.* **25** 1
[4] Istomin P V, Nadutkin A V and Grass V E 2011 Forced SHS compaction of layered non-powder Ti–SiC compositions *Proceedings of the Komi Science Centre of the Ural Division of the Russian Academy of Sciences* **8** 12 (in Russian)
[5] Istomin P V, Nadutkin A V and Grass V E 2012 Production of Ti_3SiC_2-based materials by SHS forced compaction of layered composite Ti–SiC *Glass Ceram.* **68** 366
[6] Istomin P, Nadutkin A and Grass V 2013 Fabrication of Ti_3SiC_2-based ceramic matrix composites by a powder-free SHS technique *Ceram. Int.* **39** 3663
[7] Istomin P, Nadutkin A and Grass V 2014 Combustion synthesis of Ti_3SiC_2-based ceramic matrix composites using non-powder reactant solids *Ceram. Trans.* **248** 515
[8] Istomin P V, Grass V E and Nadutkin A V 2013 Method of producing ceramics and composite materials based on Ti_3SiC_2 *Patent RU* No. 2486164
[9] Istomin P V, Nadutkin A V and Grass V E 2017 Method of obtaining ceramic composite with a multichannel structure *Patent RU* No. 2622067

Refractory and heat-insulating materials based on aluminosilicate SHS compositions

R D Kapustin

Merzhanov Institute of Structural Macrokinetics and Materials Science, Russian Academy of Sciences, Chernogolovka, 142432 Russia

Abstract. The phase composition, structure, and physical and mechanical characteristics of refractory aluminosilicate SHS materials (AS materials) were investigated taking into account the properties of initial components and the production method. Computational methods were used to reveal macrokinetics of processes and determine the possible formation of mullite structures and corundum in aluminosilicate refractory SHS materials during the synthesis reactions. Heat losses caused by the small thickness of a coating or the walls of the material between pores were found to have a significant effect on the synthesis reaction in thin coatings and porous AS materials, as well as the on the formation of various chemical compounds.

1. Introduction

In industrial processes the refractory lining of thermal generating units is often exposed not only to high temperatures, but also to abrasive wear or corrosive media. The combined effect of destructive factors on refractory lining leads to fast damage to the material of refractory lining and requires significant repair costs.

A review of research and developments of new refractory materials and products used in modern heat generating units for metallurgy and engineering demonstrates a worldwide increasing interest to unmolded refractories and their mechanical characteristics and high temperature resistance. Refractory and heat-insulating materials are more often required due to the ability to operate for a long period of time at extremely high temperatures (up to 1700°C and above), as well as due to high wear resistance and resistance to corrosive media and flue gases.

Aluminosilicate materials are of the greatest interest caused by chemical and mineral composition, since their proportion in the refractory linings of heat generating units exceeds 80% of all types of refractories used, in particular, in the engineering industry.

Among the modern methods for producing and using unmolded refractory materials, self-propagating high-temperature synthesis (SHS) is one of the most attractive methods, since it can significantly reduce the energy consumption in the production of refractory products. In most cases, the refractory lining acquires final refractory and operational properties in the first hours of operation of heat generating units that uses SHS refractories.

SHS can be used to obtain various high-purity and lightweight refractories. SHS allows materials to be obtained with high physical and mechanical characteristics, such as strength, wear resistance, corrosion resistance, as well as high stability of properties, shapes and sizes when exposed to high temperatures. In recent years, various refractory AS materials on various bases have been developed.

In 2000 a Malaysian-American team of scientists conducted much experimental and theoretical work to obtain magnesium-aluminum refractory spinel $MgAl_2O_4$ [5]. The researchers used the SHS method for obtaining magnesium aluminate powder with very high phase purity. Low-temperature melting of aluminum followed by an exothermic reaction between molten aluminum and magnesium oxide allowed a powder to be obtained with a very high proportion of spinel phase, small particle size and narrow particle size distribution. Compacts with a density of up to 92% were obtained from this powder during sintering for 4 h at 1600°C. The addition of a small amount of aluminum oxide was found to have a favorable effect on the sintering characteristics and density of the material.

Studies concerning the use of mineral and secondary raw materials for the production of AS refractories were conducted by the Kazakh Interbranch Scientific and Technical SHS Center. Alumothermite refractories "Furnon" and "Thermok" [1] were developed. The Combustion Problems Institute (Kazakhstan) also investigated the systems TiB_2–Al_2O_3, CrB_2–Al_2O_3 and ZrB_2–Al_2O_3 which were subjected to SHS with preliminary mechanical activation. The products of exothermic interaction under constant pressure in an argon atmosphere were found to be refractory compounds of metal borides and corundum which formed a dispersed phase and a ceramic binder in a ceramic composite material [4].

In the early 2000s, much work has been done at Samara State University of Architecture and Civil Engineering to study the conditions for the production of refractory composite materials from AS refractories using various industrial wastes in AS mixtures [2]. The dispersed materials, such as a spent alumochromite AM 2201 catalyst, fireclay dust, and dusty metal processing waste can be used as ballast in AS mixtures for the production of aluminothermic composite materials. A production process was also developed to obtain products from aluminothermic composite materials.

The possible obtaining heat-resistant coatings on fireclay refractories by the method of self-propagating high-temperature synthesis in the MgO–SiO_2–Al system using liquid sodium glass as a binder with the addition of sodium tripolyphosphate and titanium dioxide activating the SHS process was studied at the Physical-Technical Institute of the National Academy of Sciences of Belarus [3]. The optimal composition was determined, and the physical and mechanical properties, macro- and microstructure of the coatings were investigated.

A large number of refractory AS materials was developed at the Scientific and Production Commercial Company "MaVR", Zhukovsky [6, 7]. Synthesis of ceramic structures (corundum, sillimanite, mullite, etc.) in most of these materials is conducted after heating up to a temperature of 850–900°C and initiating self-propagating high-temperature synthesis. The synthesized ceramics is a mullite structure with the formula $mAl_2O_3\ nSiO_2$.

Despite a significant number of studies, the practical application of such materials is limited due to insufficient studying the processes that take place in the materials during the SHS process, which does not allow their physical and mechanical and operational properties to be predicted with sufficient accuracy.

The purpose of this work is to find the phase composition, structure and physical and mechanical characteristics of refractory aluminosilicate SHS materials, taking into account the properties of initial components and the production method, as well as to improve the production process and use of AS materials and products from them to protect the heat generating units against high temperatures.

2. Materials and Methods

In the work the parameters were calculated and the composition of final products was determined for all AS materials studied. The computation of the combustion parameters for the AS material (M-1 grade) and the composition of the products obtained, depending on the initial temperature for the initiation of the synthesis reaction, is given as an example.

The computation was performed in the «Combustion Regime P = Const» mode. Table 1 provides the input data for the mass ratio of the initial components in the mixture (M-1 grade). The initiation temperature varied from 800 to 1500°C with a step of 100°C. When the initiation temperature exceeds the melting point of metallic aluminum (T_m = 660°C) means that aluminum starts reacting in the molten state. The formation of the melt indicates the occurrence of synthesis reactions on the surface of solid SiO_2 particles with the melt in the mode of capillary spreading.

The experimental procedure involved the annealing of samples in the step heating mode in a laboratory electric chamber furnace, followed by an investigation of phase composition, microstructure, and material properties. The maximum annealing temperature of samples varied from 900 to 1600°C with a step of 100°C [8].

Table 1. Chemical composition of the AS material (M-1 grade) during the reaction

	Mass, g	Amount of substance in the composition, mol	aggregative state
Al	135	5.0000	Liquid
Al_2O_3	6	0.0590	Solid
SiO_2	989	16.5000	Solid
Fe_2O_3	0.7	0.0044	Solid
N_2	77.8	2.7900	Gas
Mixture	1200	24.3534	-

3. Results and Discussion

The calculated adiabatic temperature of combustion of the AS material is 1727°C. The enthalpy of products ranges from minus 13,000 kJ to minus 14,000 kJ. All computations of the thermodynamic parameters for the synthesis process, as well as the composition of the final products obtained from the mixture (M-1 grade), depending on the initiation temperature, are given in table 2.

Table 2. Thermodynamic synthesis parameters and the composition of the final reaction products in the AS material (M-1 grade), and computations performed using the code ISMAN-THERMO

Initiation temperature, °C	800	900	1000	1100	1200	1300	1400	1500
Gas pressure, atm	1	1	1	1	1	1	1	1
Volume of gaseous products, liter	355	394	434	474	513	554	595	637
Adiabatic temperature, °C	1727	1727	1727	1727	1727	1727	1727	1727
Moles of the gas phase	2.094	2.326	2.558	2.793	3.028	3.267	3.508	3.758
Heat capacity (Cp) of products in the mixture, J/K	1739	1720	1700	1680	1660	1639	1619	1598
Entropy of products, J/K	4144	4147	4148	4151	4153	4156	4158	4161
Enthalpy of products, kJ	−13980	−13839	−13697	−13554	−13410	−13265	−13118	−12967
Jump in enthalpy, kJ	2413	2412	2411	2412	2410	2411	2410	2412
Mass of the system, kg	1.2	1.2	1.2	1.2	1.2	1.2	1.2	1.2
Mass of the product $Al_2Si_2O_{13}$, kg	0.3165	0.2768	0.2369	0.1977	0.1572	0.1173	0.0757	0.0342
Mass of the product $Al_6Si_2O_{13}$, kg	0.1584	0.1762	0.1941	0.2116	0.2297	0.2476	0.2662	0.2847
Mass of the product SiO_2, kg	0.2741	0.3049	0.3358	0.3661	0.3975	0.4284	0.4607	0.4927
Mass of the product Al_2O_3, kg	–	–	–	–	–	–	–	–
Mass of the product Si, kg	0.1174	0.1119	0.1063	0.1008	0.0951	0.0895	0.0837	0.0778

Experimental annealed samples were analyzed using X-ray diffraction by DRON-3M diffractometer that showed that after heating to 900°C, the structure consisted of Al_2O_3, SiO_2 and Si, which is related to the nonadiabaticity of synthesis in thin coatings due to the significant heat removal to the environment and the fireclay base material. In this regard, the elements obtained in the coating at 900°C interacted during the annealing of the samples in the temperature range from 1300 to 1600°C and formed the mullite structure (sillimanite) with the chemical formula Al_2SiO_5.

This fact is confirmed by studies of the microstructure of the coating using a scanning electron microscope. The photograph of the microstructure of the coating, which is presented in figure 1a (heat treatment at 1100 C), clearly shows corundum in the form of spherical boulders, while figure 1b (heat treatment at 1600°C) shows a monolithic mullite structure.

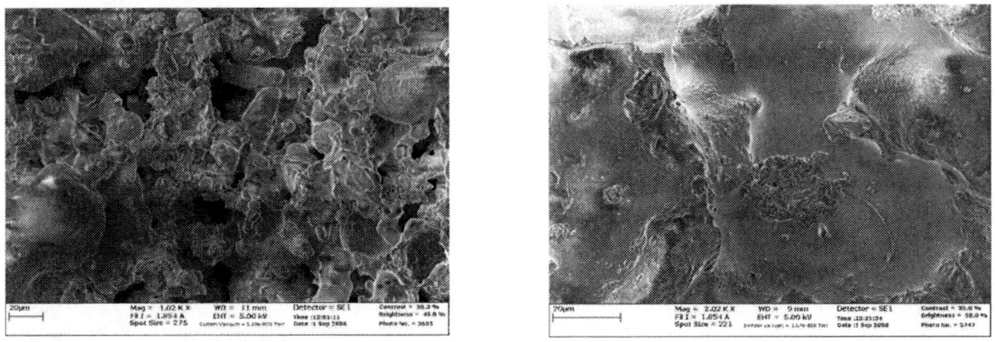

(a) (b)

Figure 1. Microstructure of the AS material (M-1) after heating at: (a) 1100°C and(b) 1500°C

Experimental studies of AS materials (VBF grade) in the form of bulk cellular samples showed that a synthesis process was initiated in the material after heating to 1100°C, and as a result, such compounds as Al_2O_3, SiO_2 (christabolite) and SiO_2 (quartz) became the main components in the structure of materials (VBF grade). Mullite structures (as well as in thin coatings) were not formed due to the high porosity of porous materials, the small thickness of the walls between pores, and significant heat losses.

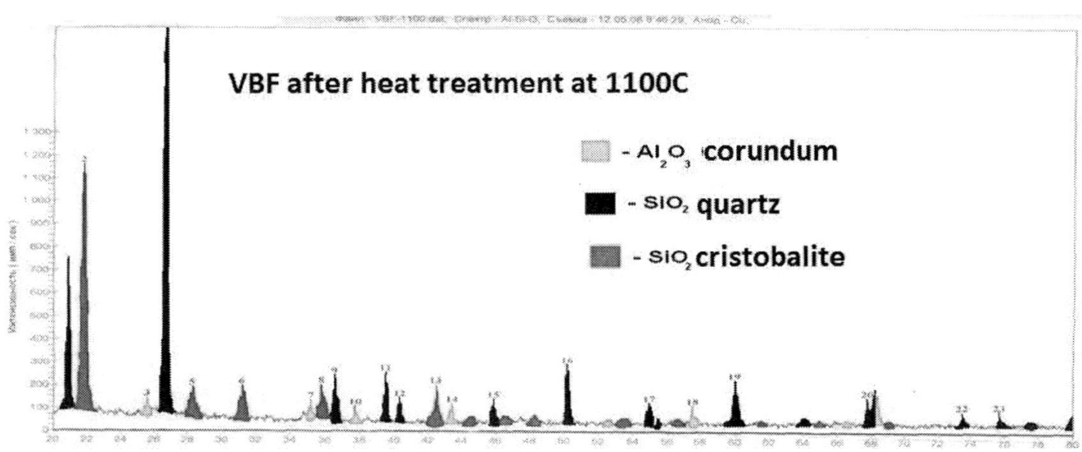

Figure 2. Structure of porous AS material (VBF grade) after annealing at the temperature of 1100°C

Thus, the following conclusions can be drawn from this study:

- macrokinetics of processes was established and the formation of mullite structures and corundum in aluminosilicate refractory SHS materials subjected to heating to different temperatures was determined during the synthesis reaction using computational methods and the specialized code ISMAN-THERMO.

- composition of compounds obtained in the material was experimentally studied to develop the application technique and evaluate structural changes in various modes of operation of refractory products based on AS materials. The computations were in good agreement with experimental data.

- heat losses caused by the small thickness of a coating or material walls between pores have a significant effect on the development of the synthesis reaction in thin coatings and porous AS materials, as well as the on the formation of various chemical compounds.

References

[1] Seidaev A P, Ismailov MB, et al 1993 *J. Eng. Phys. Thermophys.* **65** 623–626
[2] Ryazanov S A 2005 *Development of the method for the production of refractory products from aluminothermic composite materials using recycled materials* Ph.D. thesis (Samara) 163 p
[3] Hina B B, Volochko A T and Zhukov A A 2011 *Glass Ceram.* **2** 29–32
[4] Abdulkarimova R G, Seidualiyeva A J and Kamunur K 2018 *J. Mater. Sci. Eng.* B **8** 56–65 doi: 10.17265/2161-6221/2018.3-4.003
[5] Lim Rooi Ping, Abdul-Majeed Azad and Teng Wan Dung 2001 *Mater. Res. Bullet.* **36** 1417–30 https://doi.org/10.1016/S0025-5408(01)00622-5
[6] Vladimirov V S, Galagan A P, Ilyukhin M A, Karpukhin I A, Moizis S E and Moizis E S 2002 *Refrac. Indust. Ceram.* 87–88
[7] Vladimirov V S, Karpukhin I A, Moizis S E and Moizis E S 2004 *Refrac. Indust. Ceram.*
[8] Kapustin R D 2016 *Refrac. Indust. Ceram.*

IOP Conf. Series: Materials Science and Engineering **558** (2019) 012017 doi:10.1088/1757-899X/558/1/012017

Synthesis of Ti_2AlN MAX-phase by sintering in vacuum

A A Kondakov, I A Studenikin, A V Linde, N A Kondakova and V V Grachev

Merzhanov Institute of Structural Macrokinetics and Materials Science, Russian Academy of Sciences, Chernogolovka, 142432 Russia

E-mail: kondakov@ism.ac.ru

Abstract. The dependence of MAX-phase synthesis on phase composition of the initial mixture and heat treatment terms in the electric furnace was investigated. Using Ti and AlN powders with the molecular ratio 2:1 as an example, the sequence of phase composition changes with increasing exposure temperature was tracked. It was established that the maximum amount of Ti_2AlN MAX-phase in the end products is obtained at 1400°C. At this temperature a single phase product (100% wt Ti_2AlN) was obtained for the initial mixture Ti:Al:TiN = 1:1:1. The principal possibility of scaling for obtaining a single-phase product of Ti_2AlN by vacuum sintering was experimentally shown on the example of the mixture of this composition weighing 500 g at a certain mode of thermal vacuum treatment.

1. Introduction

Ti_2AlN ternary compound belongs to MAX-phase systems with the general formula $M_{n+1}AX_n$ (usually $n = 1 \div 3$), where M is a transition metal, A is an element from IIIA and IVA groups of Mendeleev periodic system, X is carbon or nitrogen [1]. MAX phase compounds are the point of the scientific interest for different research groups since they possess properties of metal and ceramic simultaneously [2, 3]. Ti_2AlN-based materials are characterized by high electric and heat conductivity, high elastic modulus, low density, high thermal stability and ability to mechanical treatment that makes them attractive for various high-temperature applications [4].

The aim of this work was to obtain a single phase product (Ti_2AlN) by sintering different mixtures in the vacuum furnace. The additional aim was to define an opportunity of the process scaling to obtain a considerable amount of the single phase product of Ti_2AlN (using the example of 500 g specimen). Literature data analysis did not reveal any application of this method for obtaining Ti_2AlN.

2. Experimental

The powders of commercial Ti (PTS-1 brand), Al (ASD-4 brand), TiAl (PT65Yu35 brand) and home-made SHS-produced AlN, TiN were used as raw materials. The following powder blends: 2Ti + AlN (blend **1**), TiAl + TiN (blend **2**), and Ti + Al + TiN (blend **3**) were intermixed in a ball mill. The prepared blends in quartz tubes were placed in a vacuum furnace (SNVE-16/16) and held at $T = 1100$, 1200, 1300, 1400, and 1500°C (pressure 5.8×10^{-6} mm Hg) for 60 min. The sintering products were characterized by XRD (DRON-3M) and SEM/EDS (Zeiss Ultra Plus, Carl Zeiss equipped with an INCA Energy 350 XT accessory, Oxford Instruments).

3. Results and discussion

Diffraction patterns of the products obtained by sintering of blend **1** are presented in figure 1. It is evident that at $T = 1100–1200$°C the obtained product is multiphase. It contains Ti_2AlN, TiN, AlN, Ti_3AlN, and Ti_3Al. With increasing temperature, the Ti_2AlN content is seen to grow from 20 to 52% (figure 2).

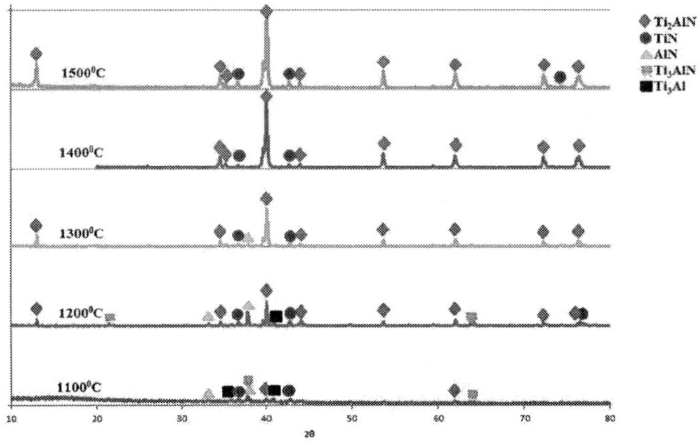

Figure 1. Diffraction patterns of products obtained by sintering of blend **1** at various temperatures.

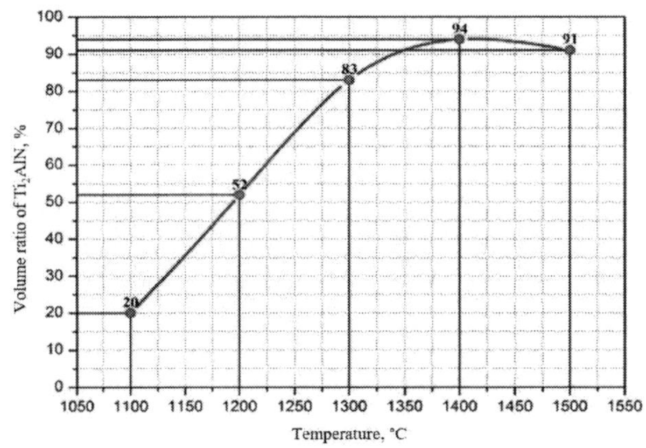

Figure 2. Volume ratio of Ti_2AlN at various sintering temperatures, %.

With an increase in the temperature to 1300°C the obtained product contains 83% Ti_2AlN, 12% TiN and 5% AlN. The product formed at 1400°C contains 94% Ti_2AlN and 6% TiN. But in the product formed at 1500°C the amount of Ti_2AlN decreases to 91% and the amount of TiN increases to 9%.

The decrease in Ti_2AlN amount coincides with the data in [5], which demonstrated that at $T > 1550$°C in dynamic vacuum, aluminum evaporated from the MAX-phase to form non-stoichiometric titanium nitride:

$$Ti_2AlN_{(s)} \rightarrow 2TiN_{0.5} + Al_{(g)}$$

That implies that the maximum amount of Ti_2AlN MAX phase (94%) can be obtained at $T = 1400$°C. That is why blends **2** and **3** were sintered at the same temperature.

After sintering of blend **2** the product formed at 1400°C contains 94% Ti_2AlN, 4% TiN, 1% Ti_3AlN and 1% Ti_3Al. The diffraction pattern is shown in figure 3. After sintering of blend **3** the single phase Ti_2AlN product was obtained (figure 4a). The material exhibits a nanolaminate structure typical of MAX compounds (figure 4b).

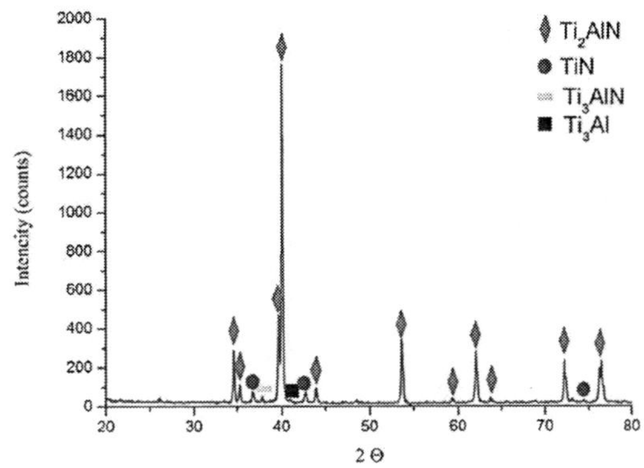

Figure 3. Diffraction pattern of the product obtained by sintering of blend **2** at 1400°C.

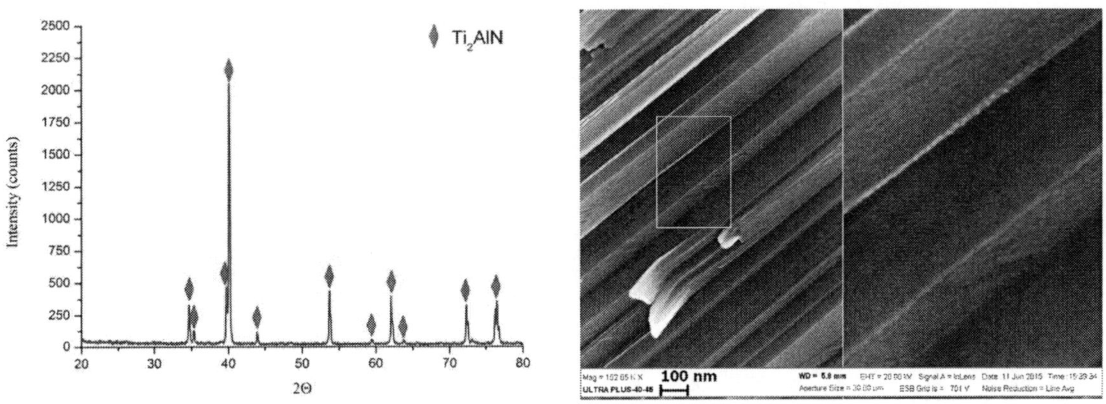

(a) (b)

Figure 4. (a) Diffraction pattern and (b) fracture surface of the product obtained by sintering of blend **3** at 1400°C.

The scale factor investigation was carried out with blend **3** since the maximum product yield of Ti_2AlN MAX-phase was obtained with it. The experimental results are presented in figure 5.

Figure 6 shows the fracture surface of the product. The EDS results for the area indicated in figure 6 gave the following elemental composition: N 23.56 at %, Al 25.68 at %, and Ti 50.76 at %; it is in good agreement with that of Ti_2AlN.

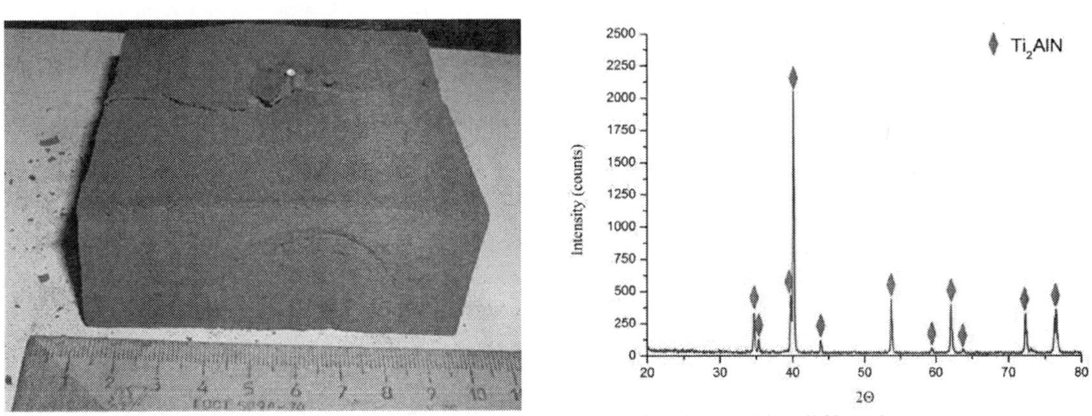

Figure 5. The product obtained from 500 g of blend **3** and its diffraction pattern.

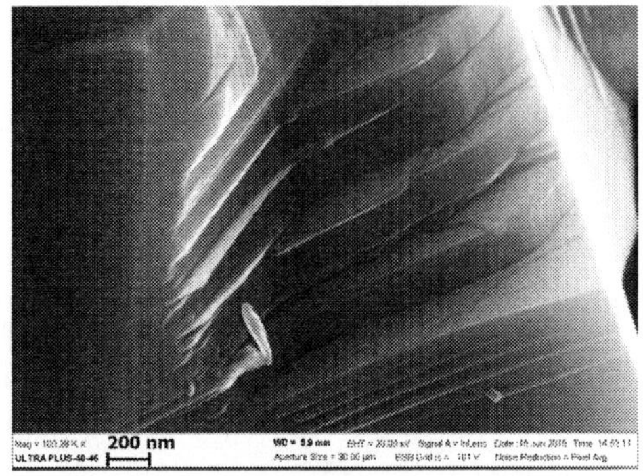

Figure 6. Fracture surface of the product obtained from 500 g of blend **3**.

4. Conclusions

Thus, as a result of the studies the exposure temperature of 1400°C was determined, at which the maximum content of Ti_2AlN MAX phase was observed in the end product. The maximum amount of Ti_2AlN MAX-phase (100%) was obtained for the mixture Ti:Al:TiN=1:1:1. Using the example of sintering of 500 g of mix 3, the principal possibility of scaling for the sintering process in vacuum and perspectives of this method for industrial application were demonstrated.

References

[1] Barsoum M W 2000 *Prog. Solid St. Chem.* **28** 201–281
[2] Radovic M and Barsoum M W 2013 *Am. Ceram. Soc. Bullet.* **92** 20–27
[3] Barsoum M W 2013 *MAX Phases: Properties of Machinable Carbides and Nitrides* (Wiley VCH)
[4] Lin Z, Zhuo M, Li M, Wang J and Zhou Y 2007 *Scripta Mater.* **56** 1115–18
[5] I M Lowa, W K Pang, S J Kennedy and R I 2011 *J. Eur. Ceram. Soc.* **31** 159–166

Theoretical and experimental study of cellular structures in restricted filtration combustion of porous media

P M Krishenik[1], S V Kostin[1], N I Ozerkovskaya[1] and K G Shkadinsky[2]

[1]Merzhanov Institute of Structural Macrokinetics and Materials Science, Russian Academy of Sciences, Chernogolovka 142432 Russia
[2]Institute of Problems of Chemical Physics, Russian Academy of Sciences, Chernogolovka 142432, Russia

E-mail: petr@ism.ac.ru

Abstract. Infiltration-mediated combustion of a thin porous cylinder in reactive gas was studied by mathematical modeling. Gas reagent is infiltrated through an annular gap left of the side surface. In conditions of instability, the self-sustained propagation of individual cellular spots of exothermic combustion was observed. The structure of cellular waves, their dynamics, and travel direction are defined by such factors as temperature profile, pressure gradients, heat release rate, porous structure, and heat losses. Cellular combustion of Ti powder in dish was also explored experimentally and the results were compared with theoretical predictions.

1. Introduction

Still back in the 60s, Zel'dovich and Barenblatt [1] have demonstrated that the process of flame propagation in gases can be spasmodically interrupted upon temperature decrease below some critical value. In the case of solid-state combustion, critical conditions for flame propagation are complicated by front instability phenomena and formation of periodic wave propagation modes [2]. According to Aldushin [3], the extent of conversion in infiltration-mediated combustion initially grows with increasing heat losses and then the blow-out of combustion takes place upon reaching complete conversion. In conditions of unsteady infiltration combustion, an initially planar combustion front was found to degenerate into several self-sustaining hot spots of cellular combustion [4–6].

This work was aimed at the theoretical and experimental study of cellular combustion of porous thin disk in conditions of restricted supply of gas reagent.

2. Mathematical model

The nonlinear dynamics of the filtration combustion front was numerically studied using a three-dimensional system of equations of filtration combustion [5, 6]. Combustion geometry is schematically presented in figure 1. Gas reagent entered through a gap at the top. After ignition at the center, an annular expanding combustion begins to propagate over a porous combustible disk (figures 1a, 1b).

The processes are described by a system of differential equations reflecting the conservation laws with the use of simplifying assumptions. We assume that the motion of the gas reagent from the outside is almost completely due to longitudinal motion along the slit, and the gas is consumed with a thin layer of porous condensed reagent in the zone of intense chemical interaction. Transverse filtration transport of the gas from the slit into the interior of the thin layer is under almost isobaric conditions. The longitudinal filtration of the gas in the porous layer is neglected. In the mathematical analysis, the parameters of the considered system were chosen so that, in the front propagation at insignificant heat losses, the conversion of the condensed phase is incomplete. The proposed mathematical model is similar to the model of thermal diffusion combustion of gases, and, consequently, the characteristics of the frontal exothermal chemical conversion (cellular structure of the flame front for gases and porous media) may have common features.

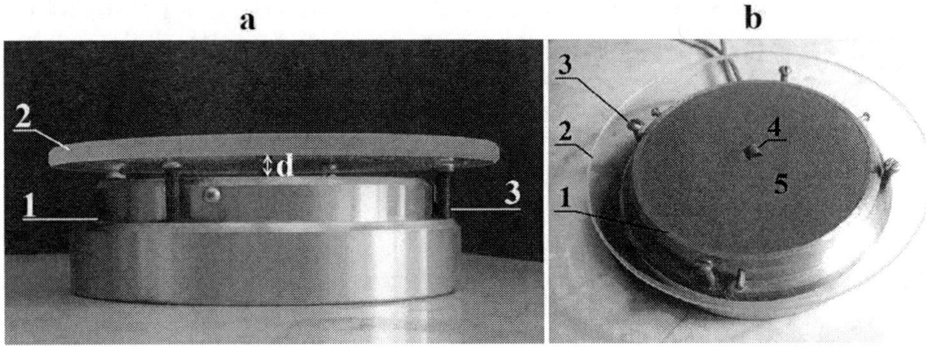

Figure 1. Experimental setup: *1* cylindrical tray (inner ∅ 70 mm), *2* silica shield, *3* supports, *d* gap width (2 mm), *4* igniting coil, and *5* layer of Ti powder (2 mm thick, mean particle size below 100 μm).

On going outward (figure 2a), the combustion loses the stability and degenerates into a pair of hot spots that move toward the incoming gas flow (figure 2b). Because of gas deficit, one of the cells disappears (figure 2c), while the remaining one continues its motion (figure 2d). The distribution of temperature in hot spots in conditions of low heat losses is shown in figure 3.

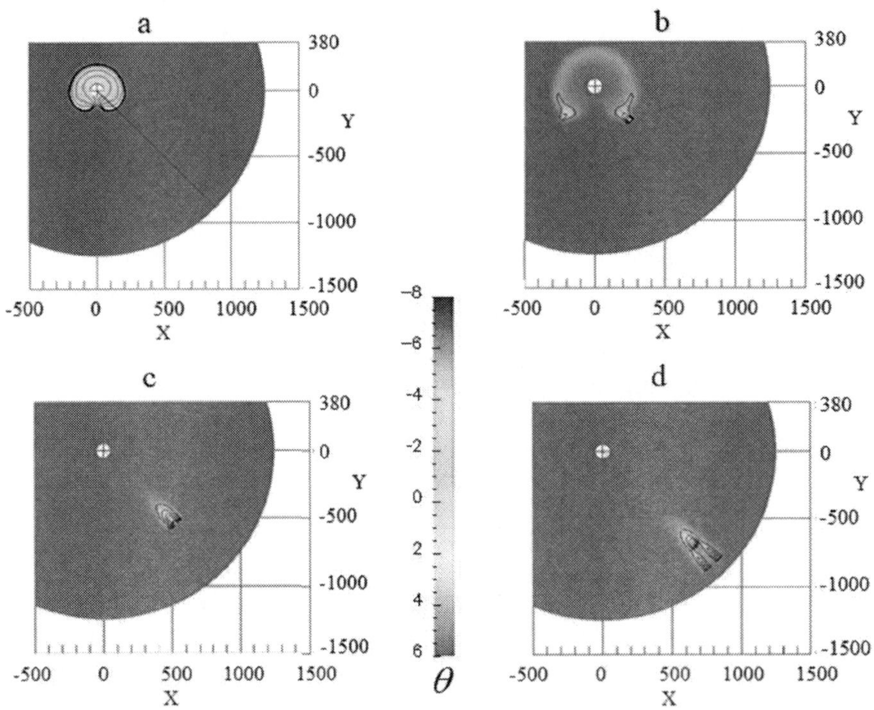

Figure 2. Evolution of cellular combustion over a combustible porous disk, θ is the temperature.

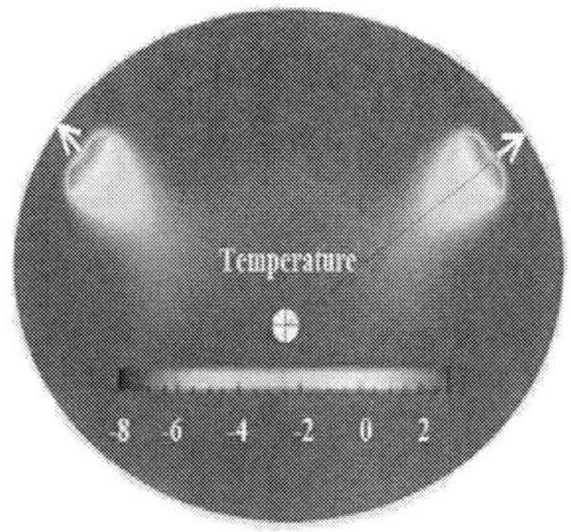

Figure 3. Temperature distribution in hot spots in conditions of low heat losses.

Upon reaching the disk edge, the hot spot divides into three independent cells: two moving into different directions while the third one toward the disk center (figure 3).

3. Experiment

To check the predictions of the model, we carried the combustion in the experimental setup shown in figure 1. Providing a visual access to combustion processes, we study the filtration combustion of a thin layer of an energetic porous composition. The layer thickness is determined by temperature uniformity and good permeability for the gas reagent in the transverse direction. The porous layer should be on an insulating inert substrate with a possibility for controlling the heat transfer process between them. It is reasonable to implement filtration of the active gas reagent from the outside through the slit between the layer of the porous condensed reagent and the transparent flat wall located parallel to the porous layer at an adjustable distance. The location and nature of the motion of cells of an intense exothermic chemical conversion depend on the conditions of the combustion process and heat and mass transfer with the environment. This is confirmed by experiments carried out in a circular Duralumin bath (inner diameter of 70 mm). The titanium powder filled the bath to the upper edge, the combustion process was initiated in the center of the circle, and the gap for the supply of the gas reagent was located at its periphery because the glass was lifted above the bath by means of small screws. Initially, we observed the formation of an expanding high-temperature source of filtration combustion provided by the active gas reagent because of its great curvature. As it expanded, the inside of the source reacted completely. Cells formed at the outer edge of the expanding combustion region separated from the front (Fig. 4a), and independently propagated toward the gas flow. The number of such cells increased, and the circular symmetry was destroyed. A set of cells of exothermic conversion moving over the titanium layer was formed. They moved to the periphery, "aiming" to provide themselves with a necessary amount of active gas. Naturally, the area of unreacted titanium layer in the bath decreased, and some of the cells die (figures 4b, 4c). After the last cell disappeared, the entire bath with reacted and unreacted titanium "islets" cooled (figure 4d). The trails of the cells can be studied in detail from changes in the color of the titanium layer.

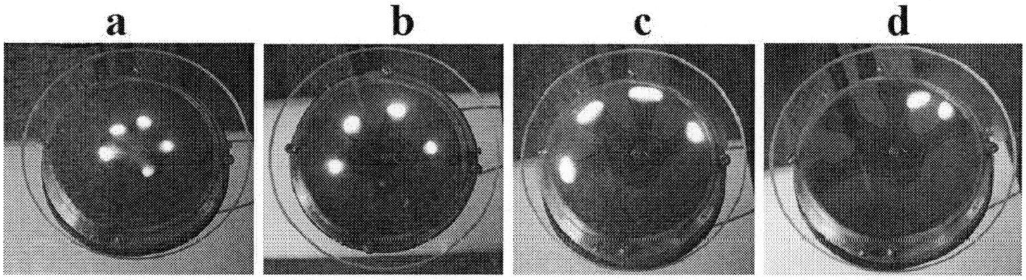

Figure 4. Still frames of combustion over a 7-mm sublayer of titania used as heat insulation; time elapsed from the onset of combustion: (a) 14, (b) 30, (c) 46, and (d) 72 s.

Figure 4 presents the still frames of Ti powder combustion in atmospheric air over a 7-mm sublayer of titania placed beneath sample *5* used as heat insulation. Comparison shows that theoretical predictions qualitatively agree with experimental results.

4. Conclusions

Evolution in a mode of combustion propagation is defined by not only infiltration of gas reactant, heat/mass exchange with environment but also by combustion geometry. Our results shed new light on the non-liner dynamics of combustion in heterogeneous systems and also on the character of critical conditions for combustibility limits.

References

[1] Zeldovich Ya B and Barenblatt G I 1959 *Combust. Flame* **3** 61
[2] Firsov A N and Shkadinsky K G 1987 *Combust. Explos. Shock Waves* **23** 288
[3] Aldushin A P 1988 Infiltration-mediated combustion *Rasprostraneniye teplovykh voln v geterogennykh sredakh* (Propagation of Thermal Waves in Heterogeneous Media) ed YuSh Matros (Novosibirsk: Nauka) chapter **2** pp 9–52
[4] Kostin S V and Shkadinsky K G 2010 *Dokl. Phys.* **55** 533
[5] Ozerkovskaya N I, Firsov A N and Shkadinsky K G 2010 *Combust. Explos. Shock Waves* **46** 515
[6] Kostin S V, Krishenik P M, Ozerkovskaya N I, Firsov A N and Shkadinsky K G 2012 *Combust. Explos. Shock Waves* **48** 1

IOP Conf. Series: Materials Science and Engineering **558** (2019) 012019 doi:10.1088/1757-899X/558/1/012019

Effect of nanosized titanium and silicon carbides on synthesis and consolidation of titanium silicon carbide during spark plasma sintering

V B Kulmet'eva, M N Kachenyuk, S A Oglezneva and A A Smetkin

Perm National Research Politechnical University, Perm, 614990 Russia

Abstract. Using spark plasma sintering method, at temperature T = 1400°C and load of 30 MPa samples of composite material Ti_3SiC_2/TiC were obtained. Before SPS powder mixtures with molar ratio 3Ti–1.25SiC–0.75C with addition of up to 4% wt of nanopowders of titanium carbide or silicon carbide were mechanically activated in a planetary mill. Material structure was studied using X-ray structural analysis, hardness and fracture toughness was evaluated on basis of Vicker's indentation results. It was established that introduction of nanocarbides up to 1% wt significantly reduces contents of Ti_3SiC_2, and further on practically has no influence on phase composition. Increase of isothermal time from 5 to 15 min leads to an increase in the Ti_xSi_y secondary phases to 11–14%. Introduction of nano-TiC and nano-SiC into mixtures increases crack fracture toughness to 7.06–7.55 MPa·m$^{1/2}$ for samples after 5 min of SPS and practically does not change it after 15 min of SPS.

1. Introduction.

Composite materials on of oxygen-free compounds are widely enough known and are used in production of wear-resistant parts for extreme (temperature, environment) operational conditions. Ti–Si–C composition is most promising system due to number of advantages over other systems. Optimal elemental contents of such composition during high-temperature consolidation provides for formation of multi-phase system, consisting mainly from titanium silicon carbide Ti_3SiC_2 and titanium carbide TiC. Carbides are widely used as basis for tungstenless hard alloys, in carbidic steels, wear-resistant materials, as they have high hardness, wear resistance, resistance to acids, alkali, temperature up to 1000°C [1, 2]. Disadvantage of traditional ceramics is low resistance to damage. In this aspect materials on basis of laminate compound – titanium silicon carbide are of interest, which has quasi-plasticity feature, i.e. is able to restrict crack propagation effectively [1, 3]. Addition of additional phase-forming elements, improving complex of properties of wear-resistant materials is a relevant objective [3–6].

The purpose of the work was to study evolution of phase composition during spark plasma sintering of Ti–SiC–C powder composition, supplemented with nanosized powders of silicon carbide and titanium carbide, and evaluation of Vickers hardness depending to contents of secondary phases.

2. Methods of experimentation and analysis.

For synthesis of titanium silicon carbide powders of titanium TPP-7, silicon carbide with fraction less 10 µm and graphite S-1. To increase hardness and wear resistance comparing to soft titanium silicon carbide obtained by reaction sintering of the source powders, 0.5–4.0% wt of nanosized powders of silicon carbide and titanium carbide were added. Powder mixtures were treated in planetary mill

SAND under following conditions: 1) ratio of drum angular velocity to cuvette angular velocity $k = -0.39$; 2) ratio of planetary rotation radius to cuvette radius $n = 2.7$; 3) drum rotation frequency 280 min^{-1}; 4) cuvette rotation frequency 620 min^{-1}. To reduce milling of foreign material from the equipment titanium cuvettes were used, to prevent oxidation of the components mechanical activation (MA) was performed under vacuum. As grinding bodies titanium cylinders diameter 10 mm, height 10 mm were used. Ratio of masses of grinding bodies and material processed was 7.5:1.

Consolidation of powder mixtures was performed by spark plasma sintering method using SPS 1050b unit (Dr. Sinter, Japan) at temperature 1400°C, holding time for 5 and 15 min [2, 3].

Phase composition was analyzed on a XRD-6000 (Shimadzu) X-ray diffractometer in λ-Cu radiation using Rietveld full-profile analysis. Structural analysis was performed using an Axiovert 40MAT (Carl Zeiss) optical microscope.

Vickers hardness was determined by indenting samples by diamond pyramid under aload of 30 kg, and calculated per formula

$$H_v = \frac{18187.74 \cdot P}{a^2} \tag{1}$$

where P is the load, kg; a is the indentation diagonal, μm.

Crack resistance was evaluated per Evans formula [7]:

$$K_{IC} = 0.16 \cdot \left(\frac{c}{a}\right)^{-1.5} \cdot \left(H_v \cdot a\right)^{0.5} \tag{2}$$

where c is the mean length from center of impression to tip of crack obtained during indentation, μm; a is the ½ length of indentation diagonal, μm; H_v is the Vicker's hardness, GPa.

3. Discussion of results.

Analysis of shrinkage during sintering shows that addition both of silicon carbide and titanium carbide in quantity up to 2% reduces shrinkage, and in greater quantity – increases it again (figure 1). Increase of holding time even greater emphasizes differences in shrinkage. For 1–2% of additives increase of holding time practically has no influence on shrinkage, while for other compositions holding time of 15 minutes increases it by 10–12%.

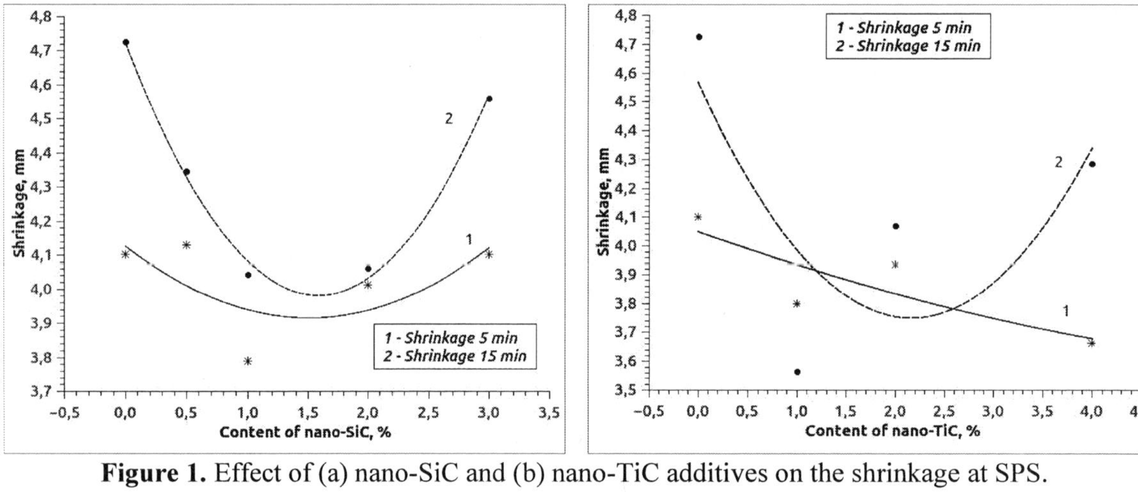

Figure 1. Effect of (a) nano-SiC and (b) nano-TiC additives on the shrinkage at SPS.

All compositions after SPS consist mainly from titanium silicon carbide, but introduction of nanosized additives somehow reduces its content: from 80 % for composition without additives to 65 for sample with 4% SiC. Probably carbides nanopowders inhibit not only shrinkage, but also synthesis of the compounds during reaction sintering. Figure 2 shows dependencies of content of Ti_3SiC_2 and TiC phases on contents of TiC and SiC nanosized powders during spark plasma sintering at

T = 1400°C. Obviously, introduction of up to 1% wt nanocarbides significantly reduces contents of ternary compound, and further on does not practically influence phase composition. Independently to what additive is doped to mixtures nano-TiC or nano-SiC, phase composition in studied range of 1–4% wt stays stable.

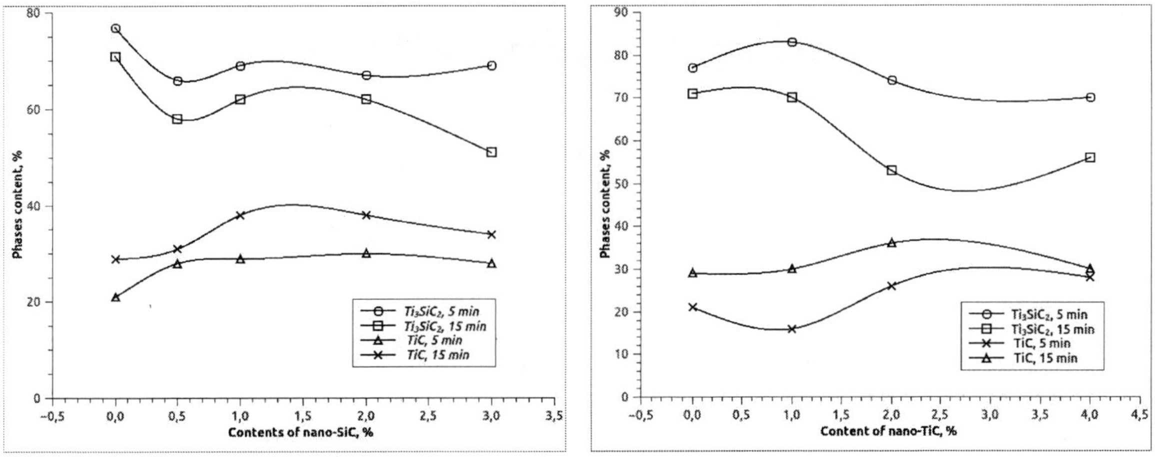

(a) (б)

Figure 2. Phase composition dependence at SPS of mixtures with different content of (a) nano-SiC and (b) nano-TiC additives.

It should be noted that during SPS, except for main phases of Ti_3SiC_2 and TiC, secondary phases of titanium silicide Ti_xSi_y are observed. In largest quantity (up to 11–14%) they are found during SPS for 15 min for all compositions. During SPS for 5 min their contents does not exceed 1–3%. Obviously, this is determined by increase of role of diffusion mass transfer at increase of high temperature consolidation time. In majority of cases the researchers showed that already at 1200°C shell structures from Ti_5Si_3 and TiC are formed on surface of titanium particles, and further on with temperature increase, due to interaction of Ti_5Si_3 and TiC titanium silicon carbide Ti_3SiC_2 is formed. On the other side, mechanism of phase composition change depending on content of additives and SPS duration at 1400°C is complex enough and is possibly determined by several important processes. During heating not only reaction sintering takes place, but at high temperature partial decomposition of titanium silicon carbide is possible, leading to increase of titanium carbide share at1400°C [8]. Wei-Kong Pang and I.M. Low [9], on basis of refinement of Ti_3SiC_2 phase transformations determined by thermal decomposition during vacuum annealing, showed that increase of TiC_x as result of Ti_3SiC_2 decomposition was observed during temperature increase from 1200 to 1500°C. That is, Ti_3SiC_2 originally decomposes very slowly to TiC_x at 1200°C through sublimation of Si and Ti, because their vapor pressure exceeds pressure in vacuum furnace. Consequently speed ofTi_3SiC_2 decomposition to TiC_x significantly increases.

All samples after SPS are characterized by 3–5% porosity. Microstructure of one of compositions is shown in figure 3. The material consists of titanium silicon carbide grains (dark phase) and titanium carbide grains (bright phase). Titanium silicon carbide grains are bigger than titanium carbide grains due to greater growth speed during spark plasma sintering.

Figure 4 shows the dependence of the Vickers hardness on the content of nano-TiC and nano-SiC additives. From the results obtained it follows that composites obtained with nano-SiC additives have slightly greater hardness, as there is a greater titanium carbide content in SPS-consolidated samples. At the same time, in samples made from mixtures with addition of nano-TiC, the TiC content is lower, and consequently Vickers hardness is lower.

Figure 3. Microstructure of Ti–SiC–C + 2% nano-TiC composition.

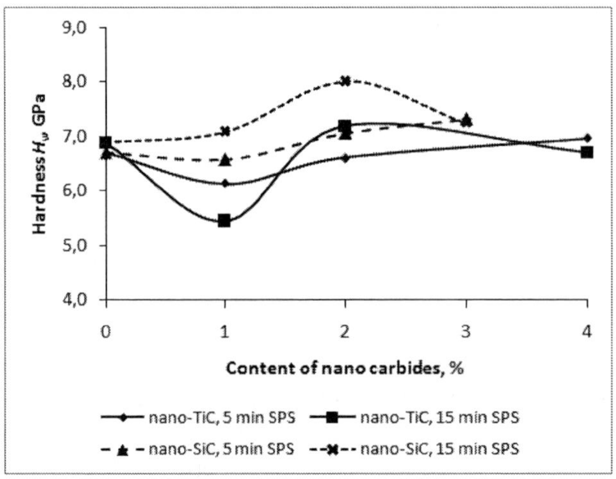

Figure 4. Effect of nano-TiC and nano-SiC additives on the Vickers hardness of the composite material.

Fracture toughness calculated per Evans for material without nanocarbide additives after SPS for 5 and 15 min was 6.58 and 6.28 MPa·m$^{1/2}$. Introduction of nano-TiC and nano-SiC into mixtures increases crack resistance to 7.06–7.55 MPa·m$^{1/2}$ for samples after 5-min SPS and practically does not change it after SPS for 15 min.

4. Conclusions

Using spark plasma sintering method at 1400°C under a load of 30 MPa, Ti_3SiC_2/TiC composite material samples were obtained. The addition of nano-TiC or nano-SiC powders up to 1% wt, prior to consolidation, significantly reduces the Ti_3SiC_2 content, and further on (up to 4% wt) practically does not affect the phase composition. The increase in the isothermal holding time from 5 to 15 min leads to an increase in the content of Ti_xSi_y secondary phases up to 11–14%. Introduction of nano-TiC and nano-SiC into mixtures increases the fracture toughness to 7.06–7.55 MPa·m$^{1/2}$ for samples after 5-min SPS and practically does not change it after 15-min SPS.

Acknowledgments

The work was supported by the Ministry of Education and Science of the Russian Federation in the framework of the basic part of the state task (no. 11.8353.2017/8.9).

References

[1] Sun Z M 2011 *Int. Mater. Rev.* **56** 143

[2] Smetkin A A and Mayorova Yu K 2015 *Bulletin of Perm National Research Polytechnic University. Mechanical Engineering Materials Science* **17** 120
[3] Barsoum M W and El-Raghy T 2001 *Am. Sci* **89** 334
[4] Zhu J and Mei B 2003 *J Mater Sci Lett* **22** 889
[5] Yang J, Gu W, Pan L M, Song K, Chen X and Qiu T 2011 *Wear* **271** 2940
[6] Zhu Y 2015 *Ceram. Int.* **41** 6950
[7] Evans A G and Charles E A 1976 *J Am Ceram Soc* **59** 371
[8] El Saeed M A, Deorsola F A and Rashad R M 2013 *Int. J. Refrac. Met. Hard Mater.* **41** 48
[9] PangW-K and Low I M 2018 *Advances in Ceramic Matrix Composites (Second Edition)* ed I M Low (Woodhead Publishing Series in Composites Science and Engineering) chapter 18 pp 429–460

Processing of Al–Zn–Mg–Cu/SiC composite prepared by mechanical alloying

E I Kurbatkina, A A Shavnev, O I Grishina and A V Gololobov

Federal State Unitary Enterprise «All-Russian Scientific Research Institute of Aviation Materials» (VIAM)

E-mail: elena.kurbatkina@mail.ru

Abstract. In the present work, using powder metallurgy method, mechanical alloying and extrusion, the metal matrix composite material Al–Zn–Mg–Cu/10% SiC was obtained. Phase transformations occurring during heat treatment were studied. After deformation aluminum solid solution, $MgZn_2$ and SiC are present in the composite material. The X-ray phase analysis and TEM results showed that when the composite material is heated for quenching, the $MgZn_2$ phase is completely dissolved. Study of the aging kinetics showed that the peak of aging of the composite material was achieved in 15 h, which is much faster than in the matrix alloy (24 h). After aging, the solid solution decomposition products appear in the form of a metastable η'-phase ($MgZn_2$), looking like rods located in three mutually perpendicular directions <100> Al.

Currently, composite materials with an aluminum matrix are widely used due to such properties as high specific strength, low density, low thermal coefficient of linear expansion and high processability of these materials [1–5]. Most often, in aluminum matrix composites, silicon carbide is used as a reinforcing component because it has high hardness and resistance to aluminum matrix. The addition of SiC increases the elastic modulus and tensile strength of the composite material [6–12]. Silicon carbide particles and features of the technological process can affect the phase transformations occurring in the composite material [13, 14]. For example, some papers concluded [15, 16] that the difference in thermal expansion coefficients between the matrix and SiC leads to an increase in the residual thermal stresses and density of dislocations. Also, the reinforcing component can affect the formation of nucleation centers of secondary phases at the interfaces, chemical reactions between the matrix and the reinforcing component, which lead to enrichment and depletion of local zones in the composite [17]. For aluminum alloys of the Al–Zn–Mg–Cu system, which belong to the group of dispersion-hardening alloys, the addition of silicon carbide can have an effect on the dissolution/precipitation of phases [18–21]. Due to the structural features of the composite material, time ranges of heat treatment, especially of aging, may change [22, 23].

The purpose of this work is the analysis of phase transformations occurring in a composite material based on the Al–Zn–Mg–Cu dispersion-strengthened alloy reinforced with SiC particles.

As the initial materials, we used a spherical powder of an Al–6% Zn–2.5% Mg–2.5% Cu–0.5% Mn aluminum alloy with a particle size of 10–100 μm, as well as silicon carbide powder with a particle size of 2–10 μm. Mass of the initial components, loaded into the mill for processing, was selected based on the content of silicon carbide in the composite material - 10%. Powders were mixed in a V-shaped mixer for 30 min, then were subjected to the mechanical doping in a planetary mill Fritsch Pulverisette 5 for

30 min in inert gas. The weight raito of balls to powder charge was 20:1. After mechanical alloying composite granules were hot extruded on a 500 ton press at a temperature of 575°C with extrusion ratio of 15:1. Microhardness was determined using a HV-1000 (Time Group) tester. Microstructure of the composite material was studied using scanning electron microscopy (SEM) on a JEOL JSM-6490 LV microscope and transmission electron microscopy (TEM) on a Tecnai G2 F20 S-TWIN microscope. X-ray phase analysis was carried out using a Rugaku D / MAX-2500 diffractometer.

Study of the structure of the composite material after extrusion using scanning electron microscope showed that it contains matrix aluminum alloy and particles of a reinforcing component uniformly distributed in it (figure 1a). The particles of silicon carbide have a size from tenths to 5 μm. The particles of silicon carbide are crushed during the process of mechanical alloying and have a size from tenths to 5 μm. X-ray phase analysis showed (figure 1b) that after extrusion structure contains aluminum solid solution, silicon carbide and $MgZn_2$ phase. Quenching according to the regime typical of aluminum alloys of the Al–Zn–Mg–Cu system: temperature of 465°C, holding time for 1 h, leads to dissolution of this phase (figure 1b) in an aluminum solid solution.

(a)

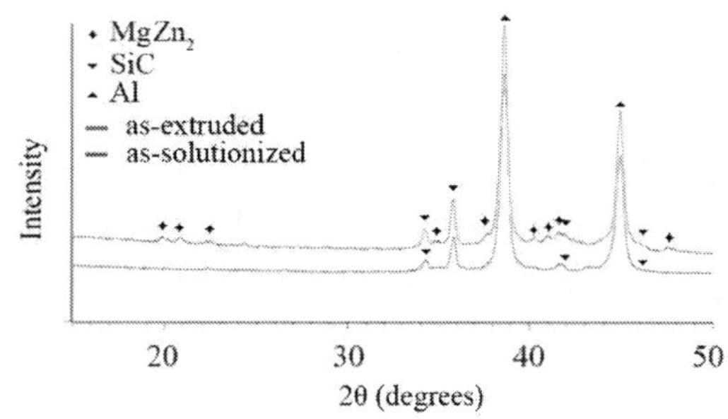

(b)

Figure 1. (a) Structure and (b) X-ray diffraction pattern of the composite material

To determine the heat treatment mode, leading to the maximum strength properties of the composite material, samples were aged in the time range from 0 to 35 h. Based on the results, the kinetic curves of aging (the dependence of the microhardness of the matrix aluminum alloy on the aging time at a temperature of 120°C) were obtained (figure 2).

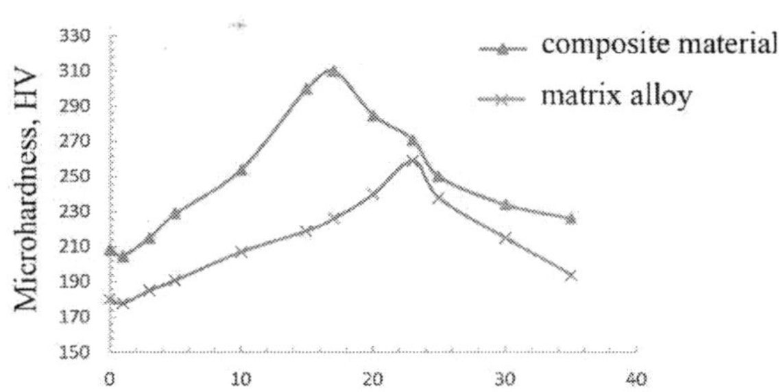

Figure 2. Kinetic curves of aging of the composite material and the matrix alloy

Peak-aging of the composite material is reached in 15 h. After the peak-aging, the microhardness decreases, which is probably associated with overaging. The comparison of kinetic curves showed that an increase in the strength of the composite material occurs much faster than that of the matrix alloy.

Analysis of the composite material aged according to the established regime (15 h, 120°C) using transmission electron microscope showed the formation of solid solution decomposition products, metastable η'-phase ($MgZn_2$), which has the form of rods located in three perpendicular directions <100> Al (figure 3).

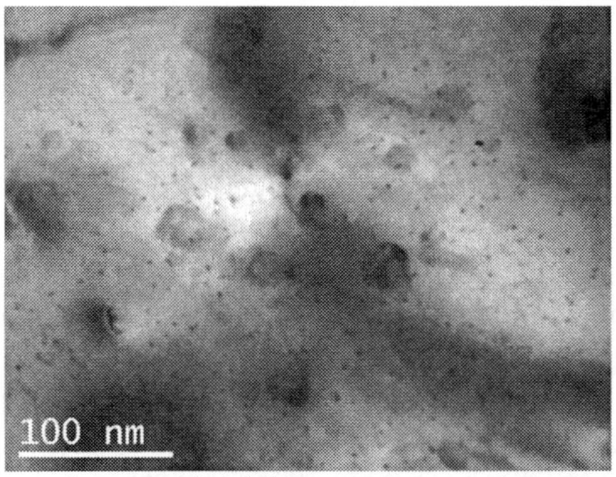

Figure 3. η'-phase in a composite material (phase rods are oriented perpendicular to the image).

Acknowledgments
The work was carried out with the financial support of the Russian Scientific Fund Agreement (no. 17-73-1032).

References

[1] Saheb N, Khalil A, Hakeem A S and Laoui T 2013 *J. Nano Res.* **21** 29–35
[2] Pal S, Mitra R and Bhanuprasad V V 2008 *Mater. Sci. Eng.* A **480** 496–505
[3] Li B, Luo B, He K, Zeng L, Fan W and Bai Z 2015 *J. Alloys Comp.* **649** 495–499
[4] Wang Z, Li C, Wang H, Zhu X and Wu M 2016 *J. Mater. Sci. Tech.* **32** 1008–12
[5] Papazian J M 1988 *Metall. Trans.* A **19** 2945–53
[6] Ludtka G M and Laughlin D E 1982 *Metall. Trans.* A **13** 411–418
[7] Oguocha I N A and Yannacopoulos S 1997 *Mater. Sci. Eng.* A **231** 25–33
[8] Hunt E, Pitcher P D and Gregson P J 1990 *Scripta Metall. Mater.* **24** 937–941
[9] Guo J and Yuan X 2009 *Mater. Sci. Eng.* A **499** 212–214
[10] Arsenault R J and Wu S B 1988 *Scripta Metall.* **22** 767–772
[11] Chawla N and Shen Y L 2001 *Adv. Eng. Mater.* **3** 357–429
[12] Ganesh V V and Chawla N 2005 *Mater. Sci. Eng.* A **391** 342–353
[13] Cottu J P, Couderc J J, Viguier B and Bernard L 1992 *J. Mater. Sci.* **27** 3068–74
[14] Badini C, Marino F and Verné E 1995 *Mater. Sci. Eng.* A **191** 185–191
[15] Hadianfard M J, Mai Y W and Healy J C 1993 *J. Mater. Sci.* **28** 3665–69
[16] Lin G, Zhen-zhu Z, Gui-song W and Bin X U 2005 *Transac. Nonferr. Met. Soc. China* **16** 387–391.
[17] Hayoune A and Hamana D 2009 *J. Alloys Comp.* **474** 118–123.
[18] Cheng N, Zeng S and Liu Z 2008 *J. Mater. Process. Tech.* **202** 27–40
[19] Gouma P I, Lloyd D J and Mills M J 2001 *Mater. Sci. Eng.* A **319–321** 439–442
[20] Liu P, Wang A, Xie J and Hao S 2015 *Transac. Nonferr. Met. Soc. China* **25** 1410–18
[21] Choi H J, Min B H, Shin J H and Bae D H 2011 *Compos. Part A: Appl. Sci. Manufac.* **42** 1438–44
[22] He Y, Wang N, Qiao B, Feng L and Chen Z 2010 *Chinese J. Nonferr. Met.* **20** 1302–08
[23] Bhaduri A, Gopinathan V, Ramakrishnan P and Miodownik A P 1996 *Mater. Sci. Eng.* A **221** 94–101

IOP Conf. Series: Materials Science and Engineering **558** (2019) 012021 doi:10.1088/1757-899X/558/1/012021

Synthesis of nanopowders of Fe-C system via Flow-Levitation method and study of their properties

M L Kuskov[1], A N Zhigach[1], I O Leipunsky[1], E S Afanasenkova[1], O A Safronova[1], N G Berezkina[1], V V Artemov[1,2], O M Zhigalina[2,3], D N Khmelenin[2]

[1] V L Talrose Institute for Energy Problems of Chemical Physics, Russian Academy of Sciences, 38, Bld. 2, Leninsky prospect, Moscow, 119334, Russian Federation
[2.] Federal Scientific Research Centre "Crystallography and Photonics", Russian Academy of Sciences, 59, Leninsky prospect, Moscow, 119333, Russian Federation
[3.] Bauman Moscow State Technical University, ul. Baumanskaya 2-ya, 5/1, Moscow, 105005, Russian Federation

E-mail: mkuskov72@yandex.ru

Abstract. Nanopowders of Fe-C system are synthesized via modified Guen-Miller Flow-Levitation method by *in situ* reaction of nascent iron nanoparticles with acetylene. Morphology, internal structure, chemical and phase composition are studied by electron microscopy methods, including electron diffraction and element analysis, X-ray phase analysis and CNHS chemical analysis. It is shown that depending on the parameters of the reaction with acetylene (temperature, concentration), particles composition can vary from pure iron with thin carbon coating complex composition consisting predominantly of iron carbide.

1. Introduction.

Nanoparticles with iron-containing core and carbon-containing shell are used widely in biomedical research [1,2], in electrochemical technologies [3], as catalysts for olefin synthesis [4], and as a component of radio-frequency absorbing materials [5]. Plasma chemical methods [6], shock waves pyrolysis of iron carbonyl $Fe(CO)_5$ [7], mechanochemical synthesis [8], hydrothermal carbonization of iron chloride $FeCl_2$ [9], and evaporation-condensation Flow-Levitation method where nascent particles react with butane or methane [10,11], are successfully used for synthesis of nanoparticles of Fe-C system.

In the present work synthesis of nanoparticles of Fe-C system via Flow-levitation method, using argon as carrier gas and acetylene as reactant, is described. The aim of this work is to study the influence of acetylene/iron nanoparticles reaction parameters on morphology, structure, element and phase composition of the resulting nanoparticles of Fe-C system.

2. Experimental.

Samples of Fe-C system nanoparticles were synthesized using the modified Guen-Miller Flow-Levitation equipment described in [12,13]. The scheme of the method and operating reactor are presented in Figure 1.

Iron drop is suspended and heated up to the temperature higher than melting point in high-frequency electromagnetic field (440 kHz) of counter-current inductor inside the quartz tube (14-15 mm inner diameter). The drop evaporates in argon flow blown over it downstream. Iron vapor condenses homogeneously in argon flow thus forming nascent iron nanoparticles. Hot nascent Fe nanoparticles react with acetylene injected into argon flow below the drop. Parameters of reaction of acetylene with Fe nanoparticles are controlled by position of the injector with regard to the drop (determining the reaction temperature) and by acetylene flow rate (determining its concentration in Ar flow carrying nascent nanoparticles).

Iron loss due to evaporation is continuously replenished by Fe wire feeding at constant rate (5.9 g/h in this work).

Synthesized Fe-C nanoparticles are cooled down in the gas path of the equipment, then collected on a fabric filter and subsequently dropped into a special container. After the synthesis process completion, the particles are subjected to a passivation with atmospheric air using at least two cycles "slow inflow followed by pumping out", for safety.

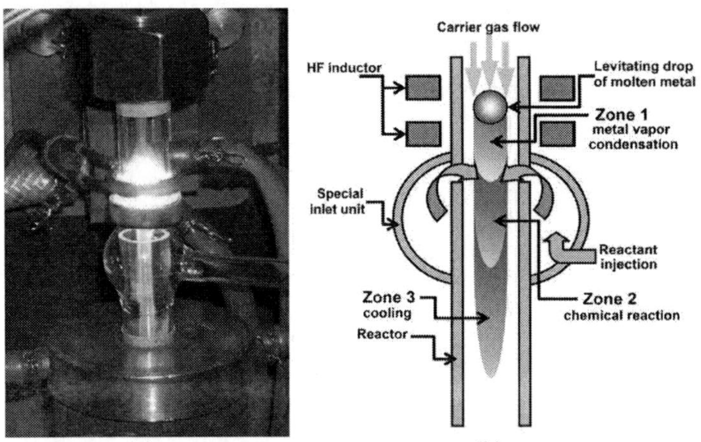

(a) (b)

Figure 1. Synthesis of Fe-C system nanoparticles via Flow-Levitation method (a) and scheme of the method with nascent particles chemical treatment (b).

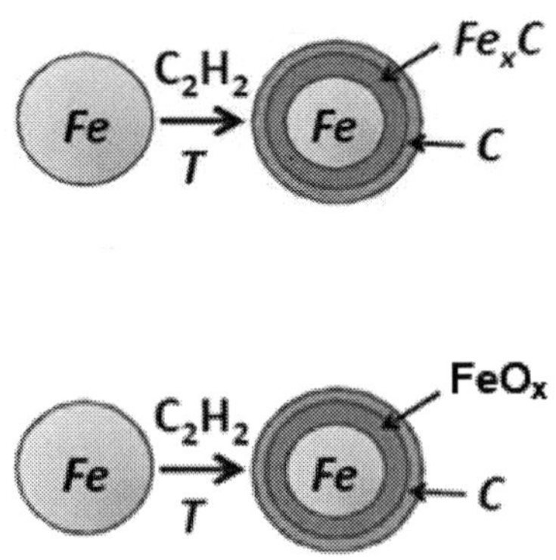

Figure 2. Schematic representation of chemical reactions in Fe + C_2H_2 system.

As a result of the chemical reaction

$$\mathbf{Fe + C_2H_2 \rightarrow Fe_xC_y + C + H_2}$$

particles of complex internal structure are formed, containing metallic iron, iron-carbon compounds, elemental carbon, and passivating layer of iron oxide as a result of the reaction of iron in particles with atmospheric air (figure 2).

All experiments are fulfilled at argon pressure of 0,2 ata and the flow rate of 0.18 st. m^3/h. To vary the conditions of chemical reaction between acetylene and nascent Fe nanoparticles, the point of acetylene injection was shifted within the range from the hotter zone at 8-10 mm below the drop (synthesis of Fe1) to the colder zone at approximately 30 mm (synthesis of Fe2, Fe3). Acetylene concentration in the reaction zone was varied by more than 2 orders of magnitude: from 5.5 to 0.02 vol. %.

The samples of Fe-C system nanoparticles are studied by scanning (SEM) and transmission (TEM) electron microscopy, including high-resolution microscopy (HRTEM) and scanning transmission electron microscopy (STEM) and electron diffraction analysis (EDA) (FEI Tecnai Osiris, LEO 912 AB OMEGA), energy-dispersive X-ray spectroscopy (EDX) with TEM mapping (JEOL JSM 7401F, FEI Quanta 200 3D FIB). X-Ray diffraction analysis (XRDA) (ADP-1), catalytic burning element CHNS-analysis (Euro Vector EA 3000).

Particle sizes are determined by direct measurements of their TEM images (at least 900 particles per a sample).

3. Results and discussion.

Synthesis parameters, particle sizes and crystalline phase composition of nanopowders of Fe-C system are presented in Table 1.

Table 1. Synthesis parameters and characteristics of nanopowders of Fe-C system.

Sample Fe-C	Acetylene flow rate (st.m^3/h)	Mean <D> and mass-mean D_m particle sizes	Crystalline phases	
			XRDA	Electron diffraction
			Phase composition (w. %)	Phase
Fe1	$1{,}0*10^{-2}$	<D> = 38 nm D_m = 84 nm	Fe_3C (Pnma) - 56.1 %, α-Fe (Im3m) - 16.4 %, $Fe_{2.4}C$ (P63 2 2) - 27.5%	α-Fe, C Fe_3C, Fe_2C
Fe2	$1{,}6*10^{-3}$	<D> = 38 nm D_m = 78 nm	Fe_3C (Pnma) - 55.6 % α-Fe (Im3m) - 44.4 %	α-Fe, C, Fe_3C FeO, Fe_3O_4
Fe3	$4*10^{-5}$	<D> = 45 nm D_m = 109 nm	α-Fe (Im3m) -100 %	α-Fe

Results of XRD analysis are shown in Figure 3.

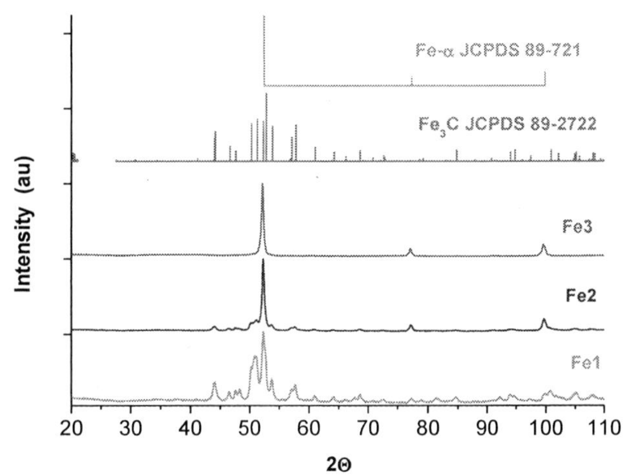

Figure 3. XRDA patterns of the Fe-C samples.

Depending on the synthesis conditions, nanoparticles contain crystalline iron, iron carbides of various compositions (total iron carbides content ranging from 0 to 83.6 w. %) and elemental carbon. The presence of iron oxides can be attributed to chemical reaction of air oxygen with metallic iron in the particles during the powder passivation.

Apparently, the synthesized particles possess a multi-layered core-shell structure. Particles of sample Fe3 shown in figure 4 exhibit spherical shapes and possess a visible core-shell structure. The particles size distribution is presented in figure 4 (c). Since the only crystalline phase revealed in the sample Fe3 by XRDA is metallic iron (Figure 3), it can be assumed safely, that clearly visible outer shells on the particles are amorphous. However, there are crystalline sections of shells revealed on the HR-TEM images due to lattice fringes resolution. The measured magnitude of the inter-planar spacing corresponds to the (111) plane of crystalline phase of Fe_3O_4 (figure 4 (d) and (e)).

Study of Fe3 sample by STEM-EDX method (figure 5) revealed at least triple-layer core-shell structure, where metallic iron core is coated with iron oxide layer 5-7 nm thick and outer elemental carbon shell 4-15 nm thick.

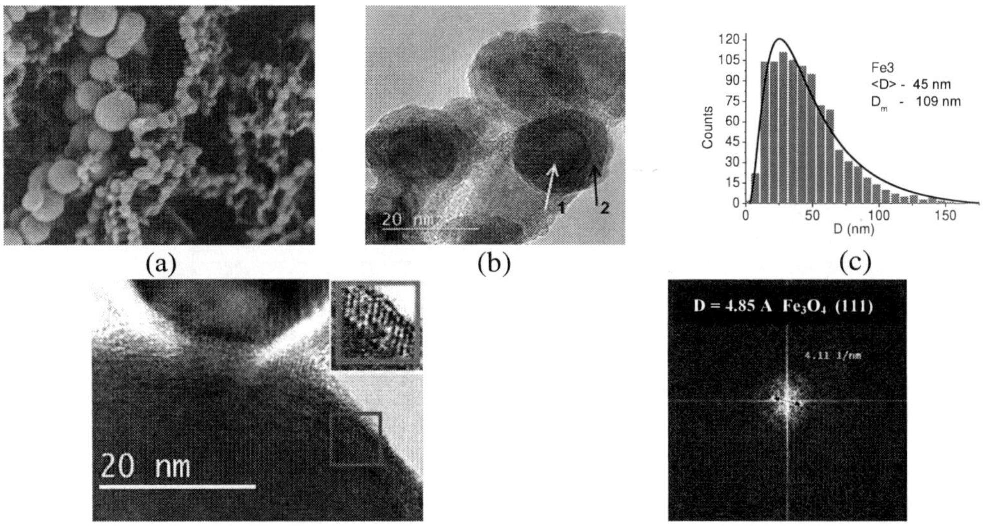

(a) (b) (c)

(d) (e)

Figure 4. SEM image of sample Fe3 (a); HR-TEM image (b), where 1 is crystalline core, 2 is shell; particles size distribution histogram (c) where black curve is log-normal fit; HR-TEM image with crystalline section of the shell (d) where inset is the enlarged view of the selected area; Fast Fourier transform pattern of the selected area (e) where inter-planar spacing corresponds to Fe_3O_4 (111).

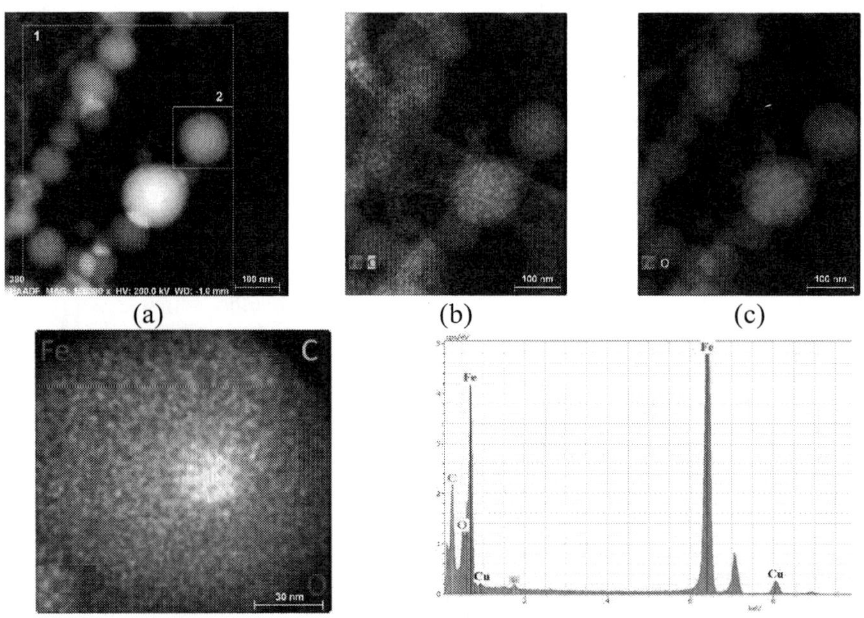

(a) (b) (c)

(d) (e)

Figure 5. STEM image (a), Fe and C elements distribution map of the particles array within the frame 1 in STEM image (b); Fe and C elements distribution map of the same array (c); Fe, C and O elements distribution map of the particle enclosed within the frame 2 in STEM image (d); EDS spectrum of the particles array (e), where peak of Cu resulted from the copper TEM grid.

Electron diffraction analysis confirms the presence of metallic iron and iron oxides in sample Fe3 (Figure 6).

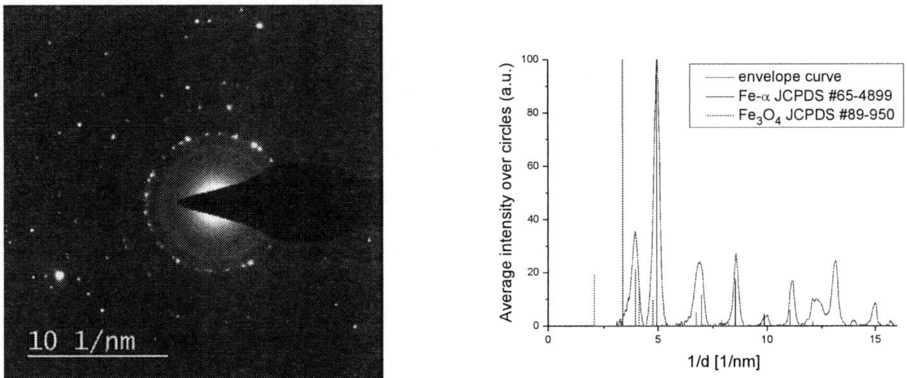

(a) (b)

Figure 6. EDA pattern of sample Fe3 (a); EDA net pattern (b)

No differences in morphology of particles of all studied samples were found.

The total carbon content of in the samples Fe2 and Fe3 was obtained using CHNS-analysis method (catalytic sample oxidation in helium flow with oxygen addition).

Weigh content of total carbon in Fe3 sample is 0.4 w. % that corresponds to mean carbon shell thickness of 1.3 nm. Total carbon content in Fe2 sample is 6.7 w. %. Comparison of this value with XRDA data enables to conclude that 3.7 w. % corresponds to Fe_3C, while residual 3 w. % belongs to outer elemental carbon shell with calculated mean thickness of about 10 nm.

Conclusion.

Nanoparticles of Fe-C system with mean size of 38-45 nm are synthesized via Flow-Levitation method using the *in situ* reaction of nascent iron nanoparticles with acetylene. Depending on the acetylene flow rate value, the method enables to synthesize core-shell nanoparticles with high content of iron carbides (about 84 w. %) or iron-core particles (α-Fe) without crystalline iron carbides. Nanoparticles of sample Fe3 have shell with the inner oxide layer of 3-5 nm thick and the outer carbon layer of 4-12 nm thick.

Acknowledgements.

The work was developed within the framework of the Program for Basic Research of the Russian State Academy of Sciences for 2013-2020 and supported by the Ministry of Sciences and Higher Education (agreement GZ №007-00269-18-00), and the Federal Agency for Scientific Organizations (agreement No. 007-GZ / Ch3363 / 26) in terms of electron microscopy

References

[1] Yushkov B G, Artashyan O S, Medvedeva S Yu, Byzov I V, Uimin M A, Ermakov A E, Bykova M Yu and Petrova I M 2015 *Bull. Exp. Biol. Med.* **159** 498

[2] Surgutskaya N S, Trusova M E, Slepchenko G B, Minin A S, Pershina A G, Uimin M A, Yermakov A E and Postnikov P S 2017 *Anal. Methods* **9** 2433

[3] Nowicka A M, Kowalczyk A, Donten M L, Donten N, Bystrzejewski M, Stojek Z 2014 *Electrochim. Acta* **126** 115

[4] Yongwu Lu, Qiangu Yan, Jun Han, Baobao Cao, Jason Street, Fei Yu 2017 Fuel **193** 369

[5] Omid Khani, Morteza Zargar Shoushtari, Mansoor Farbod 2015 Physica B **477** 33

[6] Reza Sanaee M, Enric Bertran 2015 *J Nanomater* **10** ArtID 450183

[7] Eremin A V, Gurentsov E V, Musikhin S A 2017 *J Alloy Compd* **727** 711

[8] Kasprzak A, Bystrzejewski M, Koszytkowska-Stawinska M, Poplawska M 2017 *Green Chem.* **19** 3510
[9] Qiangu Yan, Jason Street, Fei Yu 2015 Biomass Bioenerg **83** 85
[10] Galakhov V R, Shkvarin A S, Semenova A S, Uimin M A, Mysik A A , Shchegoleva N N, Yermakov A Ye, Kurmaev E Z 2010 *J. Phys. Chem. C* **114 (51)** 22413
[11] Tsurin V A, Yermakov A Ye, Uimin M A, Mysik A A, Shchegoleva N N, Gaviko V S and Maikov V V 2014 *Phys of Solid State* **56.2** 287
[12] Zhigach A N, Kuskov M L, Leipunsky I O, Stoenko N I and Storozhev V B 2012 *Nanotechnol. in Russia* **7** 28 (in Russian)
[13] Leipunsky I O, Zhigach A N, Kuskov M L, Berezkina N G, Afanasenkova E S, Lopez G W, Vorobjeva G A, Naumkin A V 2019 *J Alloy Compd* **778** 271

IOP Conf. Series: Materials Science and Engineering **558** (2019) 012022 doi:10.1088/1757-899X/558/1/012022

Combined equipment for synthesis of ultrafine metals and metal compounds powders via Flow-Levitation and crucible methods.

M L Kuskov, A N Zhigach, I O Leipunsky, A N Gorbachev, E S Afanasenkova, O A Safronova

V L Talrose Institute for Energy Problems of Chemical Physics, Russian Academy of Sciences, 38, Bld. 2, Leninsky prospect, Moscow, 119334, Russian Federation

E-mail: mkuskov72@yandex.ru

Abstract. Laboratory automated equipment for synthesis of ultrafine particles of metals, alloys and metal compounds via evaporation-condensation Flow-Levitation Guen-Miller method combined with crucible method using high-frequency (440 kHz) electromagnetic field for heating is described. The equipment synthesizes the said particles and multilayer core-shell structures with mean size ranging from tens to hundreds of nanometers with output from grams to tens of grams per hour. Synthesized particles may be collected into container with inert gas, or into non-volatile liquid, or *in situ* passivated with air for further handling. Examples of synthesized particles are presented.

1. Introduction.

Ultrafine (sub-micron and nano-sized) particles (UFP) of metals and metal compounds are of significant interest and in high demand for scientific research. They have an ample spectrum of applications ranging from production of high-energy nanocomposites to a wide variety of materials used in agriculture and biomedicine. The synthesis of metal and metal compound UFPs at nanoscale magnitudes ($<D>$=5-1000 nm) is a proven complex exertion that requires meticulous consideration to countless parameters and mode adjustments to obtain the product at optimum conditions. These conditions often include chemical, phase, and granulometric characteristics and other requirements such as custom composition gradient, and non-agglomerative properties of the UFPs in order to achieve a more uniform distribution of the particles in a medium. Therefore, equipment to synthesize nanoparticulate must possess the necessary versatility and adaptability to accommodate the vast and often uncounted number of manufacturing requirements. The equipment should, for instance, be capable to manufacture nanoparticles using a wide assortment of metals, synthesize UFPs with a wide range of properties, facilitate a wide throughput spectrum, adapt to synthesis-technique modalities, uncomplicated and inexpensive maintenance, etc.

2. MIGEN Equipment description.

MIGEN equipment is based on the Flow-Levitation (FL) evaporation-condensation method invented by Guen and Miller [1-3].

Figure 1 demonstrates the principle of the FL-method. A drop of molten metal levitates inside a quartz tube reactor where it is heated by the electromagnetic field of a high-frequency (~440 kHz) countercurrent inductor. The drop temperature is high enough as to provide an acceptable evaporation rate of the metal. Evaporated material is continuously replenished using feed-wire at preset rates ranging from 1 to 100 g/h, depending on the metal and particle size required.

To synthesize alloys of desired composition, the apparatus is equipped with two independent mechanisms for simultaneous wire (0.4-1.0 mm in diameter) feeding. Wire feeding rates are set for each mechanism separately to provide mass feeding in the range 0.1...100 g/h with an accuracy of 0.1%. Coils of wires are placed inside a vacuum chamber of the apparatus. Wire is fed to the evaporating drop with stepper motor mechanism, placed outside of the said vacuum chamber to avoid undesirable contamination of the synthesized powder with the substances released by operating motors. Wire feeding mechanisms rotating shafts with driving rollers are passed from the motors into the vacuum chamber through vacuum-proof oil-free Wilson-type seals.

Inert gas or inert-reactive gas mixture of preset composition is continuously blown downstream onto the molten metal drop from the top.

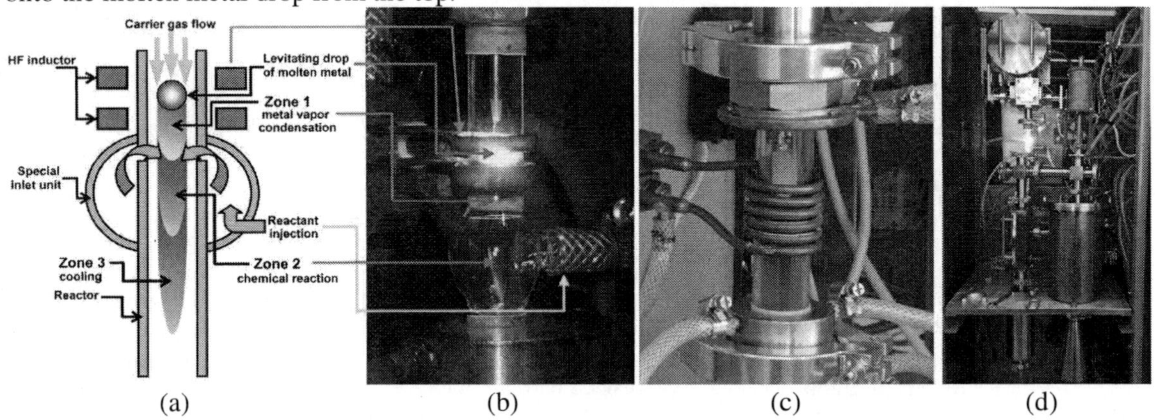

Figure 1. Flow-Levitation method for synthesis of ultrafine particles: (a) – scheme of nanoparticles synthesis process, (b) – appearance of a FL-mode reactor with side reactant input; (c) – appearance of a crucible mode reactor, (d) view of actual equipment in operation

This method is known in general in other research groups. It is used in equipment assembled independently in Russia [4,5], Korea [6,7], Iran [8,9], India [10] and China [11-13].

Metal vapor condenses homogeneously within the gas flow below the drop, then fresh-formed nanoparticles are cooled and collected by a fabric filter or a bubbler filled with non-volatile liquid. The concentration of metal vapor in the gas, and its cooling rate determines the final size of the nanoparticles in this section. These parameters can be controlled by presetting the drop temperature (dependent on the wire feed rate), the type of inert gas (argon, helium, hydrogen etc.) and the pressure and velocity of the gas flow over the molten drop. Whereas diffusion coefficients, viscosities, thermal conduction and capacity of the said inert gases are strongly different from each other, the composition of the carrier gas influences not only on particle size and size distribution, but also on the morphology and crystallinity of the formed particles.

Desired physicochemical characteristics of nanoparticles and/or their surface are achieved by treatment of nascent particles with gaseous reactants injected into the gas flow carrying the condensing nanoparticles. Depending on chemical treatment regime (reactant concentration, temperature in the point of reactant injection), one can synthesize a metal compound, core-shell-structured particles (e.g. with metal core and metal compound shell), or metal nanoparticles with thin surface coatings of desired compositions.

Since a temperature gradient below the drop amounts to tens or hundreds of degrees per centimeter, it is possible to set the required temperature in the reaction zone from practically the melting point to the ambient temperature, by placing the intake of reactant gas at the proper distance L from the drop.

IOP Conf. Series: Materials Science and Engineering **558** (2019) 012022 doi:10.1088/1757-899X/558/1/012022

The reactant is blown in a chosen zone of the reactor through the pathways 1, 2, 3 or 4 (figure 2) at a distance L from the drop with the proper concentration to create the required conditions for chemical reaction with metal vapor or nascent nanoparticles.

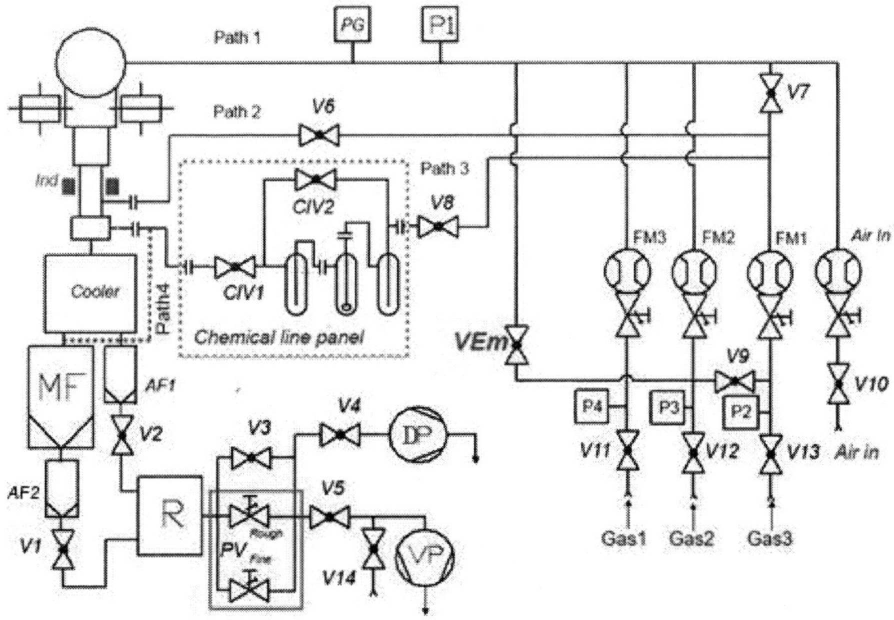

Figure 2. MIGEN equipment gas scheme. V1-V14, CLV1, CLV2, VEm – vacuum ball valves, FM1…FM4 – flowmeters with gas flow control multiturn valves. MF – main fabric filter, AF2 – additional filter for use in transition regime, AF1 – emergency filter, P1…P4 – vacuum-meters, PG – membrane gas pressure gauge, R – receiver, PV – multiturn flow rate and pressure control valves, VP – forevacuum pump, DP – diafragm pump

During synthesis of metal compounds, it is possible to admix simultaneously up to 3 reactants into the flowing gas. The reactant can be supplied at different points of the flowing gas: into the main gas flow, blowing around the molten metal drop (Path1, applicable only to synthesize metal compounds unstable at the drop temperature), downstream the gas flow at distances from 0.5 to ~20 cm from the evaporating metal (Path 2 and 3), into the outlet of the cooler (Path4). These reactant injection points provide the reaction temperature from that of the molten drop to practically room temperature. This allows variation of particles composition and synthesis of uniform compound particles as well as multilayer core-shell structures.

Both gaseous reactants and mixtures of liquid reactant vapors with inert gas may be used. In the last case, a desired evaporation rate of the reactant may be achieved by placement of the reactant containing bubbler into container with controlled temperature.

In instances where the brittleness, low melting point and poor electric conductivity of a material limits the usefulness of the FL method, MIGEN can be easily switched to crucible-based high-volume evaporation with HF heating, and no metal feed-wire. Brittle metals of interest, such as magnesium and manganese, are not available in the form of wire. This hampers the use of the Flow-Levitation method which relies on replenishment of material via feed-wire. Similarly, metals with melting point less than 400 °C get overheated in the electromagnetic field at the intensity needed to levitate the molten drop. The FL method in these cases renders extremely fast evaporation of the metal. Materials with poor electric conductivity also inhibit the use of the FL method since they cannot be levitated within the electromagnetic field.

Nevertheless, the cunning adaptability of the new generation of MIGEN equipment to crucible modality easily resolves the inherent, but minor, glitches of the FL method. The equipment is commercially available for production of a wide variety of nanoparticles (Figure 1). The technique modalities used by MIGEN (FL method and Crucible-based) are attained via exchangeable reactor assemblies (Figures 1c and 1d) that use common gas, reactant and power control of the manufacturing equipment. In both modes, the electromagnetic field generated by a tunable high-frequency (440 kHz) generator is used for heating and evaporating the metal.

The Flow-levitation mode is used for a majority of regular metals with melting point T_{melt} between 400 °C and 2000 °C and high electric conductivity. Quartz tube reactors, in which heating and evaporation of the metal take place, are available in diameters of 18 mm and 21 mm. The crucible mode is used for metals unavailable in the form of wire such as (Mg, Mn) and materials with poor electric conductivity (e.g., rare earth metals). The heating of the material sample is accomplished in quartz tube reactors with diameters 18, 21 or 30 mm using a unidirectional inductor. Crucibles for the material evaporation are made of carbon, glassy carbon composite (such as Sigradur), or refractory metals, such as W, and Ta. In addition, crucible method is preferable for metals with low melting point such as tin and lead, because their confinement in electromagnetic field leads to overheating and fast boiling evaporation, that is not acceptable for nanoparticle synthesis.

Only replacement of inductor, quartz tube and the tube holding heads is required to change flow-levitation mode to the crucible one and back, while all other parts of the equipment, including gas control scheme, pump, electronics etc. remain unchanged.

Depending on the required reaction temperature, either straight tube quartz reactor (reactant input paths 3 or 4), or quartz tube reactor with side reactor input (Figure 1B, reactant input paths 2, higher reaction temperature) may be used both in flow-levitation and in crucible regimes.

Since most of materials in the form of synthesized ultrafine particles are extremely pyrophoric, they need to be passivated by slow in-leakage of atmospheric air (repeatedly, if necessary) before unloading, or poured into a container filled with inert gas.

3. Examples of synthesized nanoparticles.

Table 1. Examples of metal, alloys and metal compound nanoparticles synthesized using MIGEN equipment.

Substance	Particle mean size, nm	Substance	Particle mean size, nm
	Metals		Metal compounds
Ag	5-200	Al_2O_3	50-150
Co	20-100	NiO	50
Fe	20-100	FeO_x	50
Zn	170-1400	TiO_x	15-35
Ti	20-40	TiH_2	23-30
Al	5-300	Cu_xO	70
Cu	30-150	CoO	50
Ni	5-100	TiC_x	15-30
Mn*	80	Fe_xC	38-45
Mg*	85	TiN_x, x=0.19-1	15-35
	Alloys		Intermetallic compounds
Ni-Cu	40-100	Ni_xAl	70-120
Fe_x-Ni_y-Co_z	40	Fe-Al	200
Ag-Al	200	Fe-Cu	28-44

* -- synthesized in crucible regime.

Technical capability of the equipment covers synthesis of nanoparticles of metals with boiling point ranging from some hundreds to approximately 3300 °C with output more than 1 g/h (Table 1).

To manufacture compound or core-shell nanoparticles, practically any desirable gaseous reactant may be used. Liquid reactants with saturation vapor pressure higher than 100 Pa at 20 °C are easy to use for nascent surface chemical treatment. Table 2 presents the examples of reactants and products obtained in nanoparticles synthesis process. Depending on the reactant input conditions (gas pressure, concentration, temperature in the reaction zone), either metal with thin surface coating, or multilayer core-shell structure, or metal compound nanoparticles may be manufactured. Such synthesis influences not only the particles composition, but particle morphology, especially the morphology of the particles surface layer (see figure 3).

Figure 3. Electron microscopy images of metal nanopartiles, synthesized via MIGEN equipment:
a) Al, "thin" oxide coating, 18 mm straight tube reactor, mean size <D>=85 nm, mass mean size D_m=131 nm;
b) Al orgnic coating (acrylic acid, Path 3), 18 mm straight reactor, <D>=72 nm, D_m=228 nm;
c) Al "thick" oxide coating, 18 mm reactor, Path 2 side O_2 input, <D>=43 nm, D_m=162 nm;
d) Al with (oxy)nitride coating 18 mm reactor Path 2 side NH_3 input, <D>=93 nm, D_m =245 nm;
e) Zn, Ar as carrier gas, 18 mm straight reactor<D>=0.72 μm, D_m =3.91 μm;
f) Zn, H_2+He mixture as carrier gas, 18 mm straight tube reactor <D>=206 nm, D_m =430 nm;
g) Cu, 18 mm straight tube reactor, reactant (Ar+H_2O vapor) input through Path 3, <D>=86 nm;
h) Mn, Ar, crucible method, 30 mm straight tube reactor <D>=80nm, D_m =113nm;
i) Mg, Ar, crucible method, 30 mm straight tube reactor, <D>=85 nm, D_m=186 nm.

Table 2. Examples of reactants and products obtained in nanoparticles synthesis process

Reactant	Product
Oxygen	MeO_x (Me: Ti, Fe, Al, Cu etc.), oxide coating
Dry ammonia	MeN_x (Me: Ti, Al), (oxy)nitride coating
Hydrocarbons	MeC_x, C, MeH_x (Me: Ti, Fe)
Hydrogen	TiH_2
Carbonic acids	Organic hydrophobic surface coating
Organosilazanes	Silicon-organic hydrophobic surface coating
Perfluoorganic acids	Fluorine-organic hydrophobic surface coating
Air	Passivating surface coating

Main technical characteristics of MIGEN equipment.

Metal wire consumption 1-100 g/h
Power consumption 10 KWt;
Overall size LxBXH = 140X70X200 cm,
Net weight ~ 400 kg.

The equipment automatic control and monitoring system maintains long-term (up to 8 hours) operation without the operator interaction, and is capable to turn itself off in emergency situations.

MIGEN equipment has been delivered and it is currently in use at: The Institute of Nuclear Physics and Chemistry, China Academy of Engineering Physics, MianYang, and Shenzhou Space Biotechnology Group, China Academy of Space Technology, Beijing, China; Advanced Research Center for Powder Metallurgy and New Materials, Hyderabad, India.

Acknowledgements.
The work was developed within the framework of the Program for Basic Research of the Russian State Academy of Sciences for 2013-2020 and supported by the Ministry of Sciences and Higher Education (agreement GZ №007-00269-18-00).

References.

[1] Guen M Ya and Miller A V, Method for production of metal aerosols, 1961 SU Patent 814432.
[2] Jigatch A N, Leipunsky I O, Kuskov M L, Stoenko N I and Storozhev V B 2000 *Instr Exp Tech* **43** 839
[3] Jigatch A N, Kuskov M L, Leipunsky I O, Stoenko N I and Storozhev V B 2012 *Nanotechnol. in Russia* **7** 28.
[4] Galakhov V R, Shamin S N, Mesilov V V, Ovechkina N A, Uimin M A, Yermakov A Ye, Schneider L, Balinsky K and Senkovskiy B V 2014 *J Phys Chem C* **118** 30216
[5] Ortega D, Kuznetsov M V, Morozov Ya G, Belousova O V and Parkin I P, 2013 *J Alloy Compd* **579** 495
[6] Uhm Y R, Han B S, Lee H M and Rhee C K 2008 *Sol St Phen* **135** 123.
[7] Kim C K, Lee G J, Lee M K and Rhee C K 2014 *Powder Technol* **263** 1.
[8] Mohammadi A V and Halali M 2014 *RSC Adv.* **4** 7104
[9] Khodaei A, Hasannasab M, Amousoltani N and Kermanpur A, 2016 Mat. Res. Bull. **74** 212.
[10] Sivaprahasam D, Sriramamurthy A M, Bysakh S, Sundararajan G and Chattopadhyay K 2018 *Metall Mater Trans A* **49a** 1410
[11] Liu Wei, Yang Tian-zu, Chu Guang, Luo Jiang-shan and Tang Yong-Jian 2007 *Trans. Nonferrous Met Soc China* **17** 1347
[12] Chen S J; Chen Y; Li S, Wang Y E, Zeng H, Tang Y J and Sun W G. 2015 *Nano* **10** Art.ID 1550002
[13] Xiaoyang Zheng, Jiangshan Luo, Kai Li, Yong Yi and Kai Du. 2017 *IEEE Trans Nanotech* **16** 1106.

Synthesis of nanopowders of titanium compounds via Flow-Levitation method and study their properties

M L Kuskov[1], A N Zhigach[1], I O Leipunsky[1], E S Afanasenkova[1], O A Safronova[1], N G Berezkina[1], and G A Vorobjeva[2]

[1]V.L. Talrose Institute for Energy Problems of Chemical Physics of Russian Academy of Sciences, 38, bld.2, Leninsky prospect, Moscow, 119334, Russian Federation
[2]N.N. Semenov Institute of Chemical Physics of Russian Academy of Sciences, 4, Kosygina Street, Moscow, 119991, Russian Federation

E-mail: mkuskov72@yandex.ru

Abstract. Nanopowders of titanium compounds TiH_2 and TiC are synthesized via Flow-Levitation method using *in situ* reaction of nascent titanium nanoparticles with the proper reactants. The influence of manufacturing parameters on the composition and the internal structure of the synthesized nanoparticles is examined. Study of TiH_2 nanoparticles using thermal methods revealed hydrogen evolution within the narrow temperature range (390-510 ^{o}C), while hydrogen evolution temperature of the commercial samples is higher than 500 ^{o}C.

1. Introduction

Titanium compounds are widely used in various fields of science and technology. Titanium hydride is known as a source of hydrogen gas in manufacturing of porous metal materials, specially aluminum foams [1,2]. It is also used as material for storage, transport and recovery of hydrogen in power engineering applications [3,4]. Titanium carbide is a hard, refractory, abrasion, chemical inert material, which is used as a component of heat-resistant hard alloys, carbidized steels, reinforced alloys, abrasive materials, coatings for cutting tools, etc. [5]. Titanium carbide may be used for metal processing as an alternative to expensive tungstem. The study of physical, chemical and other properties of nanoparticulate titanium compounds is of scientific and practical interest. It is expected that due to higher specific surface area, the properties of nanoparticulate titanium compounds are different from those of “coarse” powders of the same materials. This work is devoted to study of physicochemical properties of nanoparticulate titanium hydride and titanium carbide synthesized via Guen-Miller Flow-Levitation (FL) method invented and developed at the Institute for Energy Problems of Chemical Physics of the Russian Academy of Sciences [6].

2. Experimental

2.1. Method and materials

Figure 1 illustrates the main elements of the apparatus for synthesizing of nanoparticles via the Guen-Miller FL-method. A molten titanium drop levitates inside a high frequency (440 kHz) electromagnetic field produced by a countercurrent inductor coiled around the wall of a cylindrical quartz reactor of 14-15 mm inner diameter. The inductor heats up the metal ensuring its steady evaporation. The mass lost by evaporation is continuously replenished by metal wire fed into the drop at a preset rate. An inert gas (Ar) input through the main inlet flows steadily from the top of the reactor over the levitating molten drop sweeping the metal vapor downstream. Metal vapor condenses homogeneously into nanoparticles in Zone 1. Nascent nanoparticles flow to Zone 2 where they react with gaseous reactant injected into the reactor through the mane or the special inlet at a preset rate. The resulting nanoparticles of titanium compound continue to flow to Zone 3 where they are cooled down, and subsequently collected on a filter at room temperature.

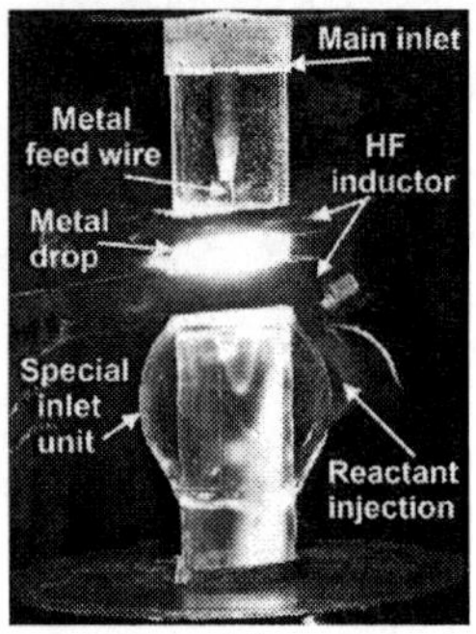

(a)

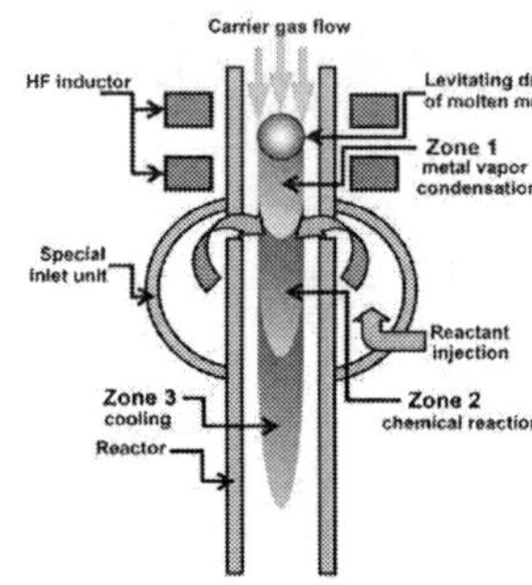

(b)

Figure 1. Appearance of the reactor in action (a) and a scheme of the synthesis process (b).

On completion of the synthesis process, nanoparticles of titanium compound were passivated *in situ* by pumping out the inert gas from the apparatus and slowly filling it with atmospheric air (~10-20 h) in order to form a protective titanium oxide layer on the surface of nanoparticles.

2.1.1. Synthesis of titanium hydride nanoparticles. Titanium hydride nanoparticles were obtained via direct interaction of titanium with hydrogen gas [7,8]. Hydrogen can be admixed to the inert gas from the top or injected through the special inlet directly into Zone 2. At this stage, the temperature of the flowing nascent titanium nanoparticles is sufficiently high (400-500 °C) as to secure the stoichiometric reaction: $Ti + H_2 = TiH_2$. Subsequently, the fast motion of the flow to a colder Zone 3 prevents the thermal decomposition of the synthesized TiH_2 particles.

The method allows synthesis of customized "core-shell" nanoparticles under highly controllable conditions. Depending on the manufacturing parameters, it is possible to synthesize nanoparticles of different composition, from pure stoichiometric titanium hydride TiH_2 to mixed composition such as TiH_2-Ti. One of the main factors determining the composition of the synthesized titanium hydride nanoparticles is the hydrogen concentration in the reaction zone. In the present work, it is demonstrated by the examples of samples TH-1 and TH-2 synthesized at different hydrogen concentrations achieved using different gas inlets (Table 1).

Table 1. Manufacturing parameters and characteristics of titanium hydride nanopowders

Sample	Feed wire rate (g/h)	Ar gas pressure (at.)	Hydrogen gas flow rate (Nl/min)	Hydrogen gas inlet	Crystal phase content (w.%)	Mean size (nm)
TH-1	1.37	0.2	0.125	special design (reaction zone)	TiH_2 – 100	25
TH-2	2.25	0.2	0.125	main (top)	TiH_2 – 94.2 α-Ti – 5.8	30

Synthesized titanium hydride nanopowder is a loose material of black color. The powder, if not properly passivated, is extremely pyrophoric and needs to be handled in an inert atmosphere with decreased oxygen and moisture content. For the present study, the synthesized samples of titanium hydride were subjected to a suitable process of passivation in order to form a protective titanium oxide layer on the surface of nanoparticles which prevents sudden ignition in atmospheric air.

2.1.2. Synthesis of titanium carbide nanoparticles. Titanium carbide nanoparticles were obtained via interaction of titanium in nascent nanoparticles with carbonic reactant gases [9]. In the present work, hydrocarbons such as acetylene (C_2H_2), methane (CH_4) and mixture of propane (C_3H_8) and butane (C_4H_{10}) were used as gaseous reactants. The reactants were injected in the reactor through the adjustable inlet of special design, which location determines Zone 2, depending on the required temperature of the reaction of titanium with the chosen hydrocarbon: $Ti + C_xH_y \rightarrow TiC, TiH_2, C + H_2$.

The reaction conditions and a type of hydrocarbon used determine the composition of the resulting nanopowder. It is possible to synthesize pure crystalline TiC nanoparticles (with small amount of crystalline carbon), as well as nanoparticles of mixed composition such as $TiC+TiH_2$, TiC+Ti. In the present work, it is demonstrated by the examples of samples TC-1, TC-2, TC-3 (Table 2).

Table 2. Manufacturing parameters and characteristics of titanium carbide nanopowders

Sample	Ar gas pressure (at.)	Ar gas flow rate (m^3/h)	Reactant gas (hydrocarbon)	Reactant gas flow rate (m^3/h)	Crystal phase content (w.%)	Mean size (nm)
TC-1	0.2	0.15	C_2H_2	0.15	TiC – 95,2 C – 4,8	21
TC-2	0.24	0.15	CH_4	$3.6 \cdot 10^{-3}$	TiC – 62,9 TiH_2 – 37,1	27
TC-3	0.5	0.25	$C_3H_8 + C_4H_{10}$	$4.4 \cdot 10^{-2}$	TiC – 86,6 Ti – 13,4	26

Synthesized titanium carbide nanopowder is a loose inert material of black color. However, the synthesized nanopowder of mixed composition can exhibit pyrophoric properties, depending on the content of titanium hydride or residual titanium metal.

3. Results and discussion

3.1. Titanium hydride

Characterization of the titanium hydride nanoparticles was performed via complex of methods of transmission electron microscopy (TEM/HRTEM/EDA/EDX), X-Ray diffraction (XRD), thermal analysis (TGA/DTA/DSC/MS).

X-ray diffraction patterns of titanium hydride samples TH-1 and TH-2 synthesized under conditions indicated in Table 1 are shown in Figure 2.

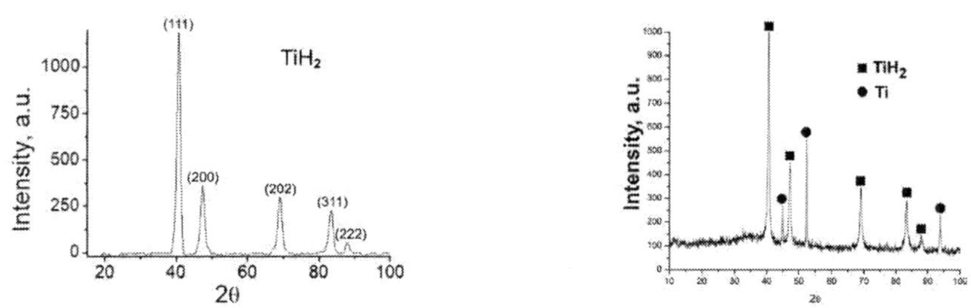

(a) (b)

Figure 2. X-ray diffraction patterns of titanium hydride samples TH-1 (a) and TH-2 (b)

The fact that XRD analysis revealed no crystalline phase of titanium oxide formed on the surface of nanoparticles during passivation indicates its amorphous state.

Typical TEM/HRTEM images of samples TH-1 and TH-2 are shown in figure 3 and figure 4, respectively. TEM study of samples TH-1 and TH-2 revealed non-agglomerated arrays of particles of various sizes. The diameters of particles in both samples measured from TEM images ranged from 5 nm to 70 nm (nanoparticles mean size is presented in Table 1). The TEM/HRTEM examination of nanoparticles revealed that the structure of the particles depends on their size. Generally, the smallest particles of the array (up to about 10 nm) look equiaxial and homogeneous, while larger particles

IOP Conf. Series: Materials Science and Engineering **558** (2019) 012023 doi:10.1088/1757-899X/558/1/012023

exhibit a core-shell structure with a pronounced facetted shape. Amplified views of HRTEM images expose lattice fringes corresponding to the atomic (inter-planar) spacing. These lattice fringes are found only on the images of the particle cores, and the measured magnitude of the inter-planar spacing d ≈ 0.26 nm corresponds to the (111) plane of TiH_2 lattice (Figure 3 (b) - inset). No signs of crystallinity of the particle shells (such as lattice fringes, moiré patterns) are observed on the HRTEM images of both samples, which is consistent with the results of XRD analysis. Measurements of shell thicknesses of the HRTEM images of about 20 nanoparticles ranged between 3.5-5.0 nm.

Results of electron diffraction analysis of sample TH-1 and TH-2 are consistent with XRD data.

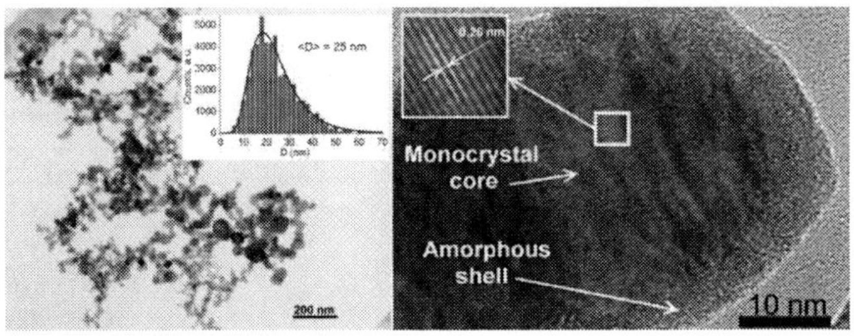

(a) (b)

Figure 3. TEM image of sample TH-1 where the inset is the particles size distribution histogram (a) and HRTEM image of an individual particle of the same sample where the inset is the enlarged view of the selected area (b).

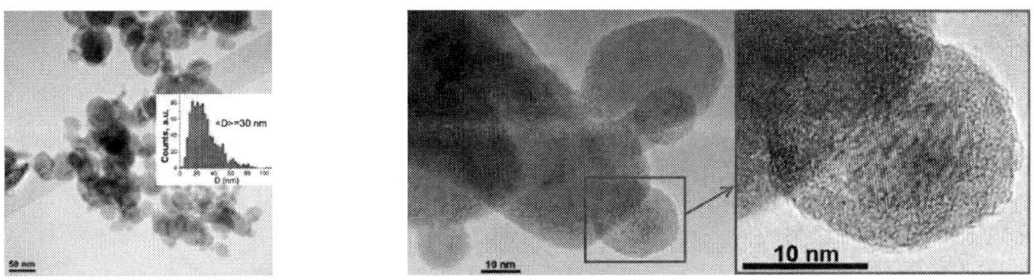

(a) (b)

Figure 4. Typical TEM image of sample TH-2 where the inset is the particles size distribution histogram (a) and HRTEM image of the same sample where the enlarged view of the selected particle is shown (b).

Hydrogen release from TiH_2 nanoparticles was studied using thermal methods of analysis. Decomposition behavior of the synthesized sample TH-1 of TiH_2 nanopowder (NP) and commercial titanium hydride powders as the references were studied using simultaneous TG-DTA/DSC and MS analysis. Reference Ref1 is a commercial titanium hydride powder with particle mean size $\langle D_1 \rangle \approx 15$ μm, and reference Ref2 is a coarse titanium hydride powder of particle size $\langle D_2 \rangle \approx 0.5$ mm.

The studies were conducted in high-purity helium gas flow (40 ml/min) at heating rate of 10 ^{0} C/min within the temperature range of 50-700 ^{0} C. TG/DSC analysis of all samples included two successive heating cycles for each sample with no exposure to air during the intermediate cooling. Typical results are shown in Figure 5.

It is evident from the data presented, that a significant feature of TiH_2 nanopowder synthesized via the Guen-Miller FL method is the hydrogen evolution within the narrow temperature range (390-510 ^{0}C), which contrasts markedly with commercial samples of larger particle size (with a typical temperature of hydrogen evolution higher than 500 ^{0}C). It is also evident that the TiH_2 NP requires less energy to start the decomposition process and that the diffusion of hydrogen is faster in smaller particles than in larger particles.

IOP Conf. Series: Materials Science and Engineering **558** (2019) 012023 doi:10.1088/1757-899X/558/1/012023

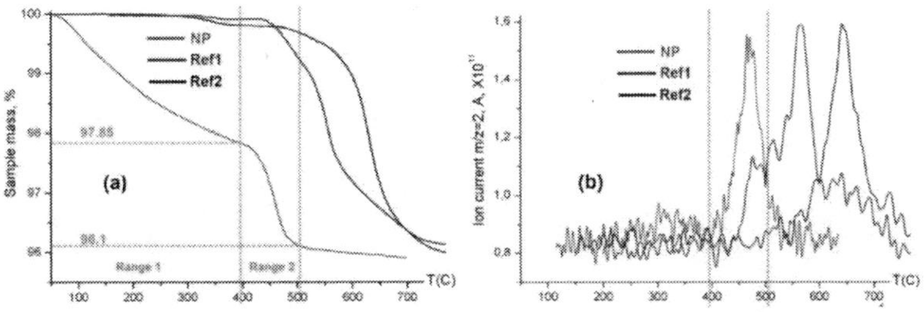

(a) (b)

Figure 5. Results of TG-DTA/DSC-MS during the heating cycle showing mass loss of titanium hydride (a) and hydrogen release intensity (b), where NP is synthesized TiH_2 nanopowder, Ref1 and Ref2 are commercial titanium hydride powders.

3.2. Titanium carbide

Characterization of the titanium carbide nanoparticles was performed via complex of methods of scanning and transmission electron microscopy (SEM/TEM/HRTEM/EDX) and X-Ray diffraction.

X-ray diffraction patterns of titanium carbide samples TC-1, TC-2 and TC-3 synthesized under conditions indicated in Table 2 are shown in Figure 6.

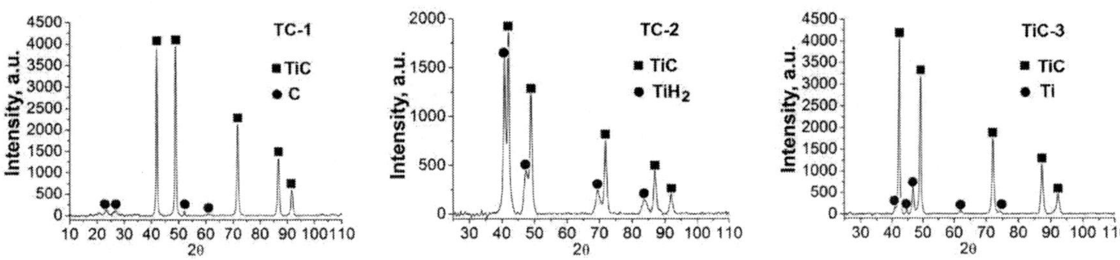

(a) (b) (c)

Figure 6. X-ray diffraction patterns of titanium carbide samples TC-1 (a), TC-2 (b) and TC-3 (c)

SEM and TEM analysis of titanium carbide samples revealed close similarity of nanoparticles morphology (Figure 7). All particles, irrespectively of the manufacturing conditions, exhibit equiaxial facetted shape. It is evident from SEM study that nanoparticles are equiaxial polyhedrons. Mean size of particles of all studied titanium carbide samples do not exceed 30 nm.

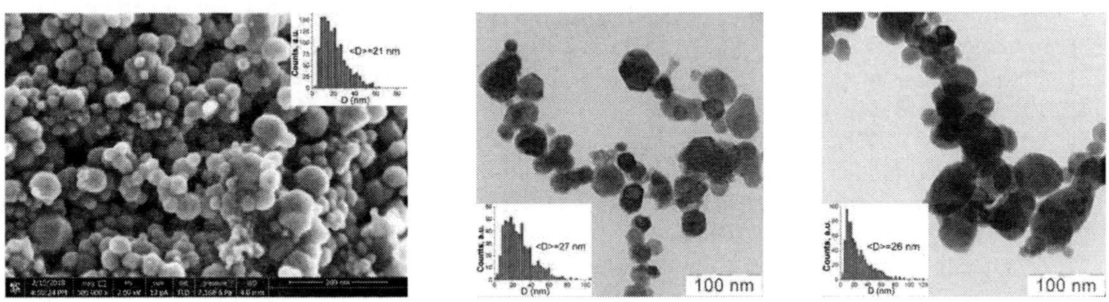

(a) (b) (c)

Figure 7. Typical SEM image of sample TC-1 (a) and TEM images of samples TC-2 (b) and TC-3 (c). Insets are the particles size distribution histograms.

HRTEM examination revealed a difference in the internal structure of titanium carbide nanoparticles synthesized under different conditions. Examples are shown in Figure 8. Nanoparticles of the sample TC-1 are pure titanium carbide monocrystals, the measured magnitude of the inter-planar spacing d $\approx$ 0.251 nm corresponds to the (111) plane of TiC lattice (Figure 8 (a) - inset). Nanoparticles of the sample TC-2 exhibit the evident core-shell structure, where the shell thickness is about 2.6-2.7 nm. The measured magnitudes of inter-planar spacings at the image of the core

IOP Conf. Series: Materials Science and Engineering **558** (2019) 012023 doi:10.1088/1757-899X/558/1/012023

correspond to the (111) planes of TiC (d ≈ 0.251 nm) and TiH_2 (d ≈ 0.265 nm) which shows complex crystalline composition of nanoparticles core (Figure 8 (b) - insets). The shell presence can be attributed to formation amorphous titanium oxide layer, as a result of TiH_2 oxidation during the passivation process.

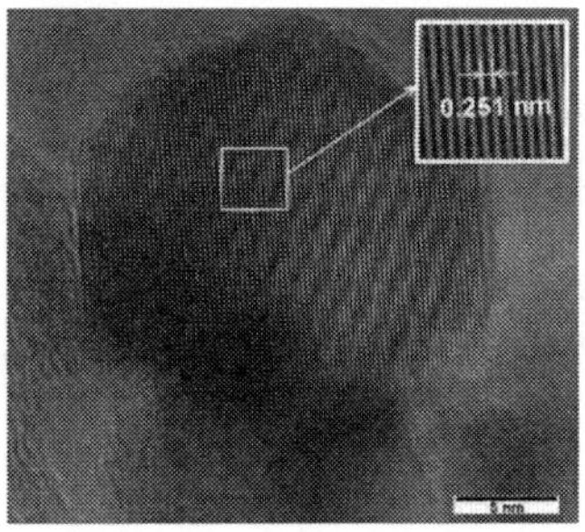

(a)

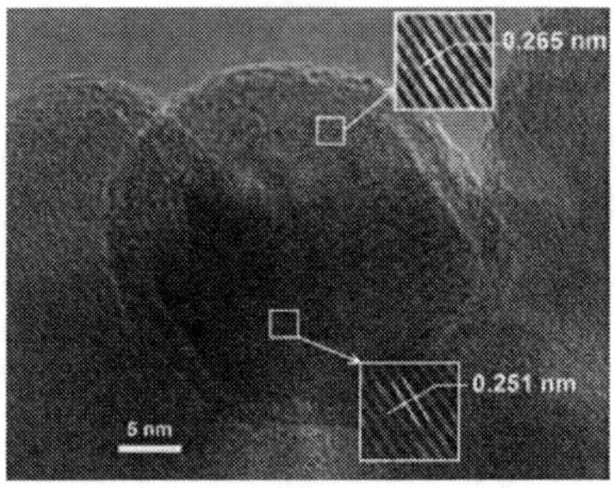

(b)

Figure 8. HRTEM images of nanoparticles of the samples TC-1 (a) and TC-2 (b), where the insets are the enlarged views of the selected areas.

4. Conclusion

The results obtained demonstrate the capability of Flow-Levitation method for synthesis of nanoparticles of titanium compounds of customized composition with mean size of less than 30 nm.

Titanium hydride nanoparticles exhibit core-shell structure and possess the promising characteristics for hydrogen recovery as compared to coarse-grained commercial powders.

Titanium carbide nanoparticles characterization displayed the relation between their composition and internal structure. Pure TiC nanoparticles are monocrystalline polyhedrons, whereas nanoparticles of complex composition exhibit core-shell structure.

Acknowledgements.
This work was developed within the framework of the Program for Basic Research of the Russian State Academy of Sciences for 2013-2020 and supported by the Ministry of Sciences and Higher Education (agreement GZ №007-00269-18-00).

References

[1] Kennedy A 2002 *Scr. Mater.* **47** 763.
[2] Rasooli A, Divandari M, Shahverdi H and Boutorabi M 2012 *Int. J. Min. Met. Mater.* **19** 165
[3] Hosoe M, Kanoya I, Sakai H and Furuta T 2012 *US Patent 8258077*
[4] Rönnebro E C E, Whyatt G, Powell M, Westman M, Zheng F and Fang Z 2015 *Energies* **8** 8406
[5] Kiparisov S S, Levinskii Yu V and Petrov A P 1987 *Titanium carbide: synthesis, properties, application* (Moscow, Metallurgiya Publ.) (in Russian)
[6] Kuskov M L, Zhigach A N, Leipunsky I O, Gorbachev A N, Afanasenkova E S and Safronova O A 2018 *Instr Exp Tech* **4** 1
[7] Zhigach A N, Lejpunskij I O, Berezkina N G, Kuskov M L and Afanasenkova E S 2017 *RU Patent 2616920*
[8] Leipunsky I O, Zhigach A N, Kuskov M L, Berezkina N G, Afanasenkova E S, Kudrov B V, Lopez G W, Vorobjeva G A and Naumkin A V 2019 *J Alloy Compd* **778** 271
[9] Zhigach A N, Leipunsky I O, Berezkina N G, Kuskov M L, Afanasenkova E S and Safronova O A 2018 *RU Patent Application 2018111411 A* (in Russian)

Study of structure of copper-based composite materials during the spark plasma sintering

K V Kuskov[1], I N Volkov[1], N F Skodich[1,2], A A Nepapushev[1], D I Arkhipov[1], and D O Moskovskikh[1]

[1]Center of Functional Nano-Ceramics, National University of Science and Technology "MISiS", Moscow, 119049 Russia
[2]Merzhanov Institute of Structural Macrokinetics and Materials Science, Russian Academy of Sciences, Chernogolovka, 142432 Russia

E-mail: kkuskov@misis.ru

Abstract. This paper is devoted to the study of copper-based composite materials structure formation processes in the conditions of SPS. The structures of Cu–Cr, Cu–Mo, Cu–W, Cu–Cr–Mo, Cu–Cr–W, Cu–Mo–W, Cu–Cr–SiC, and Cu–Zr–CNTs two-component and three-component systems were analyzed. The possibility of the formation of non-equilibrium phases under SPS conditions is shown, the effect of chemical transformations on the structure of bulk composite materials is considered.

1. Introduction

The spark plasma sintering (SPS) method has discovered new ways for producing materials with different properties due to the possibility of rapid and uniform heating combined with the application of pressure on the sample. For example, according to [1] from the 1960s, composite materials made of immiscible metals began to be used as contacts for vacuum interrupters. Conventional methods of powder metallurgy do not allow introducing something new into the structure and properties of such materials. The SPS method, due to the ability to preserve the nanostructure in the bulk material, again raised interest in contact materials, allowing one to create new composites with enhanced properties [2, 3]. However, the fundamental basis of the immiscible metals consolidation in the conditions of SPS is not yet fully understood.

2. Experimental

Commercial copper, chromium, tungsten, molybdenum, silicon carbide, carbon nanotubes, zirconium were used as initial materials. Composite powders were obtained in planetary ball mill "Activator 2S" (Activator, Novosibirsk, Russia). Pre-mixed powders were subjected to high-energy ball milling (HEBM) in an argon atmosphere in steel water-cooled vials containing 7 mm steel balls. The mass ratio of the balls to the powder was 20:1. The rotational speed of the sunwheel during process was 694 rpm with a vial-to-sunwheel rotation speed ratio (K coefficient) was equal to 1. Composite powders were consolidated in a spark plasma sintering device (SPS, Labox 650, SinterLand, Japan). The maximum load was 50 MPa.

The structure of the sintered materials was analyzed using a JSM7600F (JEOL, Japan) scanning electron microscopy (SEM). X-ray diffraction analysis was performed using a Difray 401 diffractometer

(Scientific instruments, Russia). The density of the samples after sintering was evaluated using hydrostatic weighing. Electrical resistivity was measured at KRIOTEL (Russia) device using the four-contact method. The hardness of the sintered samples was determined on polished samples by the Vickers method according to GOST 2999-75 using a HVS-50 digital hardness tester.

The effect of the SPS parameters on the structure of materials obtained from composite particles of immiscible metals was studied. It was shown that during consolidation the structure of the particles changes: in binary systems (Cu–Cr, Cu–W, and Cu–Mo) heating leads to the growth of particles of refractory components from 5–10 to 200–300 nm with copper precipitation on the grain boundaries of sintered powders (figure 1).

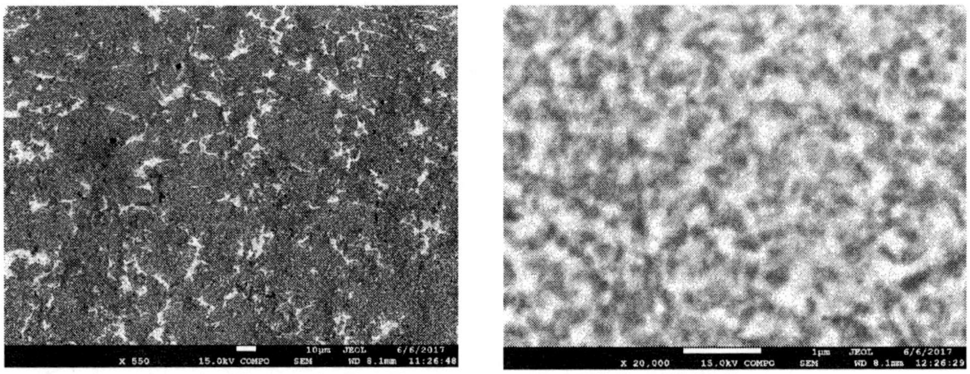

Figure 1. Cu–Cr structure after SPS at 700°C.

The structure of the Cu–Cr–W ternary composite also consists of homogeneous regions with inclusions between them. However, in some places small particles of tungsten are visible. The composition of homogeneous regions differs from the targeted, what can be explained by the copper precipitation along the boundaries, as well as the extrusion of plastic copper onto the sample surface due to overheating.

The structure of heterophase regions represents a matrix with tungsten and chromium inclusions. Cr and W powders disintegrated to the submicron sizes were overheating at the places of their contact during passing a current, forming a solid solution.

The structure of Cu–Mo–W samples also represents a heterophase region separated by a copper phase. However, in contrast to the Cu–Cr–W sample, the heterophase regions are uniform; they do not contain large inclusions of individual components. These areas consist of solid solution particles of refractory components in a copper matrix.

The microstructure of the Cu–Cr–Mo sample is completely similar and comprises heterophase region of a solid Cr–Mo solution in a copper matrix surrounded by copper inclusions formed during sintering.

The 50Cu–25Cr–25SiC and 50Cu–40Cr–10SiC materials were also investigated (the numbers indicate the volume percentage of the components).

In contrast to previous systems, the nanocomposite structure was destroyed in samples containing silicon carbide (figure 2). However, it is worth noting the presence of several phases in the back scattered electrons. The performed X-ray analysis showed a change in the phase composition relatively to the powder material, and this change is different for different SiC contents.

Thermodynamic analysis of the Cu–Cr–SiC system with a given compositions at the sintering temperature showed that in the presence of 10 vol % SiC in the initial mixture, the equilibrium composition of the solid phase will be Cu–Cr–Cr_3Si–$Cr_{23}C_6$. For the system initially containing 25 vol % SiC, the composition of the solid phase will be Cu–SiC–Cr_5Si_3–Cr_3C_2. Unfortunately, due to the similarity of the structures, the quantitative analysis of the phase composition is difficult, but it can be concluded that the reactions do not reach equilibrium under sintering conditions (heating to 900 °C at a rate of 100°/min and holding for 15 min). This is evidenced by the peaks of the intermediate phases Cr_7C_3 for the composition 50/40/10 and Cr_3Si for the composition 50/25/25.

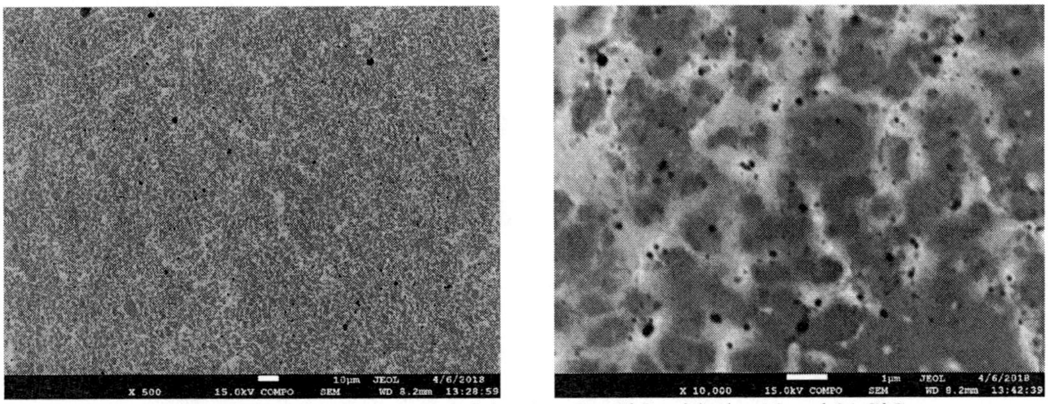

Figure 2. Microstructure of the Cu–Cr–SiC with the 10 vol % SiC.

Thus, the interaction of chromium with silicon carbide leads to the destruction of the nanocomposite structure of the powders during consolidation. The causes of this phenomenon could be the diffusion effects due to composition change, as well as a sharp increase in electrical resistance (change in the lattice type, the appearance of microstresses and other defects), leading to local overheating in the structure and accelerating mass transfer processes.

The study of the Cu–Cr–CNT sample using a scanning electron microscope showed that the structure of the material represents heterophase regions of various sizes, separated by copper inclusions, as in above-mentioned cases of binary and ternary metal systems. Energy dispersive analysis showed that heterophase regions consists of chromium, copper and carbon. However, X-ray diffraction analysis showed only two phases in the sample composition: copper and chromium carbide (II). As a result, we can conclude that long-term high-energy ball milling leads to the destruction of CNT and the intensification of the interaction processes between carbon and chromium.

Similar results were obtained for the Cu–Zr–CNT system. Powders for sintering with 50 at % Cu, 12.5 at % Zr, and 37.5 at % C were prepared in two ways: mixing in the roller mill and in the planetary ball mill.

The structure of the sample after consolidation of the roller milled powders represents copper and zirconium grains separated by the CNT agglomerates (figure 3). Zirconium concentration in the grains is not uniform; it mainly located at the edges of particles (figure 3b, 3c). X-ray diffraction analysis of the roller milled powders before and after sintering shows that a small peak of zirconium carbide appears in the structure of the sample, i.e. sintering conditions contribute to the formation of a compound between the components, but at a low rate.

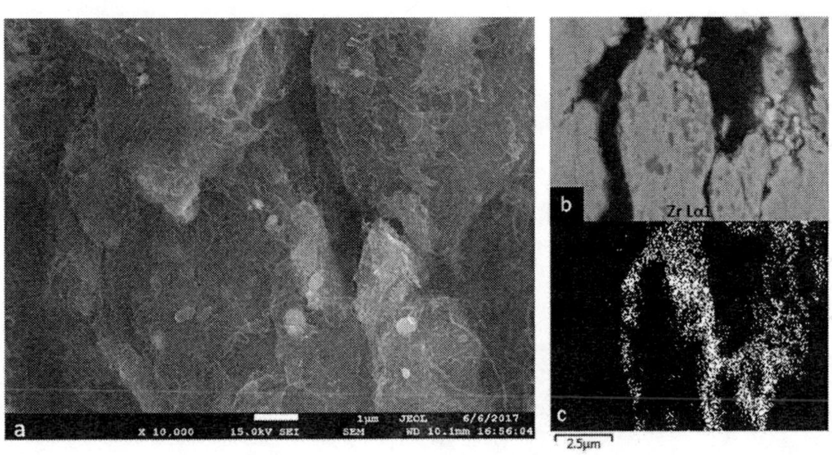

Figure 3. Microstructure of Cu–Zr–CNT powder after mixing in the roller mill.

IOP Conf. Series: Materials Science and Engineering **558** (2019) 012024 doi:10.1088/1757-899X/558/1/012024

Structure of the fractured sample obtained by the HEBM + SPS method differ from the sample after roller mill. The morphology of the initial particles is clearly visible on the surface of the fracture: there are rather large areas, to which small areas have been sintered. However, chemical interaction between zirconium and carbon occurred leading to the formation of the carbide phase.

3. Conclusion

By combining the HEBM and SPS methods, it is possible to obtain a structure consisting of heterophase regions of submicron (up to 300 nm) particles of refractory materials uniformly distributed in a copper matrix. Adding a third component does not lead to significant changes in the microstructure of the material, if the refractory components form a solid solution. Chemical transformations, on the contrary, lead to significant changes in the microstructure: to an increase in the size of the structural components and the formation of non-equilibrium phases.

Acknowledgments

This work was supported by the Russian Foundation for Basic Research (project no. 18-33-01087\18 dated March 26, 2018).

References

[1] Slade P G 2007 *The Vacuum Interrupter: Theory, Design and Application (*CRC Press, Boca Raton)
[2] Fang Q, Kang Z, Gan Y and Long Y 2015 *Mater. Des.* **88** 8–15
[3] Shkodich N F, Rogachev A S, Vadchenko S G, Moskovskikh D O, Sachkova N V, Rouvimov S and Mukasyan A S 2014 *J. Alloys Comp.* **617** 39–46

Prospective SHS composites for high-temperature applications

E A Levashov*, V V Kurbatkina, S Vorotilo, Yu S Pogozhev and E I Patsera

National University of Science and Technology "MISIS", Leninskii 4, Moscow, 119049, Russia

*E-mail: levashov@shs.misis.ru

Abstract In this article, a succinct review of the SHS ceramics prospective for the high-temperature applications is provided. Macrokinetic features and mechanisms of combustion and structure formation were studied for the systems Ta–Zr–C, Ta–Hf–C, Zr–Ta–B, Hf–Ta–B, Zr–B–Si–C, Ta–Si–C, Mo–Hf-Si–B, and Zr–Ta-B–Si. Materials were produced by the combination of SHS with hot pressing (HP) and spark plasma sintering (SPS), and their structure and properties were investigated.

1. Introduction

The development of advanced materials resistant against the oxidation and ablation in the flow of high-enthalpy gas containing abrasive particles is a crucial area of modern materials science. Such materials are employed for the high-speed aircraft as sharp wing edges and nose tips [1] as well as for the spacecraft as the nozzles for solid-fuel thrusters. Prospective materials and coatings for these applications include the composites based on HfB_2 (T_m = 3653 K), ZrB_2 (3473 K), TaB_2 (3473 K), NbB_2 (3323 K) which are characterized by the melting point above 3000 K [2], high strength, fracture toughness, wear resistance and thermal shock resistance. An essential advantage of the boride-based composites is their high thermal conductivity, which ensures the enhanced heat sink from the surfaces which are in contact with the high-temperature oxidative gas flow.

Development of the efficient production technologies for ultra-high temperature ceramics (UHTCs) based on carbides HfC (T_m = 3900°C), TaC (3800°C), borides and their solid solutions is a significant problem. Solid solutions possess higher melting points as compared to simple compounds. For example, the single-phase carbide (Ta,Hf)C with 20 % HfC has the melting point of approx. 4223 K. Moreover, the dependences of hardness, the coefficient of thermal expansion (CTE) and thermal conductivity on the composition of solid solutions are non-linear and have maxima at optimal compositions. Combination of high thermal conductivity and low CTE defines the resistance of material towards the thermal shocks.

The synthesis of single-phase compounds (Ta,Zr)C, (Ta,Hf)C, $(Ta,Zr)B_2$, $(Ta,Hf)B_2$ is another important problem. Various methods of production of complex carbides and borides are known, including the heating of the mixture of powders, co-reduction of the mixture of oxides by carbon, precipitation from solutions, microwave synthesis, sol-gel synthesis, hot pressing, etc. SHS provides unique possibilities for the production of UHTCs. Hybrid technologies SHS + HP, SHS + SPS, reactive HP and SPS demonstrated their perspectivity for the production of the dense high-temperature materials and cathodes-targets with unique structure and properties [3, 4].

2. Results and discussion

Kinetics and mechanisms of combustion and structure formation were investigated for the systems Ta–Zr–C, Ta–Hf–C, Zr–Ta–B, Hf–Ta–B, Zr–B–Si–C, Ta–Si–C, Mo–Hf-Si–B, Zr–Ta-B–Si [3–15, 18–21]. Mechanical activation (MA) of the reaction mixtures and optimization of the synthesis parameters play a crucial role in the production of the solid solutions (Ta,Hf)C, (Ta,Zr)C. MA SHS allows the formation of single-phase (Ta,Hf)C with the lattice parameter a = 0.4487 nm, which corresponds to the 18 at % dissolved HfC, while the HfO_2 impurities do not exceed 1 %. Solid solution powders are outstanding raw material for the subsequent sintering of dense ceramics by HP and SPS. For example, (Ta,Hf)C ceramics has the relative density of 98%, hardness of 27 GPa, Young modulus of 484 GPa, elastic recovery of 46%.

The possibility of SHS of single-phase solid solution on the base of diborides of tantalum and zirconium $(Zr,Ta)B_2$ as well as diborides of hafnium and tantalum $(Hf,Ta)B_2$ was demonstrated [14, 18, 19]. By thermodynamic calculations and analysis of the composition and structure of the characteristic areas of the quenched combustion fronts by SEM and EDS, it was established that in the preheating zone the volatile suboxide B_2O_2 forms, chemosorbs on the metallic particles and produces needle-like borides of Ta and Zr. At temperatures above the melting point of zirconium and boron, a melt forms and wets the tantalum particles and borides formed in the preheating zone. Main reactions take place, and ZrB_2 phase precipitates from the oversaturated melt. Borides formed in the preheating zone interact with the melt, and solid solutions are formed (figure 1). On the last stage in the zones of post-combustion and structure formation, a homogenization of the solid solutions takes place. Depending on the composition of the reactive mixture, either single-phase solid solution $(Zr,Ta)B_2$ or two-phase composite $(Zr,Ta)B_2 + TaB_2$ are formed [14].

In the Hf–Ta–B system, the phase formation occurs differently. In the Hf–B binary system, at temperature 2153 K and boron content B $\leq$ 0.8%, contact melting due to the eutectic reaction β-Hf + HfB → Liq takes place and forms a melt containing solid particles HfB + β-Hf. At 2365 K, boron melts, and almost simultaneously the first crystallites of HfB_2 and TaB_2 precipitate. Then, melting of Hf occurs (2505 K), and the active interaction of the melt with the formation of the solid solutions based on the diborides of Ta and Hf takes place [18]. Due to the high combustion temperature and considerable thermal losses, the post-combustion zone is relatively narrow. Concentration homogenization takes place in the post-combustion zone. However, the high porosity of the synthesis products and low diffusion rates do not allow the formation of equilibrium phases to complete. A significant difference in grain composition and size is retained in the synthesis products. However, the produced powders are successfully used for the subsequent consolidation by HP and SPS. A chemical homogeneity on the grain level is achieved [19].

For the Ta–Si–C system, the sequence of chemical and structural transformations in the combustion wave was studied [12]. Thermodynamic analysis, investigation of the quenched combustion fronts and time-resolved XRD revealed the sequence of phase formation during the combustion synthesis of $TaSi_2$ + SiC ceramics for application in magnetron sputtering of high-temperature tribological coatings. In the post-combustion zone, a significant refinement of the microstructure takes place due to the formation of SiC crystallites (20–40 nm) within the TaC, Ta_5Si_3 and $TaSi_2$ grains. Since one part of the SiC and $TaSi_2$ grains crystallizes from the Si melt and another part forms due to the solid-state reactions.

Submicron-sized composite $TaSi_2$–xSiC powders (x = 10, 30, 50, and 70%) consisting of $TaSi_2$ and SiC grains 50–100 nm in size were fabricated by SHS. The optimal compositions and hot pressing regimes for SHS powders were determined, ensuring the production of the $TaSi_2$–SiC ceramics with a hierarchical structure, the relative density of 96–97%, hardness of ~19 GPa, and fracture toughness of 6.5–6.7 MPa·m$^{1/2}$. Double-layered composite targets 120 mm in diameter, with the $TaSi_2$–xSiC working layer (x=10 and 30%) and molybdenum sublayer, were produced by hot pressing and used for coatings deposition. Si–Ta–C– (N) nanocomposite coatings were deposited using the magnetron sputtering. Structure of as-deposited coatings consisted of fcc Ta(Si,C,N) (the $TaSi_2$–30%SiC target) or Ta_5Si_3 (the $TaSi_2$–10%SiC target) nanocrystallites less than 3 nm in size embedded in an amorphous matrix. The coatings were characterized by high thermal stability and oxidation resistance at temperatures below

800°C. A detailed SEM analysis of the coating wear tracks showed that the tribolayer formed at room temperature is dense, although the wear mode remains abrasive. At 600°C, the tribolayer on the wear track bottom is not continuous; furthermore, individual microfibers are appeared. The wear track profile indicates that the abrasive wear still remains the main wear mode. More microfibers, including small ones, are formed at 800°C. These microfibers cover the wear track surface more densely. This leads to a combination of low coefficient of friction and high wear resistance. The fine structure of the microfibers was additionally studied by TEM. The microfibers with 1 μm thick have a smooth surface that is non-uniformly covered with particles of wear products 30–100 nm in size. EDX analysis showed that fibers contain a large amount of oxygen, as well as tantalum and silicon. This indicates that they originate from a thin oxide layer, which was formed on the surface during high-temperature tribological tests. A detailed analysis of fiber microstructure by HRTEM demonstrated that the amorphous matrix (most likely SiO_x) contains nanoparticles 2–4 nm in size. The interplanar distances (2.2 and 2.5 Å) and the interplanar angle (~ 55°) indicate that those are nanoparticles of tantalum oxide TaO with fcc structure. Hence, it is fair to assume that the coating surface undergoes oxidation due to the transformation of Ta(C,N) nanocrystallites, with their size remaining unchanged.

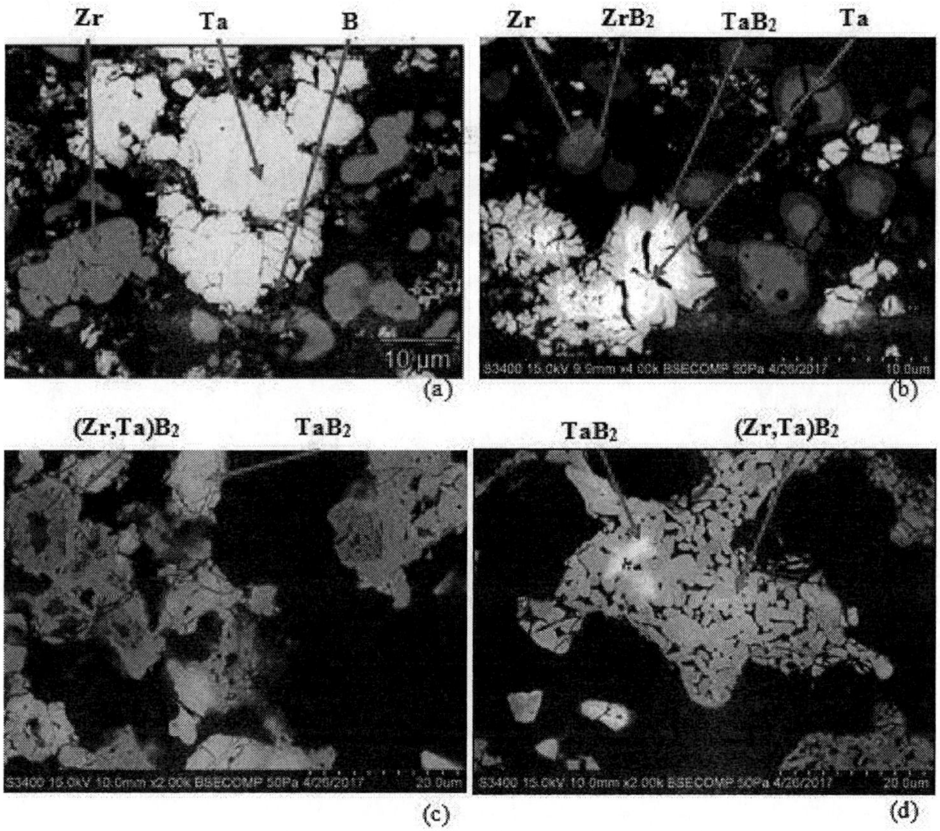

Figure 1. The microstructure of the quenched combustion front and the phase composition of the typical components: (a) initial reaction mixture; (b) the preheating zone; (c) the combustion zone; and (d) the post combustion zone [14].

The coating matrix is saturated with oxygen and retains its amorphous structure. The friction coefficient of the coatings decreases with temperature from 0.38 (25°C) to 0.28 (600°C) and 0.23 (800°C), while their wear resistance increased. A thin (~ 100 nm) oxide layer and $TaSi_xO_y$ microfibers played a key role in enhanced tribological behavior, reducing the coefficient of friction and wear due to rolling friction. Disilicides $MoSi_2$, $ZrSi_2$, and $TaSi_2$ are widely used as the alloying addition to the boride

IOP Conf. Series: Materials Science and Engineering **558** (2019) 012025 doi:10.1088/1757-899X/558/1/012025

and carbide ceramics to increase the ceramic's oxidation resistance. For example, $MoSi_2$ has an outstanding oxidation resistance and retains its properties for 2000 h at 1923 K. During the oxidation of $ZrSi_2$, glassy $ZrSiO_4/SiO_2$ forms, increasing the high-temperature oxidation resistance and providing the ability to self-heal the forming defects. Ceramics with the 1–2 μm boride grains and 2–4 μm silicide grains were produced by elemental SHS. Within the $MoSi_2$ grains, 50–300-nm eutectic colonies $Me^{IV}Si_2$–$MoSi_2$ are present.

Powders produced by the milling of combustion products were used for the fabrication of the dense samples. Physical-mechanical properties of the samples are presented in table 1. Boride compositions possess high hardness and elastic modulus, significantly surpassing the known analogous [16, 17] and reaching the level of super-hard materials. Such high values were explained by the low content of the oxides (below 1%) and high internal stresses.

Table 1. Physical-mechanical properties of UHTCs

Composition	Sintering conditions	Relative density,%	Hardness, GPa	Elastic modulus, GPa
(Ta,Zr)C	SHS + HP	96	32	520
(Ta,Zr)C	SHS + SPS	98	36	536
(Ta,Hf)C	SHS + HP	96	28	423
(Ta,Hf)C	SHS + SPS	99	32	484
70%TaSi2+30%SiC	SHS + HP	97	19	340–370
(Zr,Ta)B2	SHS + HP	97	70	594–619
80%(Zr,Ta)B2+20%TaB2	SHS + HP	96	74	647–672
(Hf,Ta)B2	SHS+SPS	98	62	570–597
80%(Hf,Ta)B2+20%TaB2	SHS+SPS	98	70	587–628

Consolidated UHTCs underwent the plasma jet tests and demonstrated the 15–20% reduction of the linear erosion rate in comparison with the commercial SiC-coated carbon-carbon composites. UHTCs on the base of $(Ta,Zr)B_2$, $(Ta,Hf)B_2$ were characterized by the enthalpy of decomposition 390 kJ/g, which is an order of magnitude higher as compared to the analogous carbides and boride-silicide composites. Solid solutions (Ta,Hf)C, (Ta,Zr,)C, $(Hf,Ta)B_2$ as well as ceramics $(Zr,Ta)B_2$ + 35% $TaSi_2$ and ZrB_2 + $ZrSi_2$ were characterized by the best high-temperature oxidation resistance.

Acknowledgment
This work was carried out with financial support from the Russian Science Foundation under project no. 14-19-00273-P.

References

[1] Paul A, Jayaseelan D D, Venugopal S, Zapata-Solvas E, Binner J, Vaidhyanathan B, Heaton A, Brown P and Lee W E 2012 *Amer. Ceram. Soc. Bulletin* **91** 22–29
[2] Wuchina E, Opila E, Opeka M, Fahrenholtz W and Talmy I 2007 *Electrochem. Soc. Inter*. 30–36
[3] Licheri R, Musa C, Orro R, Cao G, Sciti D and Silvestroni L 2016 *J. Alloys Comp.* **663** 351–359
[4] Levashov E A, Mukasyan A S, Rogachev A S and Shtansky D V 2017 *Int. Mater. Rev*. **62** 203–239.
[5] Levashov E A, Pogozhev Yu S, Potanin A Yu, Kochetov N A, Kovalev D Yu, Shvyndina N V and Sviridova T A 2014 *Ceram. Int.* **40** 6541–52.

[6] Potanin A Yu, Pogozhev Yu S, Levashov E A, Kovalev D Yu and Novikov A V 2014 *Euras. Chem. Tech. J.* **16** 53–58
[7] Pogozhev Yu S, Potanin A Yu, Levashov E A, Novikov A V, Sviridova T A and Kochetov N A 2014 *Russ. J. Non-Ferr. Met.* **55** 632–638
[8] Patsera E I, Levashov E A, Kurbatkina V V and Kovalev D Yu 2015 *Ceram. Int.* **41** 8885–93
[9] Patsera E I, Kurbatkina V V, Vorotylo S A, Levashov E A and Timofeev A N 2016 *Ceram. Int.* **42** 16491–98
[10] Pogozhev Yu S, Iatsyuk I V, Levashov E A, Novikov A V, Kochetov N A and Kovalev D Yu 2016 *Ceram. Int.* **42** 16758–65
[11] Potanin A Yu, Pogozhev Yu S, Levashov E A, Novikov A V, Shvindina N V and Sviridova T A 2017 *Ceram. Int.* **43** 10478–86
[12] Vorotylo S A, Levashov E A, Kurbatkina V V, Kovalev D Yu and Kochetov N A 2018 *J. Eur. Ceram. Soc.* **38** 433–443
[13] Kurbatkina V V, Patsera E I, Levashov E A and Timofeev A N 2018 *Ceram. Int.* **44** 4320–29
[14] Kurbatkina V V, Patsera E I, Levashov E A and Timofeev A N 2018 *J. Eur. Ceram. Soc.* **38** 1118–27
[15] Iatsyuk I V, Pogozhev Yu S, Levashov E A, Novikov A V, Kochetov N A and D Kovalev D Yu 2018 *J. Eur. Ceram. Soc.* **38** 2792–2801
[16] Guo S Q 2009 *J. Eur. Ceram. Soc.* **29** 995–1011
[17] Pierson H O 1996 *Handbook of refractory carbides and nitrides, Handbook Refractory Carbides Nitrides* pp. 8–16
[18] Kurbatkina V V, Patsera E I, and Levashov E A 2018 *Ceram. Int.* https://doi.org/10.1016/j.ceramint.2018.10.113
[19] Kurbatkina V V, Patsera E I, and Levashov E A 2018 *Ceram. Int.* https://doi.org/10.1016/j.ceramint.2018.10.165
[20] Vorotylo S A, Potanin A Yu, Pogozhev Yu S, Levashov E A, Kochetov N A and Kovalev D Yu 2018 *Ceram. Int.* https://doi.org/10.1016/j.ceramint.2018.09.138
[21] Vorotylo S A, Levashov E A, Petrzhik M I and Kovalev D Yu 2018 *Ceram. Int.* https://doi.org/10.1016/j.ceramint.2018.10.020

Processing of mineral raw materials and production wastes by combustion under atmospheric conditions

V E Loryan, A R Kachin and N Yu Khomenko

Merzhanov Institute of Structural Macrokinetics and Materials Science, Russian Academy of Sciences, Chernogolovka, 142432 Russia

Email: kachin@ism.ac.ru

Abstract. The work considers the regularities of SHS of ferrotitanium, ferrochromium and synthetic mica-based fluorophlogopite. The powder mixture of minerals (silica sand, Karelian shungite, ilmenite, chromium concentrate from Shorzha) and secondary raw materials (recirculated cryolite) with metal reducing agents and $KClO_4$ energy additive were used for the synthesis. The regularities of the initial mixture combustion under atmospheric conditions, the synthesized product composition and structure were investigated. Low-carbon ferrochromium with carbon content of 0.2%, ferrotitanium of the composition close to that of the industrial FeTi30 and synthetic mica-fluorophlogopite were synthesized. The synthesis of porous and molten-cast materials based on fluorophlogopite is shown to be possible. The composite material of $K(Na)Mg_3AlSi_3O_{10}Fe_2$–$MgAl_2O_4$–SiC was synthesized.

1. Introduction

The aim of the research is to find optimum compositions of initial mixtures which allow using ore concentrates, mineral raw materials (quartz sand, Karelian schungite) and production wastes (cryolite) [1] as components for synthesis of target products.

The following research was carried out:

- Synthesis of ferroalloys – ferrochromium and ferrotitanium [2, 3]; the annual production output of the materials is hundreds of thousands tons in Russia;
- Synthesis of fluorophlogopites ($NaMg_3AlSi_3O_{10}F_2$) [4], which is used as a material of corrosion-resistant refractories working in aggressive melts of nonferrous metals (aluminum. zinc, chlorides) at T up to 1373 K.

2. Experimental

The ferroalloys were synthesized by aluminothermic reduction of ore concentrates in a graphite pot of 40 mm in diameter and 60 mm in height in the air at P = 1 atm using the energy additive ($KClO_4$) according to the following chemical schemes:

$$FeTiO_3 + Al + KClO_4 \rightarrow FeTi + Al_2O_3 + KCl\uparrow$$

$$Cr_2O_3 + FeO + Al + KClO_4 \rightarrow CrFe + Al_2O_3 + KCl\uparrow$$

Chromium concentrate from Shorzha (Armenia) and ilmenite from the Crimea were used as initial materials (tables 1 and 2).

Table 1. Chemical composition of chromium-containing concentrate (mass. %)

Cr	Fe	O	Mg	Si	Al	Ca
35.3	13.0	27.3	9.33	4.07	3.64	0.5

Table 2. Chemical composition of titanium-containing concentrate (mass. %)

Ti	Fe	O	Cr, Mg, Mn, V, P, Si
30.3	31.0	30.4	8.3

Powder mixtures of quartz sand (SiO_2), Karelian shungite (table 3), metal reducing agents (magnesium, aluminum), fluorine-containing (recirculated cryolite Na_3AlF_6) and oxygen-containing components ($KClO_4$ is the energy additive) in accordance with the chemical schemes (a) and (b) were used to synthesize $NaMg_3AlSi_3O_{10}F_2$ mica.

$$SiO_2 + Al + Mg + Na_3AlF_6 + KClO_4 \rightarrow NaMg_3(Si_3AlO_{10})F_2 + KCl\uparrow \quad (a)$$

$$SiO_2 + Shung + Al + Mg + Na_3AlF_6 + KClO_4 \rightarrow NaMg_3(Si_3AlO_{10})F_2 + KCl\uparrow \quad (b)$$

Table 3. Chemical composition of Karelian schungite (mass. %)

1.	Si	25
2.	O_2	30
3.	C	32

3. Results and discussion

As a result of the research, the following compounds were synthesized:

- **ferrochromium** with iron and chromium content in the target product of 32% and 66%, respectively.

The burning velocity of the mixtures based on chromium-containing concentrate (table 1) was 1–2 mm/s (figure 1). According to XRD analysis, the ingot consisted mainly of Cr, Fe and Al and was a solid solution of Fe–Cr (figure 2). According to EDS data, with an increase in the aluminum content in the initial mixture (from 21.5 to 28%), its content in the product increases from 0.25 to 9.0%. The content of Cr and Fe gradually decreases from 66 to 60% and from 32 to 28%, respectively.

Metallographic analysis revealed that the structure of the ingot in the central part and at the periphery is the same. The microstructure of the alloy is represented by coarse crystals of ferrochrome solid solution separated by bright boundaries. The grain microhardness under a load of 50 g is $H_\mu = 290–410$ (kg/mm^2).

- **ferrotitanium** with 60% Fe, 28.2% Ti and 10% Al.

In the experiments ilmenite from the Crimea was used (table 2). Aluminum used in the work was characterized by various dispersion degrees; the average particle size was 30 and 300 μm. The combustion regularities of the mixtures based on the ilmenite are shown in figure 3. Figure 4 shows the XRD pattern of synthesized ferrotitanium.

Two structural components were revealed: phase based on $Fe_2Ti(Fe_2AlTi)$, $H_\mu = 750–890$ (kg/mm^2) and region which is possibly Fe–Ti solid solution, $H_\mu = 490–580$ (kg/mm^2) under a load of 50g.

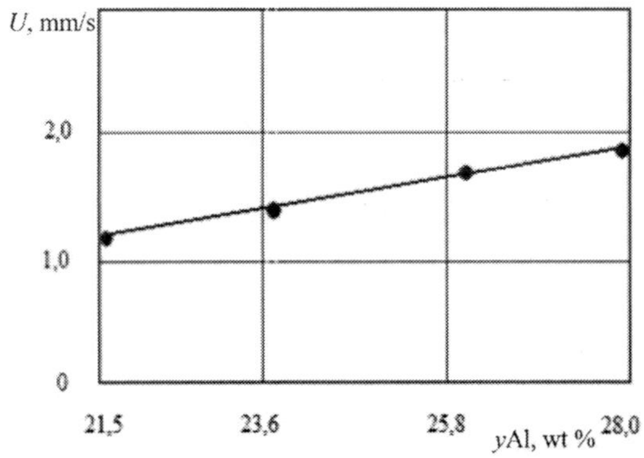

Figure 1. Influence of aluminum content on burning velocity of the mixture: $x(OC + KClO_4) + yAl \rightarrow Cr–Fe + Al_2O_3 + KCl$. OC is the ore concentrate, $x + y = 100\%$.

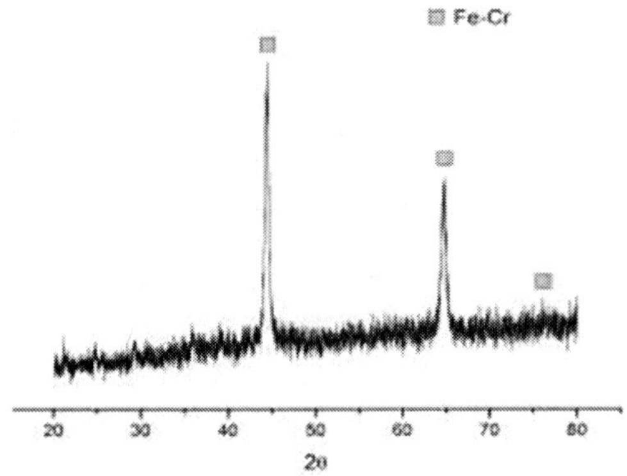

Figure 2. X-ray diffraction pattern of synthesized ferrochromium.

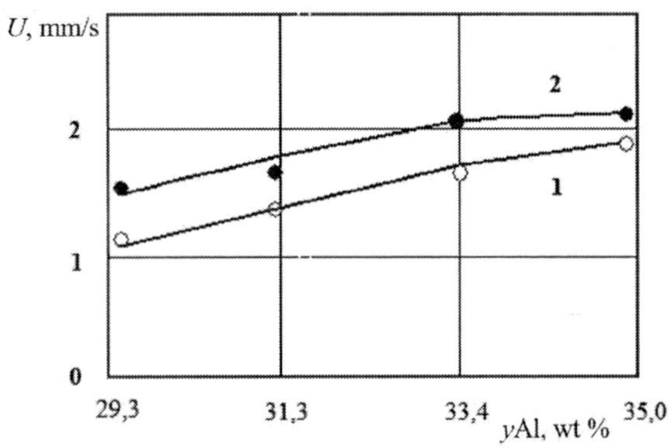

Figure 3. Influence of aluminum content on burning velocity of the mixture: $x(FeTiO_3 + KClO_4) + {}_yAl \rightarrow FeTi + Al_2O_3 + KCl$, $x + y = 100\%$, *1* Al 300 μm, *2* Al 30 μm.

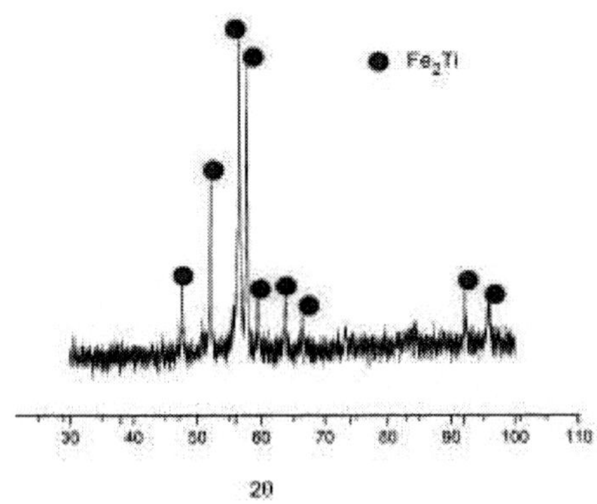

Figure 4. X-ray diffraction pattern of synthesized ferrotitanium.

- **fluorophlogopite**-based mica.

It is established that the main parameter affecting the burning velocity of the mixture for synthesizing fluoroflogopite is the content of potassium perchlorate in it (figure 5). XRD analysis shows that the synthesized product is a composite material based on monoclinic sodium fluorophlogopites $NaMg_3AlSi_3O_{10}Fe_2$ and $Na_4Mg_6Al_4Si_4O_{20}Fe_4$ and contains $MgAl_2O_4$, SiO_2 and some amount of free silicon (figure 6). The change in the initial composition allows regulating the state of aggregation of the substance in the combustion wave from the melt to an ordinary cake. It is established that it is possible to obtain porous and melt-molded products depending on potassium perchlorate content in the initial mixture (figure 7). The microhardness of the cast product matrix, which was measured under a load of 50 g, is H_μ = 140–300 (kg/mm^2). The material density reaches 2.57 g/cm^3.

The introduction of Karelian schungite into the initial composition made it possible to synthesize a composite material $K(Na)Mg_3AlSi_3O_{10}Fe_2–MgAl_2O_4–SiC$ based on fluoroflogopite. According to XRD analysis, when SiO_2 was partially replaced by schungite with a particle size of less than 50 μm, the product containing silicon carbide, K-fluoroflogopite and complex Al–Mg oxide was synthesized (figure 8).

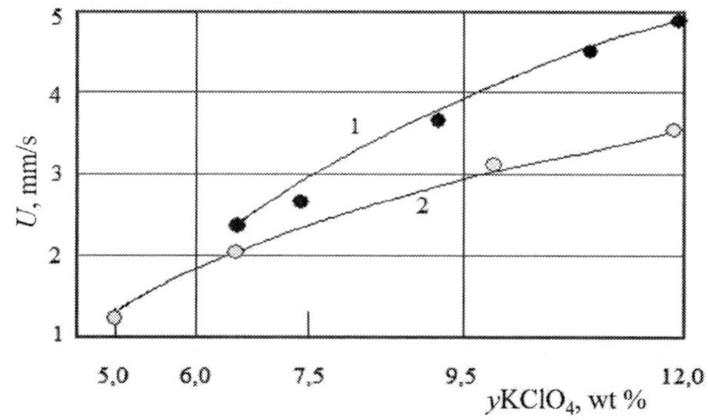

Figure 5. Dependence of burning velocity on potassium perchlorate content in the initial mixture: $x(SiO_2 + Mg + Al + Na_3AlF_6) + yKClO_4$; $x + y = 100\%$, *1* $SiO_2 < 300$ μm, *2* $SiO_2 < 50$ μm.

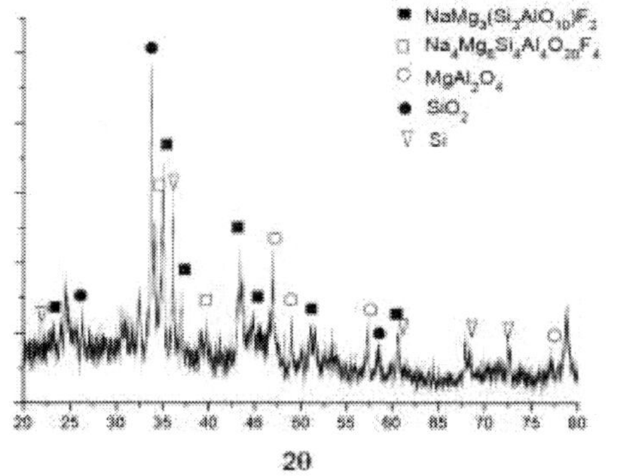

Figure 6. X-ray diffraction pattern of synthesized material.

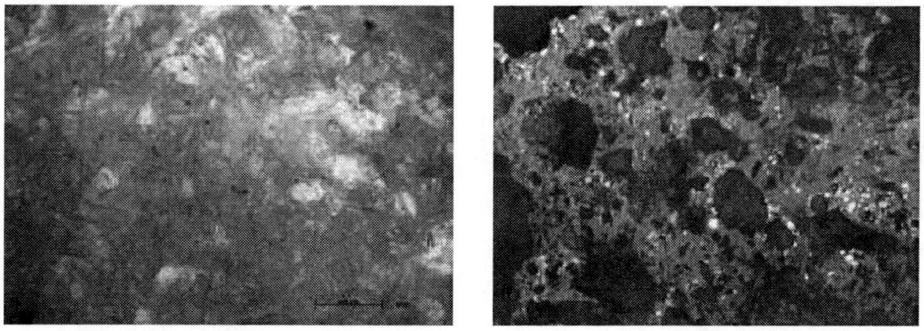

(a) (b)

Figure 7. Structure of (a) cast and (b) porous SHS-fluoroflogopite.

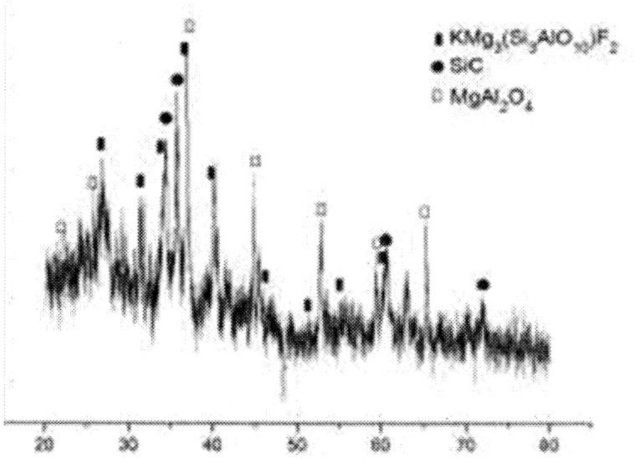

Figure 8. X-ray diffraction pattern of the synthesized composite.

4. Conclusion

According to the classification accepted in industry, the synthesized ferrochrome can be considered as a low-carbon ferrochrome with carbon content of 0.2% [5]. Ferrotitanium whose composition is close to the industrial grade FTi30, artificial mica-fluorophlogopite and $K(Na)Mg_3AlSi_3O_{10}F_2$–$MgAl_2O_4$–SiC composite material are synthesized. The possibility of obtaining porous and melt-molded materials based on fluorophlogopite is shown.

References

[1] https://studbooks.net/588604/tovarovedenie/sposoby_polucheniya_vtorichnogo_kriolita#42
[2] Kachin A R, Loryan V E and Borovinskaya I P *Int. J. Self-Propag. High-Temp. Synth.* **25** 59–61
[3] Azizov Sh T, Kachin A R, Loryan V E, Borovinskaya I P and Mnatsakanyan A S 2014 *Int. J. Self-Propag. High-Temp. Synth.* **23** 161–64
[4] Loryan V E and Kachin A R 2016 *Int. J. Self-Propag. High-Temp. Synth.* **25** 155–58.
[5] Gasik M I, Lyakishev N P and Emlin B I 1988 *Theory and technology of ferroalloy production* (Moscow, Metallurgiya)

Manufacture of products from boron-containing materials by the method of combined static-pulse compaction

V A Mironov[1], V A Glushchenkov[2], Yu S Usherenko[3], I S Belyaeva[2], P I Stankevich[1] and I I Irtisheva [1]

[1]Riga Technical University, Kalku 1, Riga LV-1658, Latvia
[2]Samara State Aerospace University, Moskovskoe 34, Samara, 443086 Russia
[3]Belarusian National Technical University, Nezavisimosti 65, Minsk, 220013 Belarus

E-mail: viktors.mironovs@rtu.lv

Abstract. In the production of long parts of powder materials, the magnetic pulse method of pressing is of great interest. In this paper a variant of the method combining the radial magnetic-pulse compaction of a powder in a metal shell and longitudinal drawing is considered. Samples were made of boron-containing materials.

1. Introduction

One of the directions of the method of magnetic-pulse compaction of powder materials (MPCP) is the production of tubular and rod products [1]. However, the main limitation in the application of the method are related to the quality of products as pressing in a rigid mold with a planar inductor does not provide the desired density and compaction of the powder material in a thin-walled shell with radial inductor does not allow to obtain a uniform density along the length [2, 3]. In this connection, a great interest is caused by the combined methods of MPCP, in particular static-dynamic compaction [4, 5]. In [6, 7] these methods were considered in connection with the utilization of brittle, solid and hazardous process waste.

This paper presents the results of experimental studies of long-length products obtained from boron-containing powder materials.

2. Experimental

Samples were made from boron-containing compositions with a boron content of 80%, 3–5% tungsten, and the rest is aluminum. The compositions are prepared by pre-grinding the boron–tungsten filaments into powder of average particle size of 20 μm and subsequently mixed with aluminum powder. The composition was filled into an electrically conductive copper shell with a diameter of 18 mm with a wall thickness of 1.0 mm. The volume of powder in each sample was 2.5 cm^3.

To obtain the samples, the scheme of the combined MPCP method combining the dynamic radial load (magnetic-pulse compaction) and the static longitudinal load (longitudinal drawing) was used (figure 1). The dynamic load was effected periodically with a repetition rate of 1–2 s and a magnetic pressure of 250–400 MPa.

During the magnetic pulse compaction of the powder composition on the MIU-10 [8] unit, the magnetic field strength was measured using an inductance sensor of sensitivity S = 1.75 × 109 [((A/m)·Hz)/V], which was located in the gap between shell and turns of the inductor. An inductor with a diameter of 20 mm was used. The sensor was connected to a 44Xs-A memory oscilloscope through a voltage divider with a factor (K = 1:10).

At discharge energy W = 2.9 kJ (charge voltage U = 11.0 kV) the discharge pulse frequency was f = 25.6 kHz (figure 2). The voltage amplitude in the sensor for the first half-wave U = 240–250 V. The calculated value of the maximum intensity of the electromagnetic field H was 16.7 ×106 A/m and the magnitude of the impulse pressure P = 250 kN.

Crimping was carried out locally along the length of the shell (figure 3).

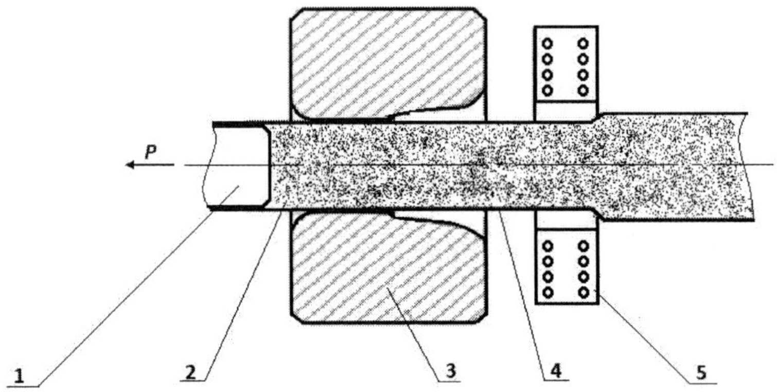

Figure1. Scheme of combined compaction of powder in a shell. *1* stub, *2* metal thin-walled shell, *3* drag, *4* powder backfill, *5* inductor.

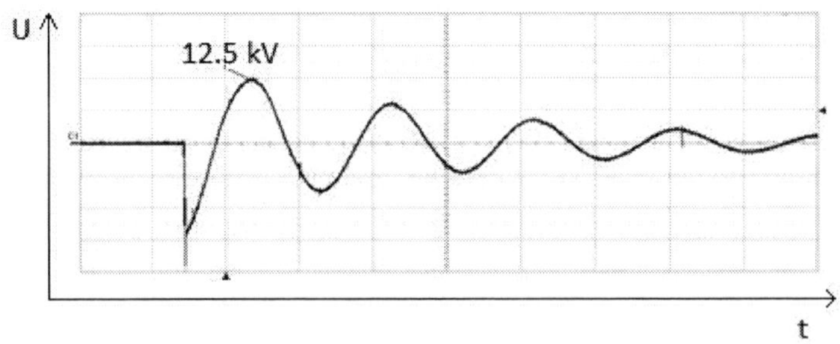

Figure 2. Amplitude of the intensity of the electromagnetic field.

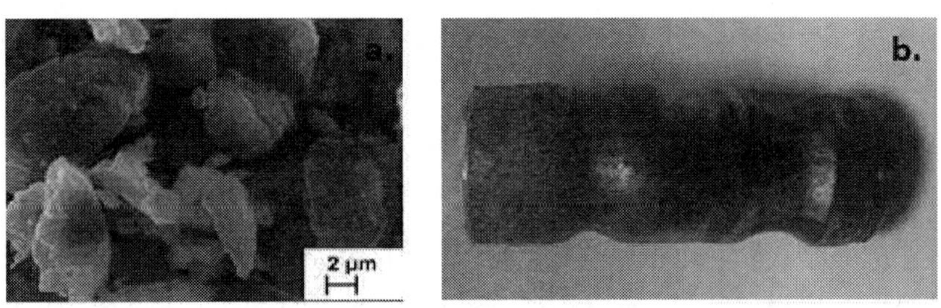

Figure 3. (a) The microstructure of the Al–W–B source material and (b) the general view of the samples in the copper shell after MPCP.

Static step is a drawing of a shell with a powder composition while moving it through a die. Static axial loading of the powder composition was carried out on a universal testing machine TSDMU-30 type with a maximum force of 30 tons-force [9]. Samples after drawing and subsequent sintering have a smooth calibrated surface, are easily machined and have a uniform dense structure (figure 4).

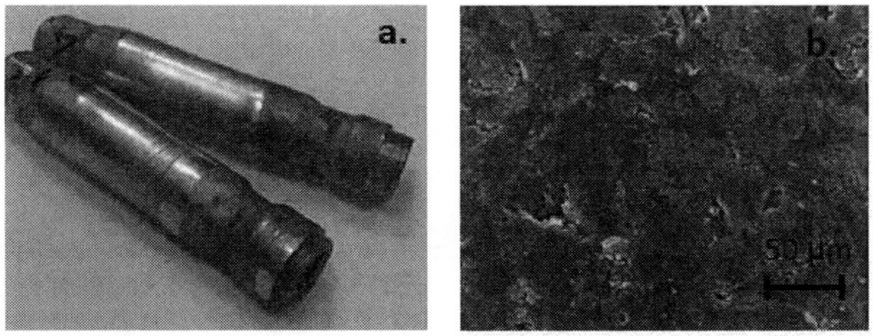

Figure 4. (a) Overall view of samples after MPCP and drawing and (b) microstructure of the material after compaction and sintering at a temperature of 900°C.

3. Conclusions

The possibility of performing combined (static-dynamic) compacting of powders encased in a shell is investigated. The method provides an improvement in the quality of the surface and increases the density of the material.

References

[1] Mironov V 1994 Magnetic pulse pressing of powders and spaping of powder products, In Proc. of the Int. Conf. *Powder metallurgy* PM-**94** Paris 2157–2160

[2] Ivanov V, Kaigorodov A, Hrustov V, Paranin S and Spirin A 2006 Hard alumina-based ceramics fabricated by the use of magnetic pulsed compaction of composite nanopowders Bd 1 (in Russian)

[3] Molochkov I, Leruma M, Mironov V and Sedmalis U 1989 *Glass and Ceramics* **10** 15–17

[4] Mironovs V, Korjakins A, at all 2017 *IOP Conf. Series: Mater. Sci. Eng.* **251** doi:10.1088/1757-899X/251/1/012020

[5] Mironovs V, Stankevich P, Belajeva I and Glushchenkov V 2016 Static–dynamic compaction of powder materials *Proceeding of the 15th International Scientific Conference "Engineering for rural development"* **15** 1128–32

[6] Glushchenkov V, Mironov V, Belyaeva A, Burmistrov E and Usherenko S 2016 *IOP Conf. Series: Mater. Sci. Eng.* **218** doi:10.1088/1757-899X/218/1/012005

[7] Mironovs V, Usherenko Y, Stankevich P and Barone E 2017 *Key Eng. Mater.* **721** 275–279

[8] Glushchenkov V and Karpuhin V 2014 *Technology of magnetic-pulsed processing* (Samara, ZAO Publishing house Fedorov) 208 p

[9] Tensile Testing-Tensile Tester http://en.labthink.com/en-us/literatures/tensile-testing-tensile-tester.html [Accessed: September 16, 2018]

IOP Conf. Series: Materials Science and Engineering **558** (2019) 012028 doi:10.1088/1757-899X/558/1/012028

The influence of electromagnetic component of the process on the structure of dissimilar alloys in high-voltage capacitor welding

S V Nescoromniy, E L Strizhakov, A A Chularis and G I Inasaridze

Don State Technical University, Rostov-on-don, 344000 Russia

E-mail: nescoromniy@mail.ru

Abstract. The article deals with the difficulties of obtaining welded structures from dissimilar alloys. It is proposed to minimize the time of thermal effect on the zone of connection of the details by using super-rigid modes in high-voltage capacitor welding. The study of the components of the process was carried out: the energy of thermal and mechanical effects on the connection zone, the ratio of these energies is recommended. The microstructure of the obtained compounds is described and the results of energy dispersion analysis are presented.

The complexity of manufacturing structures from dissimilar metals is associated with different thermal properties (melting point, thermal conductivity and electrical conductivity), chemical interaction (formation of intermetallic compounds, carbide and fusible eutectic), as well as different thicknesses of elements that affect the heat conditions and the structure of the welded joint.

To connect parts made of non-ferrous metals, it is recommended to use solid-phase impulsed welding methods characterized by relative local rates of plastic deformation $\acute{\varepsilon} = 10^{-4}–10^{-5}$ s^{-1} [1–3]. Depending on the temperature and the interaction time, diffusion processes can develop in the connection zone, sometimes reducing the quality of metal compounds with limited mutual solubility [3, 4].

To minimize the development of diffusion processes and to form a joint in the solid phase is possible by reducing the mixing of dissimilar metals in each other; reducing the residence time of metals at temperatures of formation of an unfavorable structure through the use of super-rigid modes of action.

One of the ways of welding, satisfying the above conditions is high-voltage capacitor welding (HVC) developed in DSTU [2, 5].

Features of HVC are the use of electromagnetic energy for the implementation of thermal and mechanical effects in the course of the discharge current through the welded parts and the use of induction-dynamic drive (IDD). The schematic diagram of HVC is shown in figure 1.

IDD is a set of inductor *I* and pusher 3 of different mass. The pusher 3 is electrically isolated from parts 1 and 2 through the dielectric insert 4.

The accumulated energy in the capacitor banks of the impulse current generator (ICG) when the switching device is triggered causes the flow of current I_d through the IDD, the connection zone and the welded parts 1 and 2, warming up the surface layers of metal, destroying oxide covers and other contaminants. A gap is formed equal to the height of the micro-irregularities, an impulse arc is aroused, burning in the metal vapor, which initiates the cleaning of the surfaces to be joined. When the magnetic field lines H interact with the induced currents in the pusher 3, there is a magnetic pressure P_m in the working zone of the IDD, which brings the welded parts together. When parts collide, the molten metal with oxide covers and impurities is ejected from the contact zone to the periphery of the rod element under the action of a cumulative jet.

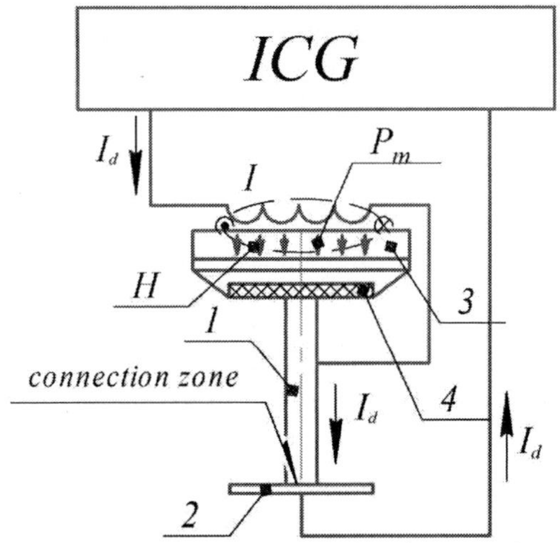

Figure 1. Schematic diagram of the implementation of HVC.

Sequential IDD connection with weldable materials allows to synchronize the process of thermomechanical effects of welded parts.

The aim of the research is to determine the influence of the controlled mass of the pusher on the structure of permanent copper-brass joints.

Experimental studies were carried out on flat samples of brass grade L63 thickness $\delta = 0.8$ mm with core elements of copper grade M01 diameter $D_r = 10$ mm. The capacity of the storage unit of ICG was $1800 \cdot 10^{-6}$ F, the frequency of the discharge current $f = 2600$ Hz. The operating voltage U varied in the range of 2.0–3.3 kV, the mass of the pusher M was 0.5; 1.0; 2.0; 3.0 and 4.0 kg.

For the criterion of the quality of the materials to be joined was chosen – the pull-off force parameter. The structure of permanent joints was investigated by metallographic and optical analysis.

The HVC process was implemented by sinusoidal damped discharge without preliminary formation of welded surfaces with their short circuit [6].

The nature of I_d, P_m changes is shown in figure 2. Discharge current every ½ of period T changes polarity, magnetic pressure decreases over time.

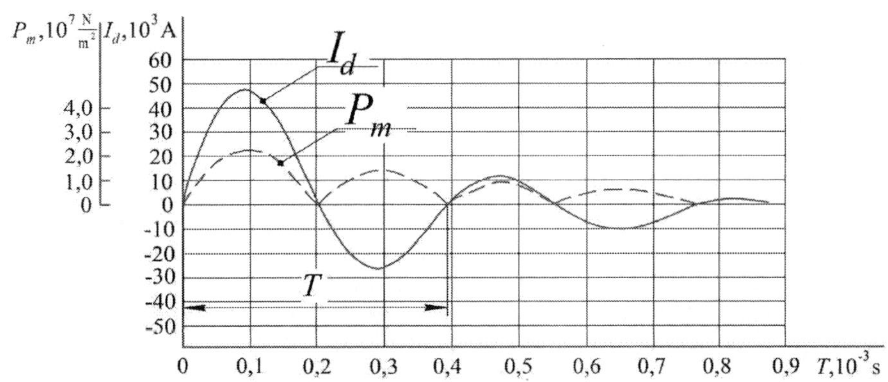

Figure 2. The nature of the change of the discharge current and magnetic pressure over time.

The energy balance at the discharge of capacitor banks at HVC can be determined by the following ratio:

$$W_{ESC} = W_h + W_m \tag{1}$$

where W_{ESC} is the energy stored in an energy storage capacitor (ESC); J; W_h is the energy spent on heat generation in the connection zone (Joule heat, impulse arc combustion), J; W_m is the energy expended on mechanical action parts to be joined, J.

When substituting all components, the energy balance at HVC will take the following form:

$$\frac{CU^2}{2} = \left(I_d^2 R t_{d.p.} + U I_d t_{a.c.}\right) + \frac{L I_d^2}{2} \tag{2}$$

where C is the capacitance ESC, F; U is the charging voltage, V; L is the inductive reactance of the discharge circuit, which is determined mainly by the spiral inductor N, H; I_d is the discharge current, A; R is the resistance of the discharge setting circuit; Ω ; $t_{d.p.}$ is the duration of the process of HVC, s; $t_{a.c.}$ is the duration of an impulse arc combustion under HVC, s.

Preliminary calculations showed that to obtain permanent connections of dissimilar metals with the strength of not lower than (0.6–0.8)σ_s, the ratio of the energies must be W_h/W_m = (60–70)%/(40–30)%.

One of the indicators of the magnetic field energy is the magnetic pressure generated by the inductor on the pusher, which can be determined experimentally by calculation [2]:

$$P_m = \frac{B}{2\mu_0} \cdot \sin^2 \omega t, \left(\frac{N}{m^2}\right) \tag{3}$$

where B is the magnetic induction generated by the inductor, T; μ_0 is the magnetic constant, H/m.

Registration of magnetic induction B in the working area of the IDD is produced by milliteslameter TPU-02 with a contactless measuring probe.

The magnitude of the specific magnetic pressure impulse $J_{m.p.}$ (mechanical component of the HVC process) was determined experimentally by calculation:

$$J_{m.p.} = P_m \cdot t_i \left(\frac{N \cdot s}{m^2}\right) \tag{4}$$

where t_i is the time of interaction between welded surfaces, s ($t_i \leq ¾\, T$).

The specific magnetic pressure impulse determines the relative degree of plastic deformation in the connection zone ε_1, which was calculated experimentally-by calculation on the macro-glyphs of welded joints. The cross-sectional structure of the welded joints and the phase composition of the joint zone were studied by micrographs using a scanning electron microscope (SEM) Carl Zeiss EVO 50.

Metallographic studies were conducted on experimental samples obtained by high-voltage capacitor welding. The microstructure of welded joints is shown in figure 3. Metallographic studies of samples at HVC showed the following. M01 grains have a polyhedral structure typical of the microstructure of annealed copper in the form of polyhedra of different geometry. In the vicinity of the connection zone (figure 3a), the zone of small grains of copper in the form of flat narrow crushed formations obtained in the solid phase under the influence of a short hard mechanical impulse is distinguished, as well as deformation doubles (crossing all fields of copper grain) characterized by intensive mechanical action at HVC. With an increase in the mass of the pusher, due to the inertia of the system, the thermal effect begins to prevail over the deformation (figures 3b, 3c). There are no oxide inclusions in the near-seam zone of copper, which does not prevent the growth of grains. The near-seam zone of brass is characterized by disoriented grains of α phase of polyhedral shape and smaller grains of β phase (dark inclusions) penetrated by small-angle boundaries of narrow doubles. With a decrease in the interaction time of the contacting surfaces of t_i (figure 3a) by reducing the mass of the pusher, the grains of α and β phases and doubles are oriented to the connection zone at angles not exceeding 90°. With increasing mass of the pusher (figure 3c) in the near-seam zone, there is a turn of grains in the flatness (parallel crystals of dark and light tones), the process of collective recrystallization develops due to crystallographic adjustment of individual verges of contacting grains and their increasing in size. Increased grain sizes indicate the intensity of the activation process of the contacting grain surfaces due to thermal deformation.

(a) (b) (c)

Figure 3. Microstructure of compounds of dissimilar alloys, brass at the top, copper at the bottom, with different weight of the pusher: (a) $M = 0.5$ kg; (b) $M = 2.0$ kg; (c) $M = 4.0$ kg.

Energy dispersion analysis of the distribution of chemical elements across the welded joint zone under low-intensity thermal effects on the transition zone metal revealed no diffusion of zinc into copper from brass, which indicates the formation of a welded joint in the solid phase (figure 4) [6]. As a result of the analysis of experimental samples, the degree of relative local plastic deformation ε_l and the width of the zone of compounds y depending on the specific magnetic pressure impulse $J_{m.p.}$ (figure 5).

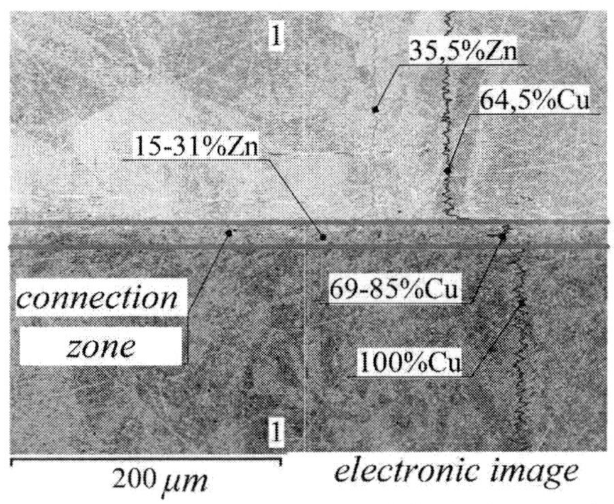

Figure 4. Elemental composition of the compound of dissimilar materials (1-1-analysis zone).

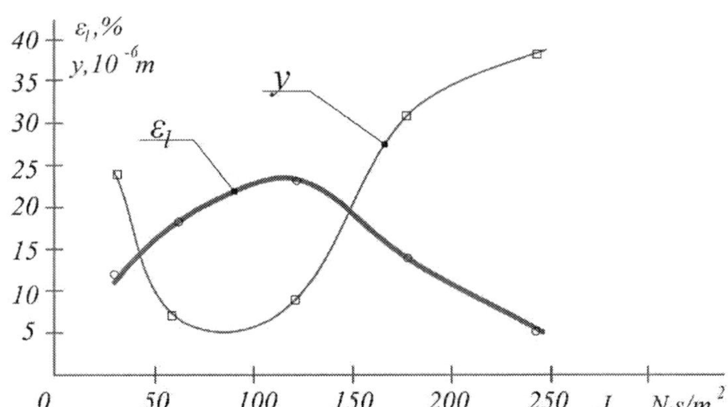

Figure 5. The dependence of the relative local plastic deformation ε_l and the width of the weld zone y on the specific magnetic pressure pulse $J_{m.p.}$ at $W = 6.5$ kJ.

With an increase in the mass of the pusher, the impulse of the specific magnetic pressure increases, but at the same time, due to the high cooling rates of the connection zone, the relative plastic deformation decreases.

By increasing the inertia of the system with an increase in the mass of the pusher, the width of the thermal impact zone increases.

Summary. One of the technological methods that affect the structure of the weld zone in high-voltage capacitor welding of dissimilar nonferrous materials is the specific magnetic pressure impulse, characterized by the mass of the pusher of the induction-dynamic drive. Varying of the mass of the pusher allows to adjust the ratio of thermal and mechanical components of the HVC process. Recommended range W_h/W_m = (60–70)%/(40–30)%. With the increase of the thermal component, in the structure of the compound grains take the form of broad formations. The prevalence of the mechanical component contributes to the formation of the structure in the form of narrow deformation doubles.

References

[1] Konyushkov G V and Musin R A 2009 *Special methods of pressure welding* (Saratov: Publ. AI PI Er Media)

[2] Strizhakov E L, Nescoromniy S V and Minko D V 2016 *Discharge-impulse processing of materials* (Rostov n/D: Publishing center DSTU)

[3] Markasheva L I, Grigorenko G M and Arsenyuk V V 2002 Proccedings of International Conference Mathematical modelling and information technologies in welding and related processes (vill. Katsiveli, Crimea, Ukraine: Institute of electric welding. E. O. Paton RAS of Ukraine) P. 72

[4] Mazanko V F, Pokoev A V, Mironov V M, et al. 2005 *Diffusion processes in metals under the action of magnetic fields and impulsed deformations* (Moscow: Mechanical engineering)

[5] Pat. 70839 Russian Federation, IPC^7 B23K 20/06. Device for shock condenser welding of rod parts with flat base / Nescoromniy S V and Strizhakov E L № 2007130908/22; appl. 13.08.07; publ. 20.02.2008, Bul. No. 5. 6 p.

[6] Strizhakov E L et al. 2015 *Welding production* **3** 42

IOP Conf. Series: Materials Science and Engineering **558** (2019) 012029 doi:10.1088/1757-899X/558/1/012029

Spark plasma sintering for high-rate diffusion welding of a UFG titanium alloy PT3V

A V Nokhrin, M S Boldin, E A Lantsev, M M Vostokov, M Yu Gryaznov, V I Kopylov and V N Chuvil'deev

National Research Lobachevsky State University of Nizhny Novgorod, Nizhny Novgorod, 603950 Russia

E-mail: nokhrin@nifti.unn.ru

Abstract. This is the first ever paper to address the prospects of using Spark Plasma Sintering (SPS) for high-rate diffusion welding of a high-strength ultrafine-grained (UFG) α-titanium alloy Ti–5A–2V. The effect of growing diffusion welding intensity in UFG titanium alloys is also described. Welds of a UFG α-titanium alloy Ti–5Al–2V obtained with SPS are characterized by high density, strength, and corrosion resistance. The rate of weld sealing in UFG alloys is shown to be in nonmonotonic (maximum) dependence on the heating rate.

1. Introduction

Besides, α- and near α-titanium alloys are widely used in petrochemistry, shipbuilding, in production of engineering structures operated at freezing temperatures, etc., requirements as to reliability and resources for which are getting stricter day by day. One of the most promising ways to improve mechanical properties of titanium alloys is to form an ultrafine-grained (UFG) structure in them using different methods of severe plastic deformation (SPD). It ensures a unique combination of high strength, ductility, corrosion resistance, superplasticity, etc.

One of the key challenges preventing extensive use of UFG materials is welding. Traditional argon-arc welding or electron-beam welding accompanied by metal melting fail to preserve a UFG structure with high mechanical properties in a weld. Spark Plasma Sintering (SPS) offers high chances of obtaining high-strength welds since the idea underlying this innovative technology involves rapid heating (up to 2500°C/min) in vacuum or inert atmosphere. This is achieved by sending high-power millisecond DC pulses through the specimen along with applying pressure [1–4]. This technology produces high-density structures in UFG materials at lower optimal sintering temperatures. High heating rates and faster diffusion at lower heating temperatures are crucial for restricting grain growth and preserving a UFG structure in the material.

This paper aims at the study of potential that SPS has for diffusion welding of high-strength corrosion-resistant UFG α-titanium alloys used in nuclear engineering.

2. Experimental

The study focuses on Ti–4.73Al–1.88V alloy (Russian industrial name PT3V). A UFG structure in the alloy was formed with ECAP.

Diffusion welding of specimens sized 7 × 7 × 3.5 mm^3 was performed using Dr. Sinter model SPS-625. Surface roughness varied as a result of polishing with diamond paste of different dispersion level.

The degree of roughness in specimens was 40–60 μm. Heating rate (V_h) varied from 10 to 350°C/min, welding temperature (T) ranged from 600 to 900°C, applied pressure (P) changed from 50 to 100 MPa. The degree of shrinkage (L_{eff}) and shrinkage rate (S_{eff}) were monitored with a dilatometer integrated in Dr. Sinter model SPS-625. Welding was performed in vacuum (6 Pa). To account for the contribution of SPS-625 thermal expansion into experimentally measured shrinkage L_{eff}, shrinkage studies were performed without specimen (L_0), afterwards true dependence of temperature on shrinkage was calculated using the formula $L(T) = L_{eff}(T)–L_0(T)$.

Structure studies were carried out using Jeol JSM-6490 SEM and Jeol JEM-2100 TEM. Microhardness (H_v) measurements were made with Duramin Struers-5. In order to study mechanical properties, relaxation tests were run to determine such values as macroelastic limit σ_0 and yield strength σ_y [5]. Tensile tests of flat specimens shaped as a 'double blade' with working part dimensions of 2 × 2 × 3 mm were performed using Tinius Olsen H25K-S stress machine.

Electrochemical studies were carried out in an aqueous solution of 10%HNO_3 + 0.2%HF with R-8 potentiostat (Russia). Ahead of tests, the surface of the specimen was covered with corrosion-resistant coating except for a 1 × 10 mm^2 spot (corrosive) located in the center of a cross-section, in the weld area. Prior to electrochemical tests, the specimen was kept in an electrochemical cell in an aqueous solution of 0.2%HF+10%HNO_3 until a stationary potential value was reached (holding time no less than 2 h), after which $E_{cor}(i_{cor})$ dependence was surveyed at the scanning rate of 0.5 mV/s.

3. Results and discussion

3.1. Experimental results

The alloy in the CG state is characterized by irregular grain size distribution. The average grain size varies from 10–20 μm to 50–100 μm. The average grain size after N = 4 ECAP cycles is 0.2–0.5 μm. EDS analysis proves that variation in local concentration of aluminum and vanadium along different grain boundaries is insignificant [6].

In-situ TEM studies performed during stage heating show that recrystallization during annealing of a UFG alloy is observed at 500°C and above. Along with the onset of grain growth, titanium carbide nanoparticles with an average size of 5–15 nm are observed in the structure of a UFG alloy. After heating to 700–800°C, the average size of titanium carbide particles reaches 30–50 nm, the average grain size in a UFG alloy is 7–9 μm. Any significant difference in average sizes and volume fractions of carbide particles Ti–C in coarse-grained and UFG alloys is not observed.

Research into mechanical properties proves that formation of a UFG structure in alloy with ECAP increases macroelastic limit (σ_0) from 400–420 to 730–750 MPa, yield stress (σ_y) from 600–620 MPa to 1020–1050 MPa, and microhardness limit (H_v) from 2.0–2.1 GPa to 3.1–3.2 GPa, respectively. Research into thermal stability of mechanical properties shows that softening of a UFG alloy starts after heating to 500–550°C, which corresponds to recrystallization temperature. A slight increase in macroelastic limit and yield stress of PT3V alloy at low temperatures ($T \leq$ 450–500°C) apparently results from precipitation of titanium carbide nanoparticles.

Comparative analysis of mechanical tensile tests at high temperatures shows that UFG alloy specimens at high temperatures are characterized by higher plasticity and lower flow stress. At deformation temperatures rising from 600 to 800°C, the flow stress in a CG alloy goes down from 355–360 to 125–130 MPa, while plasticity (δ) grows from 230 to 360%. The results of testing UFG alloy specimens in similar conditions show that σ_s goes down from 165 to 70 MPa, while plasticity (δ) goes up from 190 to 470–475% at deformation temperatures growing from 600 to 800°C.

Figure 1a shows dependences of shrinkage on heating time $L(t)$ for CG and UFG alloy specimens, while figure 1b presents dependences of shrinkage and shrinkage rate on welding temperature (continuous heating) for CG and UFG alloys. (Figure 1a also shows the temperature curve during heating.) The analysis of $L(T)$ dependences proves that during continuous heating, a monotonic shrinkage increase is observed, while dependences of shrinkage rate on heating temperature $S(T)$ have the usual two stages with a maximum that are typical of *S*(*T*) dependences observed during sintering UFG materials. As can be seen from the figures, shrinkage (diffusion welding) of UFG specimens under similar heating conditions starts at lower temperatures than shrinkage (diffusion welding) of CG specimens. The degree of shrinkage for UFG specimens is visibly higher than that for CG alloys.

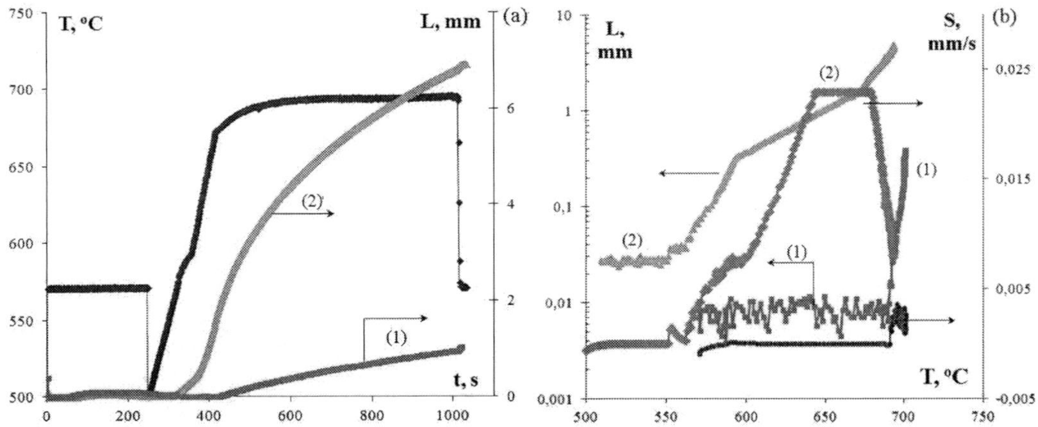

Figure 1. Shrinkage curves (welding diagrams) for coarse-grained and UFG alloy specimens ($V = 100$°C/min, $P = 100$ MPa): (a) dependence of temperature and shrinkage on heating time for coarse-grained (1) and UFG (2) alloy specimens; (b) dependences of shrinkage and shrinkage rate on welding temperature for coarse-grained (1) and UFG (2) alloy.

Generalized results of structural studies corroborate the observed effect of faster weldability in UFG alloys. Electron-microscopic studies show that the average size and volume fraction of pores in UFG specimens are much smaller than in CG materials (figure 2). Note that SPS allows for very thin welds (figure 2b) that account for much higher operational reliability of welded structures. Analysis of histograms presented in figures 2c, 2d that show pore size distribution (corresponding to figures 2a, 2b) proves that a coarse-grained metal weld has bigger pores with an average size of ~ 1.5 μm, while a UFG metal weld has pores with an average size of ~ 0.25–0.5 μm. Studies of mechanical properties suggest that the hardness of welds obtained with SPS at temperatures below recrystallization temperature corresponds to the hardness of the base metal measured at a distance of no less than three widths of a weld wall from the seam edge.

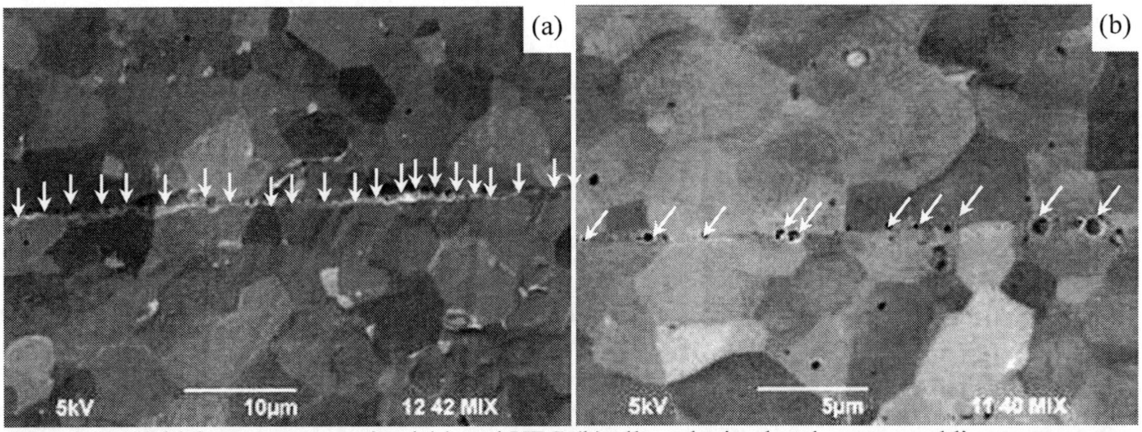

Figure 2. Weld from coarse-grained (a) and UFG (b) alloy obtained at the same welding temperature ($T = 850$°C), heating rate ($V_h = 50$°C/min) and pressure ($P = 50$ MPa). The white arrows indicate pores.

Let us analyze the impact of key factors (temperature, pressure, heating rate, holding time) on microstructural parameters of a welded joint obtained with SPS during high-rate welding of specimens obtained from UFG alloy PT3V. Dependences of shrinkage and shrinkage rate on the above parameters are shown in figure 3. As can be seen from $L(T)$ and $S(T)$ dependences, within the range of welding temperatures below $\alpha\leftrightarrow\beta$ phase transition temperature, maximum shrinkage (L_{max}) and maximum shrinkage rate (S_{max}) in UFG alloy specimens are several times higher than L_{max} and S_{max} in specimens produced from a CG alloy PT3V. Within the temperature range of over 1000°C exceeding the temperature of $\alpha\leftrightarrow\beta$ phase transition, there is no visible difference between L_{max} and S_{max} for a coarse-grained and UFG alloy specimens. Thus, it can be assumed that at low heating temperatures, the welding intensity in UFG alloy specimens is much higher than that in CG alloy specimens.

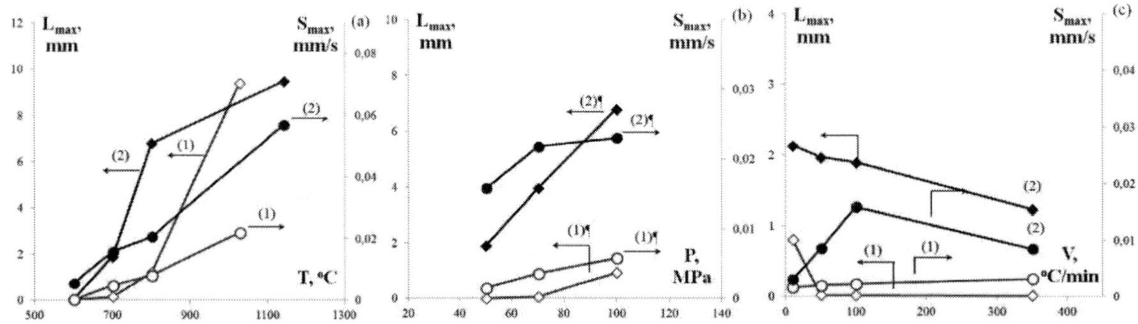

Figure 3. Dependences of maximum shrinkage L_{max} and maximum shrinkage rate S_{max} during high-rate diffusion welding of coarse-grained (1) and UFG (2) titanium alloy specimens: (a) impact of welding temperature (100°C/min, 50 MPa); (b) impact of applied pressure (700°C, 100°C/min); (c) impact of heating rate (700°C, 50 MPa).

This is accompanied by a decrease in the typical scale of diffusion mass transfer x that is proportionate to grain size $x = d/2$, and therefore, by a decrease in the standard time of diffusion mass transfer $\tau_{diff} = \delta D_b/x^3$, where δ is the grain boundary width and D_b is the grain boundary diffusion coefficient. A fine grain size in a UFG alloy dramatically reduces standard time of diffusion mass transfer τ_{diff} required for diffusion controlled dissolution of pores located near non-equilibrium grain boundaries.

Figure 3b shows dependences of shrinkage L_{max} and shrinkage rate S_{max} on applied pressure (P). As can be seen from $L_{max}(P)$ and $S_{max}(P)$ dependences, an increase in applied pressure from 50 to 100 MPa results in an increase in shrinkage rate S_{max} from $1.5\cdot10^{-3}$ to $5.8\cdot10^{-3}$ mm/s for a CG alloy and from $1.6\cdot10^{-3}$ to $23\cdot10^{-3}$ mm/s for a UFG alloy. We reckon that the results obtained testify to a big role of plastic deformation processes in diffusion welding of a CG and UFG alloy (for more details see Discussion). This assumption is also corroborated by significant shrinkage of specimens ($L_{max} \sim 1$–8 mm) exceeding surface roughness of specimens after their mechanical polishing.

Analysis of $S_{max}(V)$ and $L_{max}(V)$ dependences presented in figure 3c shows that the dependence of shrinkage and shrinkage rate on heating rate for coarse-grained alloy specimens in monotonically descending: with the heating rate rising from 10 to 350°C/min, shrinkage L_{max} goes down from 0.80 to 0.01 mm, while shrinkage rate S_{max} falls from $4.0\cdot10^{-2}$ to $1.5\cdot10^{-2}$ mm/s. This effect is obviously attributable to a decrease in standard time τ_{diff}, within which diffusion welding of specimens takes place, and consequently, to a decrease in time required for diffusion controlled dissolution of pores in a weld. $S_{max}(V)$ and $L_{max}(V)$ dependences for a UFG alloy are more complex. As can be seen from figure 3c, with heating rate rising from 10 to 350°C/min, shrinkage L_{max} goes down from 2.14 to 1.23 mm, while shrinkage rate S_{max} nonmonotonically changes with growing heating rate: with V growing from 10 to 100°C/min, S_{max} increases from $0.3\cdot10^{-2}$ to $1.6\cdot10^{-2}$ mm/s, with heating rate further rising to 350°C/min, S_{max} goes down to $0.8\cdot10^{-2}$ mm/s. We assume that such an unexpected result can be explained as follows:

with heating rate growing, the size of recrystallized grain d drops leading to a decline in the typical scale of diffusion mass transfer $x \sim d/2$ and consequently, to a decline in the standard time required to complete metal sealing and dissolution of pores in a weld. With the heating rate further rising, the factor associated with reduced welding time becomes crucial again and the sealing rate decreases.

As shown above, during high-rate low-temperature diffusion welding of UFG alloys, weld hardness is rather high (~3.2–3.3 GPa) and approximates weld hardness of a UFG alloy after ECAP (~3.5–3.6 GPa). Besides, it exceeds weld hardness achieved with argon arc welding (~2.35–2.40 GPa) or electric fusion welding (~2.40–2.45 GPa). Microhardness of weld joints in coarse-grained specimens is ~2.4–2.6 GPa and hardly depends on diffusion welding modes. It is noteworthy that weld microhardness during welding UFG alloys at low temperatures (600–700°C) is 0.4–0.6 GPa less than microhardness of a base metal away from the weld: after diffusion welding of a UFG alloy at 600°C (100°C/min, 50 MPa, holding time t = 10 min), weld microhardness is 2.7 GPa, while microhardness away from the weld is 3.0–3.1 GPa. Note also that with temperature rising to 800°C, weld H_v goes down to 2.5 GPa, while metal microhardness away from the weld declines to 2.6–2.7 GPa. Thus, it is safe to say that the weld metal obtained after welding UFG alloys is characterized by lower hardness than the base metal away from the weld. Similar results were obtained while studying the impact that the heating rate and applied pressure have on weld microhardness in UFG metals. Analysis of these results shows that with heating rate rising from 50 to 350°C/min, weld microhardness goes slightly up from 2.4 to 2.6 GPa, while microhardness of a base metal away from the weld grows from 2.7–2.75 to 2.9 GPa. An increase in applied pressure from 50 to 100 MPa leads to a slight increase in hardness from 2.4 to 2.5 GPa and to an increase in microhardness of a metal away from the weld from 2.7–2.8 to 3.0–3.1 GPa.

Microhardness of welds in CG specimens equals microhardness of a metal away from the weld or (at low welding temperatures (700°C and below) and low heating rates (not more than 50°C/min)) is 0.1–0.15 GPa more than microhardness of a metal away from the weld. We reckon that increased microhardness of a weld in coarse-grained alloys can be caused by plastic deformation in surface layers of a coarse-grained alloy and consequently their strain hardening (work hardening). Since the defect structure recovery rate in a coarse-grained alloy during high-rate heating is rather low, the hardened layer lasts during low temperature sintering[1].

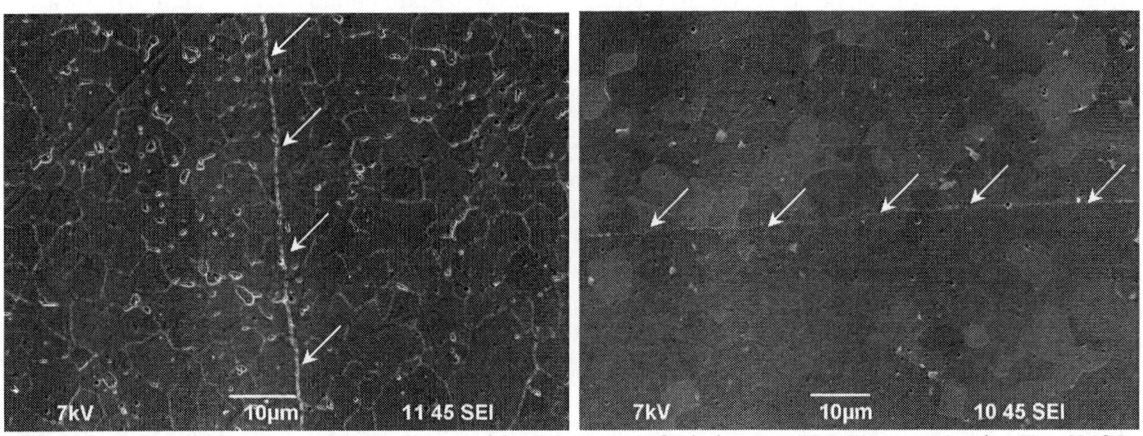

Figure 4. Microstructure of UFG alloy welds: (a) V = 100°C/min, P = 70 MPa, t = 10 min, T = 600°C; (b) V = 10°C/min, P = 50 MPa, t = 10 min, T = 700°C (in figure 4, arrows indicate a weld).

[1] Since the average grain size in the weld in UFG alloys hardly differs from the average grain size of a metal away from the weld (figure 4), the reasons for reduced weld microhardness in UFG alloys are still vague. We reckon that the most probable reason for reduced weld microhardness in UFG alloys is grain boundary recovery leading to reduced density of defects in UFG alloy grain boundaries.

Analysis of weld grain structure parameters shows that welding temperature and heating rate have the biggest impact on the average grain size in a weld joint. As can be seen from grain size distribution histograms, an increase in welding temperature from 600 to 800°C increases an average grain size in a UFG alloy weld from 3 to ~ 5 μm, while an increase in heating temperature from 10 to 350°C/min reduces an average grain size in a UFG alloy weld from 10–15 to 4–6 μm. An average grain size in weld joints in CG specimens within the experimental accuracy corresponds to an average grain size in a coarse-grained alloy.

Let us analyze the level of corrosion resistance in the weld joints obtained.

When comparing potentiodynamic dependences for coarse-grained and UFG alloys after diffusion welding, we noted that corrosion current density (i_{cor}) of welds in a UFG alloy appears to be lower than that in a coarse-grained alloy obtained under similar welding temperature and rate conditions. Note that at higher welding temperatures, differences in corrosion resistance between coarse-grained alloy welds and UFG alloy welds grow – after welding at 600, 700, and 800°C, the corrosion current density ratio for UFG alloy welds and coarse-grained alloy welds is 1.6, 1.7, and 2.1, respectively. The analysis of potentiodynamic dependencies shows that faster heating rates result in bigger corrosion resistance of welds. Corrosion current density i_{cor} in UFG alloy welds obtained at V_h = 50°C/min is 0.98 mA/cm^2, while at V_h = 200°C/min, i_{cor} = 0.5 mA/cm^2. Corrosion potential thus grows by ~ 40 mV from –490 mV to –450 mV. Current density i_{cor} for coarse-grained alloy welds obtained at the heating rates of 50 and 200°C/min is 2.5 and 2.8 mA/cm^2, respectively. In this case, corrosion potential is essentially independent of the heating rate and remains constant within the experimental error of ±10 mV caused by a specimen-to-specimen spread of properties. Analysis of the dependence of potential on holding time shows that steady potential for UFG titanium alloy welds appears to be somewhat bigger than for coarse-grained alloy welds, which also proves increased corrosion resistance of UFG alloy welds. Thus, it is fair to say that corrosion resistance of UFG alloy welds appears to be higher than that of coarse-grained alloy welds.

Figure 5 shows images of the surface of CG alloy specimens (figure 5a) and UFG alloy specimens (figure 5b) after corrosion electrochemical investigation. As can be seen from figure 5, rather active corrosion decay occurs in the weld during electrochemical investigation. As can be seen from the figures, etching (corrosion decay) is observed during testing in grain boundaries of titanium alloys. This suggests that any changes observed in corrosion properties are primarily due to the processes that occur along grain boundaries during high-rate diffusion welding of specimens.

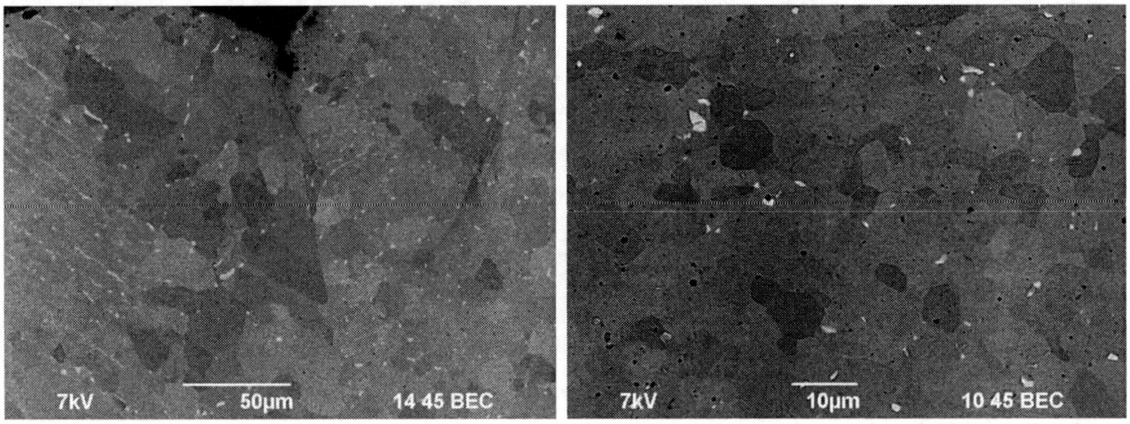

Figure 5. Surface of CG (a) and UFG (b) alloy welds after corrosion electrochemical investigation. Welding modes (10°C/min, 50 MPa, 10 min, 700°C).

We reckon that the reason for increased corrosion resistance of a UFG alloy is the change in the structural-phase composition of grain boundaries in a alloy during ECAP. As mentioned above, ECAP performed at elevated temperatures (450–475°C) leads to reduced concentrations of detrimental corrosive elements along grain boundaries (vanadium) and reduced difference in vanadium (ΔC_V) and aluminum (ΔC_{Al}) concentrations between their concentration in the crystal lattice (C_v) and in the grain boundary (C_b). The difference in aluminum and vanadium concentration between the crystal lattice and the grain boundary in a CG alloy may reach ΔC_{Al} = 5.6 at % and ΔC_V = 12.9 at %, respectively [5]. Studies of the grain boundary composition in a UFG alloy after ECAP show that ΔC_{Al} and ΔC_V approximate ~0.35–0.40 at % and slightly exceed the range of experimental error in determining the concentration of aluminum and vanadium with EDS (±0.2–0.3 at.%). (see also [5]).

While heating a UFG alloy to the temperature exceeding the recrystallization initiation temperature, fast migrating grain boundaries cover atoms of doping elements (aluminum, vanadium) uniformly distributed in the crystal lattice, which again leads to an increase in ΔC_{Al} and ΔC_V. In this case, the concentration of doping elements in the grain boundary will obviously be proportionate to a standard distance over which a grain boundary moves and consequently, to the size of a recrystallized grain. Thus, a decrease in the migration rate of a grain boundary due to an increase in the heating rate or decrease in the welding temperature is expected to boost corrosion resistance of a titanium alloy as demonstrated experimentally.

4. Conclusions

SPS is shown to allow efficient high-rate diffusion welding of titanium alloys and at the same time to preserve a UFG structure, high hardness and corrosion resistance in a weld. Pores in UFG alloy welds are found to diffuse faster due to highly intensive diffusion processes taking place along non-equilibrium grain boundaries and a small-scale diffusion mass transfer that is proportionate to grain size.

The dependence of weld sealing rate (welding rate) on heating rate is shown to be different for coarse-grained and UFG alloys. An increase in the heating rate in coarse-grained alloys leads to a monotonic decrease in the sealing rate caused by reduced welding time at high heating rates. The weld sealing rate in UFG alloys is in nonmonotonic (maximum) dependence on the heating rate. We reckon that it is due to the fact that with the heating rate growing, the size of recrystallized grain d decreases leading to a decline in the typical scale of diffusion mass transfer $x \sim d/2$ and consequently, to a decline in the standard time required to complete metal sealing and dissolution of pores in a weld. With the heating rate rising up, the factor associated with reduced welding time becomes crucial again and the sealing rate decreases.

Acknowledgments

This study was performed with the support of the Russian Science Foundation (Grant no. 16-13-00066).

References

[1] Tokita M 2011 *Spark Plasma Sintering (SPS) Method, Systems, and Applications* (Elsevier)
[2] Munir Z A, et al 2006 (New York: Wiley–Interscience)
[3] Olevsky E and Dudina D 2018 *Field-Assisted Sintering* (Springer Int. Publ.)
[4] Chuvil'deev V N, et al 2019 *J. Alloys Comp.* **773** 666
[5] Chuvil'deev V N, et al 2017 *J. Alloys Comp.* **723** 354

IOP Conf. Series: Materials Science and Engineering **558** (2019) 012030 doi:10.1088/1757-899X/558/1/012030

Synthesis and investigation of highly transparent ceramics

V V Osipov[1], V A Shitov[1], V V Platonov[1], R N Maksimov[1,2] and K E Lukyashin[1]

[1] Institute of Electrophysics UrB RAS, 106 Amundsen St., Ekaterinburg 620016, Russia

[2] Ural Federal Unversity named after the first President of Russia B.N. Yeltsin, 19 Mira St., Ekaterinbirg 620002, Russia

E-mail: osipov@iep.uran.ru; vlad@iep.uran.ru

Abstract. A short review of research works carried out in the Institute of Electrophysics UrB RAS on the development of synthesis technology for laser ceramics and the analysis of their characteristics is presented. These ceramics are widely used as active media of solid-state lasers, transparent armor, envelopes of high intensity discharge lamps, heat-resistant and mechanically strong windows etc.

1. Preparation of nanopowders

Fabrication technology of laser ceramics can be conveniently divided into three stages: synthesis of nanopowders, their compaction and sintering of compacts. The main requirements to nanopowders are small particle size, low degree of agglomeration, high purity and cubic phase. Nanopowders synthesized in a laser plume formed under the action of radiation onto target are more completely respond these requirements. We have created the laser synthesis technology of high purity nanopowders with an average particle size of ~10 nm, a size distribution of 5–45 nm (Figure 1) and an output rate up to 80 g/h. The main characteristics of a laser plume have been investigated and the reasons for appearance of large ~100-μm-sized shapeless particles and spherical particles with a diameter of ~1 mm have been determined as well as the methods of nanopowders separation and collecting have been developed [1].

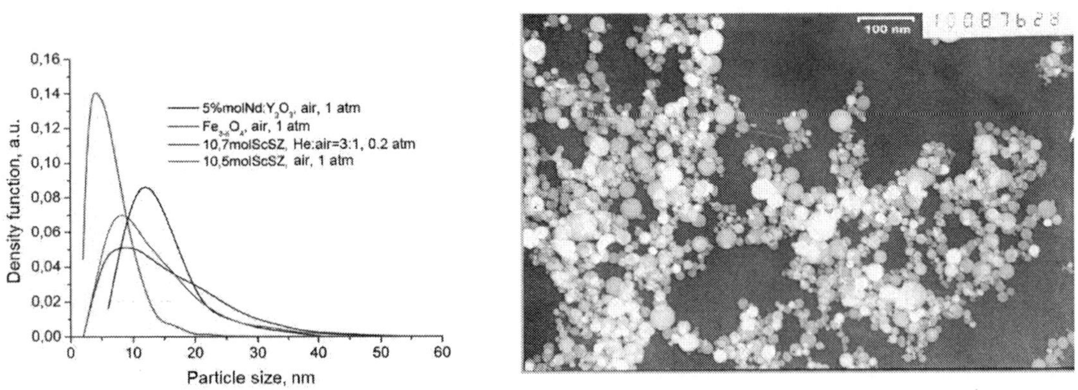

Figure 1. Size distribution and TEM image of nanoparticles synthesized in a laser plume

IOP Conf. Series: Materials Science and Engineering **558** (2019) 012030 doi:10.1088/1757-899X/558/1/012030

2. Synthesis and characteristics of highly transparent ceramics

It has been shown that the relative density of compacts is practically independent of compaction method (uniaxial static pressing with or without ultrasonic action, pulsed magnetic compaction, cold isostatic pressing) and is generally determined by the compaction pressure [2].

Synthesis methods of highly transparent ceramics have been developed on the basis of calcination of nanopowders and compacts (Figure 2). The morphology, spectral and luminescence characteristics of laser ceramics, lifetimes of most important laser levels as well as the reasons of lasing absence under the doping of Nd:Y_2O_3 ceramic by Hf^{4+} and Zr^{4+} have been investigated [3].

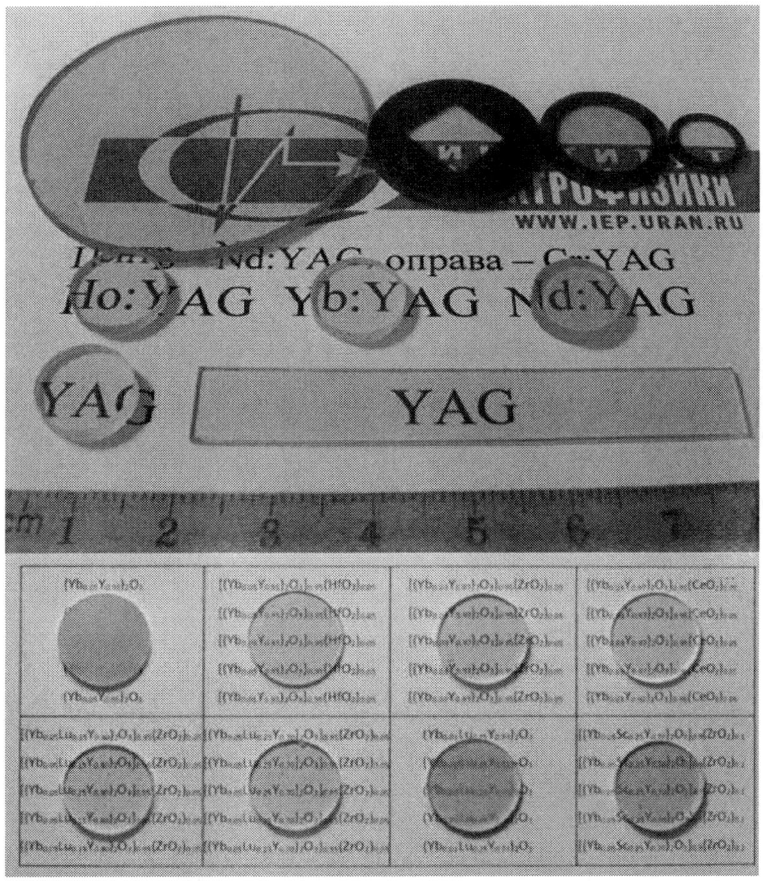

Figure 2. Photograph of ceramics produced in IEP UrB RAS

It has been shown that substitution of Nd for Yb enabled the first highly efficient CW laser oscillation with a slope efficiency of η=29% (in collaboration with the Institute of Laser Physics SB RAS, ILP SB RAS) in a strongly disordered ceramic based on Y_2O_3 [4] and also with η=51.2% in a pulsed mode of operation (in collaboration with National Institute of Optics, INO CNR, Florence, Italy) [5]. Besides that, laser oscillations in Nd:YAG with η=51.7% (Figure 3, in collaboration with INO CNR) [6] and in Ho:YAG ceramics with η=40% (in collaboration with ILP SB RAS) [7] have been demonstrated.

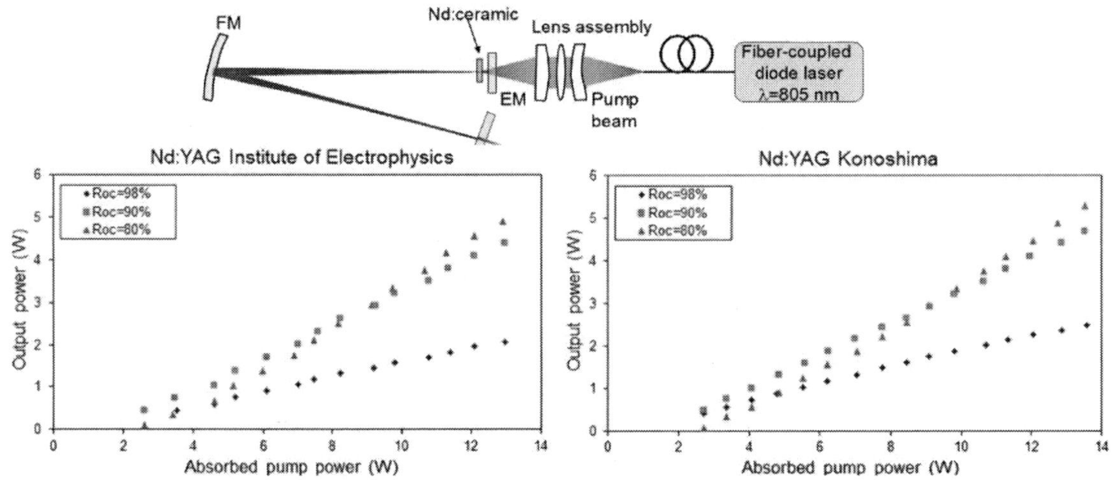

Figure 3. Scheme of laser cavity and the dependences of the output power on the absorbed pump power for Nd:YAG ceramics fabricated by IEP UrB RAS (on the left) and by Konoshima Chemical Co. (on the right)

Research works on fabricating of magneto-optical ceramic for creation of devices based on Faraday effect such as optical circuits, mirrors, isolators, laser gyroscopes are carrying out. In recent years magneto-optical single crystals and glasses are synthesized using Ce, Pr, Dy and Tb oxides. Tb_2O_3 is the best among them and thus terbium gallium garnet (TGG) single crystals with the Verdet constant of 40 rad T^{-1} m^{-1} have found extensive application. Synthesis of magneto-optical ceramics has been described elsewhere [8,9]. We have succeeded in fabrication of highly transparent ceramics based on Tb_2O_3 (Figure 4). The obtained samples exhibit highest optical transmittance reported to date (82.5% at a wavelength of 1064 nm) and the Verdet constant of 120 rad T^{-1} m^{-1}, which is three times higher than that of commercial TGG crystal. However the measured Verdet constant is inferior to the value obtained in [9].

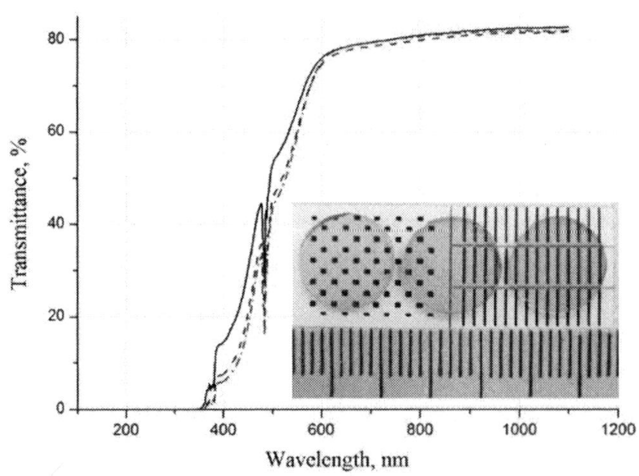

Figure 4. Photograph and transmission spectra of ceramics based on Tb_2O_3

Synthesis technology of composite Nd:YAG/Cr,Ca:YAG ceramics for thin disk lasers and high power laser systems has been described elsewhere [10].

Diffusion bonding method for YAG ceramics has been developed. According to our investigations, the bonding interface has been disappeared owing to recrystallization. Measurements of extinction coefficient have shown that extinction coefficient remained at the same level despite an increase in the thickness of bonded samples [11].

Ceramic Ce:YAG scintillator with good time-response characteristics and high light yield has been fabricated in collaboration with Ural Federal University [12].

Initial works on creating of new Fe^{2+}:$MgAl_2O_4$ ceramics for active elements of mid-IR solid-state lasers, which are important for numerous civil and military applications, have been started. Figure 5 shows the photographs and transmission spectra of Fe^{2+}:$MgAl_2O_4$ ceramics. It can be seen that with an increase in iron content the color of the samples changes from colorless to brown-red.

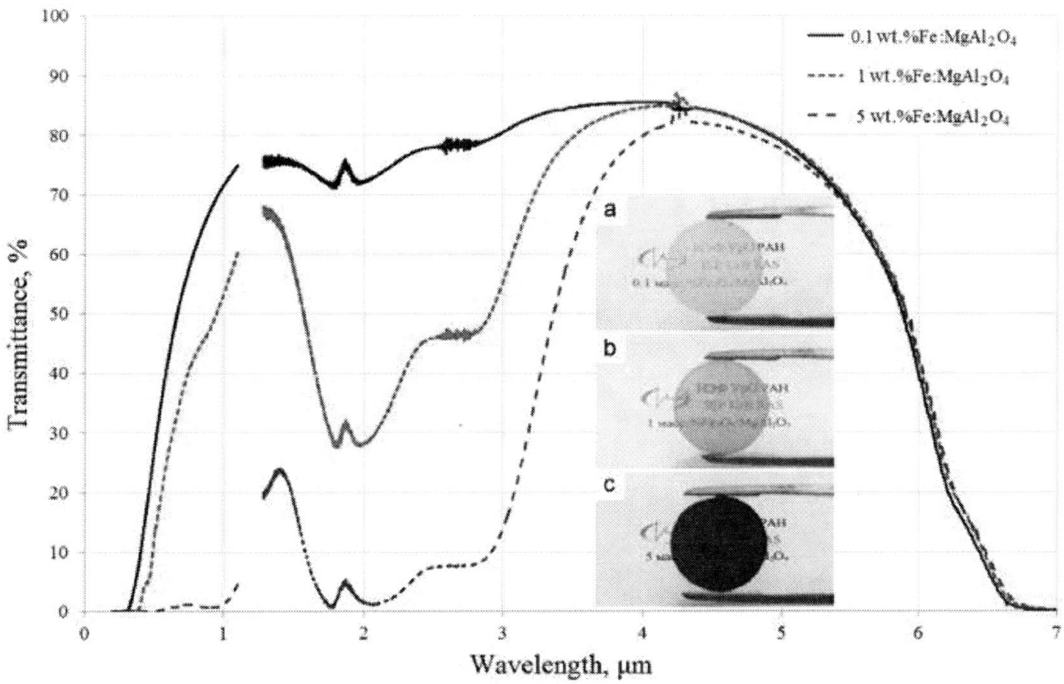

Figure 5. Photographs and transmission spectra of ceramics with various iron content: a – 0.1 wt.%Fe_3O_4:$MgAl_2O_4$, b – 1 wt.%Fe_3O_4:$MgAl_2O_4$, c – 5 wt.%Fe_3O_4:$MgAl_2O_4$

Transparency of samples is relatively low is the visible range, which is attributed to the presence of secondary phase $(MgO)_{0.9}(FeO)_{0.1}$ in ceramics with a content of several wt.% and crystallite sizes of ~120 nm. With an increase in wavelength the transparency increases and reaches 85.6% at a wavelength of 4 μm, which is close to the theoretical value of 86%. Further work in this direction will be associated with the elimination of secondary phase by selecting of nanopowders composition and sintering method.

The obtained results could serve as a basis for synthesis technology of various laser ceramics

Acknowledgements

The reported study was funded by RFBR according to the research project No. 17-08-00064 A.

References

[1] Osipov V V, Kotov Yu A, Ivanov M G, Samatov O M, Lisenkov V V, Platonov V V, Murzakaev A M, Medvedev A I and Azarkevich E I 2006 *Las. Phys.* **16** 116–25.
[2] Osipov V V, Platonov V V, Shitov V A and Maksimov R N 2017 *Photonics* 7 52–70.

IOP Conf. Series: Materials Science and Engineering **558** (2019) 012030 doi:10.1088/1757-899X/558/1/012030

[3] Osipov V V, Shitov V A, Maksimov R N and Solomonov V I 2015 *Opt. Mater.* **50** 65–70.
[4] Bagayev S N, Osipov V V, Shitov V A, Pestryakov E V, Kijko V S, Maksimov R N, Lukyashin K E, Orlov A N, Polyakov K V and Petrov V V 2012 *J. Eur. Ceram. Soc.* **32** 4257–62.
[5] Toci G, Pirri A, Patrizi B, Maksimov R N, Osipov V V, Shitov V A, Yurovskikh A S and Vannini M 2018 *Opt. Mater.* **83** 182–6.
[6] Osipov V V, Maksimov R N, Shitov V A, Lukyashin K E, Toci G, Vannini M, Ciofini M and Lapucci A 2017 *Opt. Mater.* **71** 45–9.
[7] Bagayev S N, Osipov V V, Vatnik S M, Shitov V A, Vedin I A, Platonov V V, Steinberg I Sh and Maksimov R N 2015 *Opt. Mater.* **50** 47–51.
[8] Snetkov I L, Permin D A, Balabanov S S and Palashov O V 2016 *Appl. Phys. Lett.* **108** 161905.
[9] Ikesue A, Aung Y L, Makikawa S and Yahagi A 2017 *Opt. Lett.* **42** 4399–401.
[10] Osipov V V, Shitov V A, Solomonov V I, Lukyashin K E, Spirina A V and Maksimov R N 2015 *Ceram. Int.* **41** 13277–80.
[11] Osipov V V, Lukyashin K E, Shitov V A and Maksimov R N 2016 *Mater. Lett.* **167** 81–4.
[12] Osipov V V *et al.* 2017 *Opt. Mater.* **71** 98–102.

IOP Conf. Series: Materials Science and Engineering **558** (2019) 012031 doi:10.1088/1757-899X/558/1/012031

Features of interaction of powder high-velocity particles with the surface layer of the target

E V Petrov

Merzhanov Institute of Structural Macrokinetics and Materials Science, Russian Academy of Sciences, Chernogolovka, 142432 Russia

Abstract. The interaction of a stream of high-velocity particles with the surface layer of a target was investigated. The impact of discrete particles with the sample surface was shown to lead to the formation of a coating consisting of these particles and to the changes in the physical and mechanical properties of the surface layer of the target. The obtained coatings are a composite material consisting of sintered (not melted) powder particles. The impact of high-velocity particles was shown to lead to the hardening of the surface layer of samples exposed to the stream of powder particles of tungsten, nickel and titanium nitride. The maximum increase in the microhardness of targets was observed at a depth of 2 and 4 mm from the surface treated with a stream of particles.

1. Introduction

Dynamic methods that use explosion energy occupy a special place among a large number of different methods for the processing of materials. One of these methods is shock-wave interaction of the stream of discrete particles with targets, when the material of these particles can penetrate into a target, at a depth exceeding the size of initial particles by hundred times [1–5]. This effect was discovered by Soviet scientists S. M. Usherenko in the experiments conducted for the hardening of metals subjected to explosives in the 70s of the last century. This effect is reached when a target is exposed to a stream of high-velocity particles by two main schemes: location of the powder particles in the cumulative cavity of the explosive [6] and frontal throwing of particles [7].

Shock-wave loading and impact of high-velocity particles provide improved physical and mechanical properties of both the surface and the entire volume of the materials treated. Researchers focus on the near-surface zone, where the main processes take place in the study of materials treated with a stream of high-velocity particles [8–11]. When a stream of high-velocity particles interacts with a target, most of the particles remain in the near-surface zone and form a coating. At the same time, the formed layer is stable in thickness, structure and phase composition. The formation of a special structure due to the high-velocity interaction of powder particles with a steel target allows materials to be obtained with new properties.

The purpose of this work is to study the effect of various powder particles accelerated by the energy of an explosion on the surface layer of a steel target.

2. Materials and Methods

Targets made of structural steel (grade St.3) were subjected to shock-wave treatment with a stream of discrete high-velocity particles of the powders such as tungsten with a particle size of 10–16 μm, nickel with a particle size of 6–16 μm and titanium nitride with a particle size of 45–60 μm accelerated by an explosive to study the structure and properties of the near-surface layer of these targets. To determine the particle size of the powders, a Micro Sizer 201 laser particle analyzer was used.

The scheme of the experiment was as follows: a cylindrical sample was placed in a guide channel, on top of which was a ring with a powder material with a bulk density of 3 g. An explosive with a detonator was placed on the ring [12]. There was a gap between the powder particles and the explosive. The gap provided a longer powder loading which is necessary to create a rectangular shock pulse. During detonation, the test sample was subjected to shock wave and explosion products, as well as to the treatment with a powder accelerated by them. Hexogen was used as an explosive. In all experiments, mass, type of explosives and the scheme of treatment were constant.

Vickers hardness of the near-surface zone of the samples was measured after treatment with a stream of high-velocity particles and in the initial state using a PMT-3 hardness meter under a load of 100 g. The SEM study of the samples was conducted using a Zeiss Ultra plus scanning electron microscope and a metallographic inverted microscope METAM LV-34 with a camera MC-5.3.

3. Results and Discussion

A shock-wave treatment of a steel sample with a stream of particles leads to the formation of a transition zone and a coating on the sample surface. Figure 1 shows that the thickness of the coatings formed by powder particles is distributed nonuniformly over the surface of the target and is not continuous, and there are cavities as well. This confirms that the particles of the powders are in the unmelted state while interacting with the surface of the samples, and the coating consists of sintered particles.

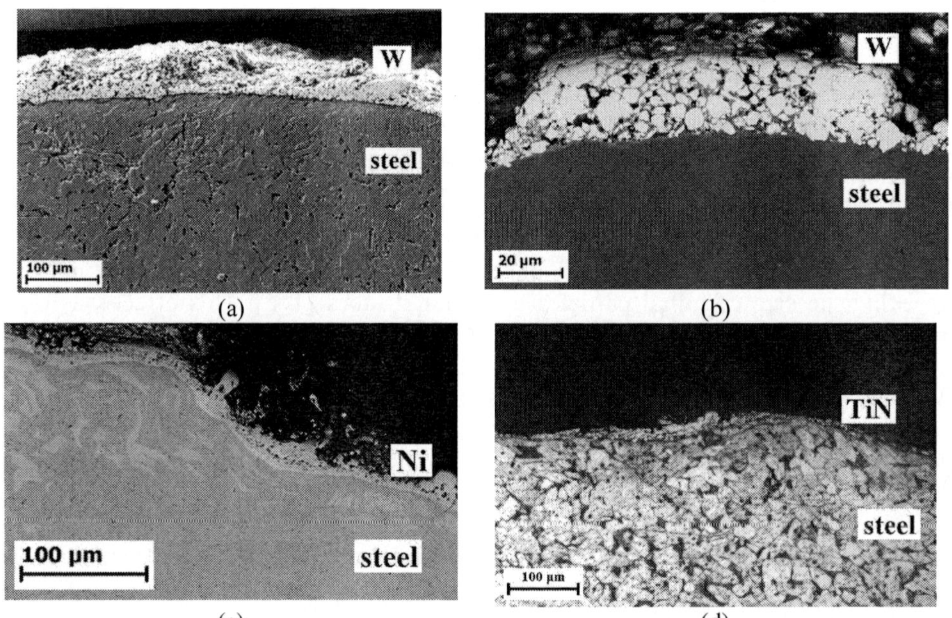

Figure 1. Microstructure of the samples exposed to powder high-velocity particles of (a, b) tungsten, (c) nickel, and (d) titanium nitride.

A similar result was obtained in the estimated calculations. Calculations of the thermal effect of the detonation products on the powder particles flying with them and the temperatures of particles reached during a high-velocity interaction with a target showed that the time during which the powder particles interact with the detonation products is not sufficient to reach the melting temperature [13].

The SEM study of the near-surface zone of the etched sample (figure 1a) with the use of the electron microscope shows that a clear boundary is observed between the steel target and the coating created from powder particles. This means that after the shock-wave interaction of powder particles with the steel target, a 2–3 μm-thick transition zone is formed, which can be seen after etching (figure 1a) or at higher magnification (figure 1b).

The energy-dispersive analysis of the samples treated with a stream of nickel particles showed structural changes detected in the near-surface layer of the target (figure 1c).

The metallographic analysis of the microstructure of the near-surface zone in the samples treated with a stream of powder particles have shown that the shock-wave treatment of samples with particles leads to grinding, flattening and stretching of ferrite and perlite grains in the near-surface layer of the samples (figure 1d). Such changes in the structure of steel are caused by both shock wave and the thrown particles of tungsten, nickel and titanium nitride powders.

The study of the microhardness distribution in the samples showed that the microhardness of the target increased after the treatment with the stream of particles. The maximum increase in the microhardness of samples is observed in the near-surface layer (figure 2) and then it decreases beginning from the surface of the treated samples. Approximately at a depth of 15–20 mm, the microhardness is about 170 units of hardness, as for the initial material, (grade St.3), since the effect of shock wave and powder particles on the structure of the steel target decreases with the distance from the contact surface of the sample.

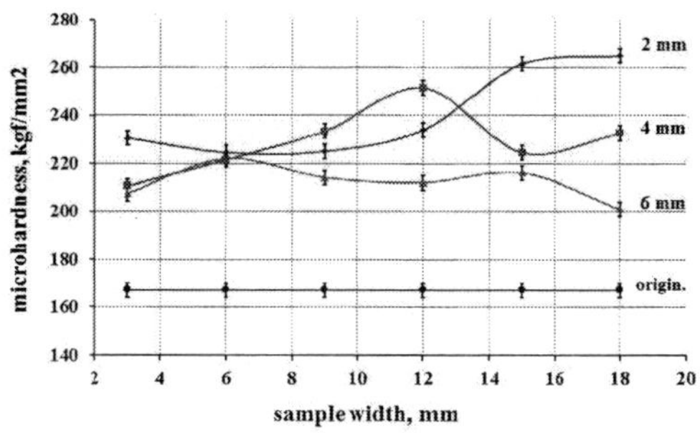

(a)

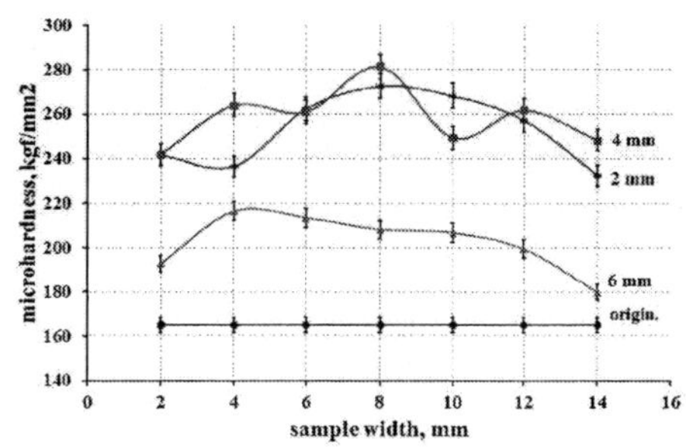

(b)

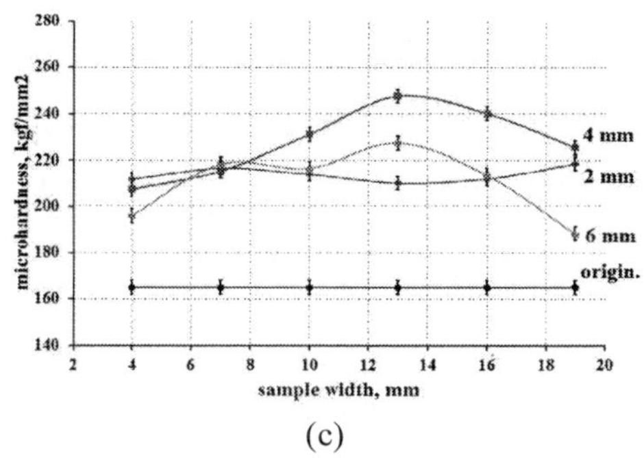

(c)

Figure 2. Distribution of microhardness across the width of the samples exposed to the particles of (a) tungsten powder, (b) nickel, and (c) titanium nitride.

Figure 2 shows the distribution of microhardness across the width of the samples exposed to the powder particles. It can be seen a wave-like distribution of microhardness across the width. The effect of high-velocity particles on the sample area is nonuniform. The samples have a maximum increase in microhardness in the near-surface layer at a depth of 2 and 4 mm from the treated surface, and mainly the center of the sample is subjected to the shock-wave treatment with the stream of particles.

The maximum increase in the microhardness of the samples treated with a stream of nickel particles (figure 2b) was observed at a depth of 2 and 4 mm compared with the samples treated with a stream of tungsten and titanium nitride particles. At a depth of 6 mm from the treated surface, all samples have the same increase in the microhardness values.

The microhardness measurements in the near-surface layer after the shock-wave treatment with the stream of the particles are given in table 1 that shows that the maximum increase in the microhardness by 51% and 54% is observed after the treatment with nickel particles at a depth of 2 and 4 mm.

Table 1. Microhardness across the depth of the sample.

	Microhardness, kgf/mm2		
	2 mm	4 mm	6 mm
W	235–240 ↑42%	230–235 ↑39%	210–215 ↑27%
Ni	250–255 ↑51%	255–260 ↑54%	200–205 ↑21%
TiN	210–215 ↑27%	225–230 ↑36%	205–210 ↑24%
Steel 3		165–170	

The interaction of high-velocity particles of tungsten, nickel and titanium nitride with the surface of the steel target forms a coating with a thickness from 10 μm to 30 μm on the surface of the treated samples. The obtained coatings according to the results of experimental studies, calculations of the effect of detonation products on flying particles and the temperatures reached during interaction with the target are a composite material consisting of sintered (not melted) powder particles.

Studying the microhardness of the surface layer of steel targets treated with a stream of powder particles accelerated by the energy of explosion have shown that there is an increase in microhardness by at least 21% after the treatment with the powder particles under study, and the maximum increase in microhardness is observed at a depth of 4 mm from the surface treated with nickel particles.

References

[1] Usherenko S M, Gushchin V I and Dybov O A 2002 *Russ. J. Phys. Chem.* B **9** 43
[2] Kiselev S P and Kiselev V P 2002 *Int. J. Impact Eng.* **27** 135
[3] Figovsky O L, Gotlib E M and Naumov S V 2010 *Bulletin of Kazan Technological University* **11** 113
[4] Aleksentseva S E 2015 *Bulletin of SamGTU. Ser. Technical Science* **1** 142
[5] Chengzhi Qi and Jianjie Chen 2014 *J. Mech. Behav. Mater.* **1-2** 21
[6] Usherenko S M 1998 *Ultra-deep penetration of particles into obstacles and the creation of composite materials* (Minsk: Research Institute of Pulse Processes Press) p 210
[7] Aleksentseva S E and Krivchenko A L 2013 *Bulletin of Sam GTU. Ser. Technical Science* **2** 71
[8] Zilberbrand E L, Rumyantsev B V, Tolkachev M D and Usherenko S M 2006 *J. Tech. Phys.* **10** 81
[9] Figovsky O L and Usherenko S M 2008 *Adv. Mater. Res.* **79-82** 1975
[10] Khristoforov B D 2010 *Engineering Transactions* **3-4** 131
[11] Ganigin S Yu, Kalashnikov V V, Ibatullin I D, Murzin A Yu , Glazunova O Yu and Grigorycv A A 2013 *News of the Samara Scientific Center of the RAS* **4-2** 339
[12] Trofimov V S, Petrov E V and Alymov M.I. 2016 *Int. J. Self-Propag. High-Temp. Synth.* **2** 125
[13] Petrov E V and Trofimov V S 2018 *Defor. Destruc. Mater.* **3** 38

Features of shock pressing and electro-impulse heating during consolidation of a powder material

R L Plomodyalo[1], E G Grigor'ev[2], L I Svistun[1] and D V Dmitrenko[1]

[1]Kuban State Technological University, Krasnodar, 350072 Russia
[2]Merzhanov Institute of Structural Macrokinetics and Materials Science, Russian Academy of Sciences, Chernogolovka, 142432 Russia

E-mail: eugengrig@mail.ru; sutk_kubstu@mail.ru

Abstract. The article discusses the method of high-speed shock consolidation of powder materials. In this method, high-speed heating of the powder is carried out by a high-voltage current pulse, and the mechanical force on the powder billet is created by a falling pile driver. A description of the experimental setup that implements this method is given. Energy and kinematics parameters of the shock consolidation process of powder materials are discussed.

Methods of consolidation of powder materials with unique performance properties that hold the metastable initial (amorphous, nanocrystalline) structure of a powder are typically characterized by short duration of thermal and force impact on the material being consolidated. These include: spark plasma sintering, magnetic pulse pressing, high-voltage consolidation [1], shock pressing [2] and a number of other methods. High-speed heating of a powder billet in these methods is created by a pulsed electromagnetic field, and force mechanical impact can be provided by various equipment: hydraulic press, pneumatic press, magnetic pulse press [1], drop-hammer [2], etc. For obtaining high performance characteristics of consolidated materials, it is necessary to determine the optimum values of the consolidation process parameters: amplitudes and durations of the electric current pulses and the mechanical pressure acting on a powder billet. Concurrently, an important aspect is the synchronism of thermal and mechanical pulse actions on the material being consolidated. High-voltage electric pulse consolidation is one of the most high-speed techniques for formation of materials and products from electrically conductive powder compositions [3, 4]. The time of input of thermal energy into the powder billet is determined by the duration of the high-voltage current pulse which is less than 300 μs, and the amplitude of the current pulse density is in the order of 1 mA/cm^2 [4]. In this case, the main release of thermal energy occurs at the points of contact between powder particles. The temperature of interparticle contacts during the current pulse action on a powder material can reach a value to the extent corresponding to the condition of an interparticle contact electric thermal explosion [3]. The average temperature in the volume of a consolidated powder billet during exposure to a high-voltage current pulse is significantly lower than the temperature of interparticle contacts. Features of the thermal processes occurring in a consolidated powder material necessitate the timely dynamic force impact on a powder billet, coordinated with the proceeding thermal processes. One of the high-voltage consolidation method variants carrying out a coordinated thermal and force impact on the consolidated material - the electric pulse pressing [5], is implemented on the experimental equipment consisting of a generator of high-voltage current pulses and a pneumatic press.

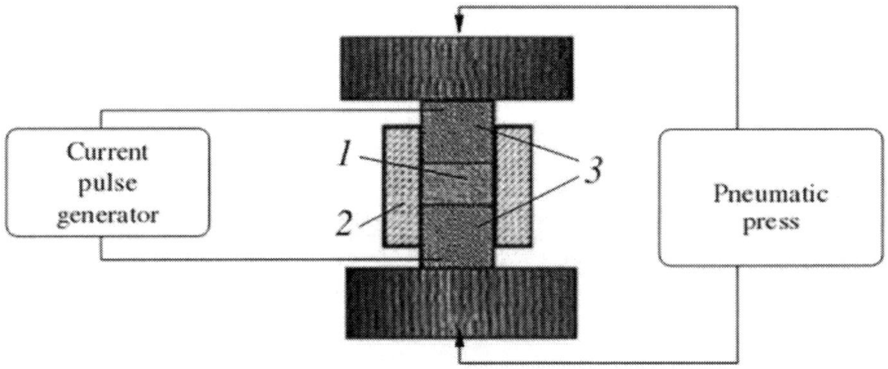

Figure 1. Schematic of the electric pulse pressing process [5].

Schematic diagram of the electric pulse pressing process is shown in figure 1. Powder material 1 is placed in die 2 made of a dielectric material. Electrodes-punches 3 transfer pressure to the powder billet from the pneumatic system, also acting as current leads from the current pulse generator to compacted powder 1. For creating powerful current pulses, a high-voltage pulse capacitor bank with stored energy of 75 kJ is used, the discharge of which provides a powerful energy release in the powder compact. The current pulse passing through the powder billet and punches intensively heats only the powder material without significant heating of the punches. This is due to the fact that the electrical resistivity of the powder material is much higher than the one of the punch material. Under intensive heating of the powder material, its resistance to plastic deformation is sharply reduced, and under the action of external mechanical pressure created by the pneumatic system, it is compacted with a characteristic speed depending on the pneumatic system. Concurrently, the powder compact is cooled by means of heat sinking into the punches and the die. It is found that the powder material compaction speed is determined by the pneumatic system and does not exceed 2 m/s [5]. This speed ensures the powder billet compaction process for 5–10 ms. It is massively longer than the powder billet heating pulse duration (< 300 μs). In this regard, there is a need to develop a more high-speed compaction system consistent with the timing parameters of powder billet electric pulse heating.

The method of shock pressing [6, 7] using experimental equipment on the basis of a drop-hammer [8] is considered as a more high-speed advanced system for compaction of a powder billet [8]. This equipment allows to increase the speed of consolidated powder compaction up to 20 m/s and to reduce the time of the compaction process by more than an order of magnitude. This will allow coordinating the time parameters of electric pulse heating and powder billet compacting.

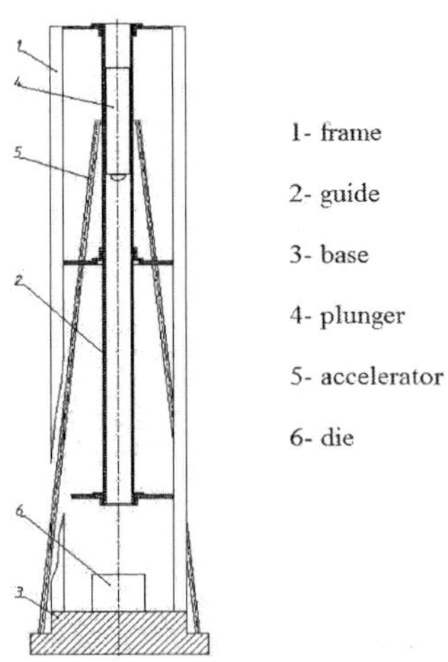

Figure 2. Drop-hammer with accelerators.

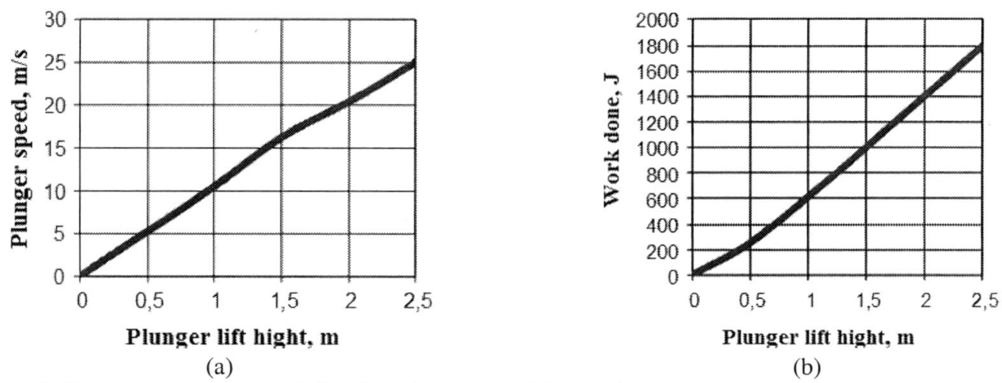

Figure 3. Energy parameters of the drop-hammer with accelerators: (a) plunger speed at different lift heights; (b) work done.

The drop-hammer with accelerators (figure 2) consists of metal welded frame *1*, pipe-shaped guide with parallel grooves *2*, cast-iron base *3* weighing 0.5 tons, mounted on a concrete foundation. Press die *6* is installed on the base. Plunger *4* moves vertically along the guiding grooves. The plunger is lifted up by the drive mechanism which includes an electric motor, a reduction gearbox and a cable wire. The plunger is dropped after releasing the dropping latch mechanism. In order to obtain loading speeds of up to 20 m/s, the drop-hammer is equipped with rubber accelerators *5*. In this case, the mass of plungers varies in the range of 0.5–40 kg. The graph of the device energy parameters is shown in figure 3. Kinematic study of the shock system is an important element of experimental studies of the process of metal powder shock pressing [9]. Punch and plunger speed measurement was carried out both under the conditions of their free collision when resistance to motion can be neglected, and in the process of pressing with progressively increasing resistance.

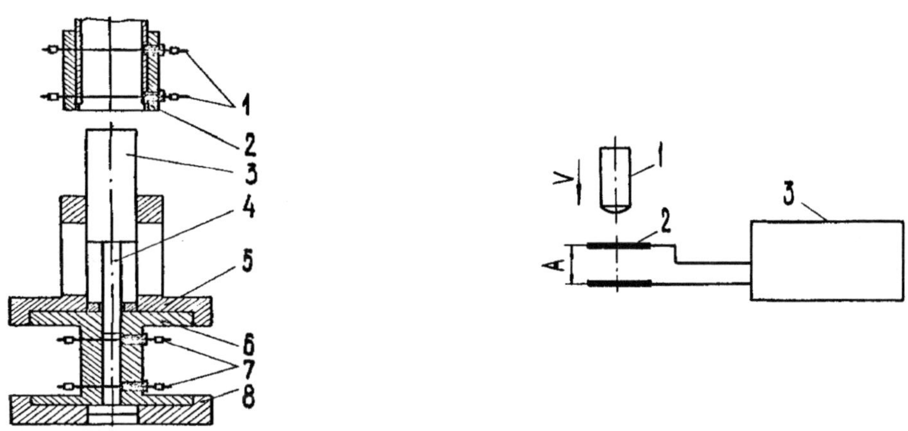

1 plunger speed measurement sensor; *2* end of the loading set-up shaft; *3* transmitting element; *4* punch; *5* muft; *6* case; *7* punch speed measurement sensors; *8* base.

1 plunger; *2* graphite rods; *3* electronic device.

Figure 4. Device and schematic for measuring kinematic parameters.

A device was used, the design of which and the speed measurement scheme are shown in figure 4. The principle of measurement was as follows. The moving plunger (or punch), meeting graphite rods on its way, destroys them, breaking the electrical circuits of the electronic device. The time of passing the basic distance by the moving body is recorded by the electronic device with an accuracy of 10^{-6} s. The average speed (V_{avg}, m/s) on the way segment is calculated according to the formula

$$V_{avg} = A/t,$$

where *A* is the basic distance, m; and *t* is the time for which the body passes distance *A*, s.

Measurement of the punch speed under the resistance force action of the pressed powder and the punch friction on the die surface was performed by determining the law of motion, i.e. the dependence of the motion of punch h on time t. True speed for each moment of time was found by differentiation of the function $h = f(t)$.

References

[1] Olevsky E A and Grigoryev E G 2016 *IOP Conf. Series: Mater. Sci. Eng.* **218** 011001
[2] Serdyuk G G, Reshetnikov V F, Svistun L I and Mikhailov O V 1985 *Powder Metall.* **12** 25–29
[3] Grigoryev E G and Olevsky E A 2012 *Scripta Mater.* **66** 662–665
[4] Grigoryev E G, Olevsky E A, Yudin A V and Yurlova M S 2015 *Comp. Mater. Sci.* **100A** 8–14
[5] Grigoryev E G 2008 *Bull. Russ. Acad. Sci. Phys.* **72** 1278–80
[6] Dmitrenko D V 2009 The shock pressing technology of carbide steel powders of the "alloy steel–carbide" type *Modern metallic materials and technologies (MMMT 2009): Proceedings of the International Scientific and Technical Conference* 131–133
[7] Dmitrenko D V, Svistun L I and Plomodyalo R L 2011 Features of the preparation of powder batch for production of carbide steels by the shock pressing method *Collection of scientific papers of the Science and Technology Conference* **13** 20–22
[8] Svistun L I and Dmitrenko D V 2009 Installation of comprehensive studies of power parameters of high-speed pressing and stamping of carbide steels *Mashinovedenie: Interuniversity collection of scientific articles* **3** 97–101
[9] Dmitrenko D V 2014 *Polythematic online scientific journal of Kuban State Agrarian University* **101** 2427–43

Features of the synthesis and consolidation of $Me^{IV}B_2$–(Me^{IV}, Mo)Si_2 ceramic powder for high-temperature applications

Yu S Pogozhev[1], M V Lemesheva[1], A Yu Potanin[1], V I Vershinnikov[2] and E A Levashov[1]

[1]National University of Science and Technology "MISiS", Moscow, 119049 Russia
[2]Merzhanov Institute of Structural Macrokinetics and Materials Science, Russian Academy of Sciences, Chernogolovka, 142432 Russia

E-mail: yspogozhev@mail.ru

Abstract. ZrB_2–$ZrSi_2$–$MoSi_2$ and HfB_2–$HfSi_2$–$MoSi_2$ composite ceramic powders were obtained by magnesiothermic SHS. The synthesized powders were consolidated by hot pressing in the temperature range of 1200–1250°C at 30 MPa. The structure and phase composition of the powders and compact ceramics were studied by X-ray powder diffraction analysis and scanning electron microscopy. ZrB_2/HfB_2, $ZrSi_2$/$HfSi_2$, ZrSi, and $MoSi_2$ were the key phases of the ceramic powders and sintered ceramics. The average grain size of the powders after grinding was 6–8 μm. Saturation of the ZrSi phase with silicon and increased $ZrSi_2$ content were observed during hot pressing. The dense ceramics had a residual porosity of 1.1–7.4%, which was largely dependent on the contents of silicides in SHS powders.

1. Introduction

Designing novel structural materials to fabricate various thermally loaded components of modern high-speed aircrafts is the primary objective of materials science. These materials are supposed to ensure constant geometry and size; stable chemical, physical and mechanical properties, as well as the heat-shielding performance of the product throughout its entire lifecycle [1]. The use of these structural materials is expected to improve the reliability and durability of thermally loaded structures, to enhance the performance of propulsion systems, and to reduce the levels of harmful emissions by increasing the temperature of combustion gases. From this perspective, the high-temperature ceramics based on HfB_2 and ZrB_2 borides shows great promise as the melting point (> 3000°C) and high thermal conductivity of these borides are comparable to those of a number of metal-based materials [2]. Furthermore, this ceramics is characterized by an elevated ablation resistance as compared to the conventionally used SiC-based materials [3, 4]. However, despite all the benefits, pure boride ceramics are not applied in practice, since low ductility of this material makes it poorly compactible and sinterable. Furthermore, its resistance to high-temperature oxidation is limited to temperatures as low as 1100°C [6] because of the formation of the B_2O_3 glassy phase (T_{melt} = 480°C), which is converted to a volatile B_2O_2 oxide at temperatures above 1100°C [7].

$TaSi_2$, $ZrSi_2$ or $MoSi_2$ silicides are added to the composition of boride ceramics to enhance its high-temperature stability and strength [8]. Oxidation of these silicides gives rise to oxide films of different compositions. Oxidation of $MoSi_2$ leads to formation of a SiO_2 protective layer on the surface, ensuring protection at temperatures up to 1800°C [9, 10]. Oxidation of $ZrSi_2$ yields the $ZrSiO_4/SiO_2$ glassy phase, which efficiently impedes diffusion of oxygen into the bulk of the ceramic material due to its barrier properties. Boron present within the ceramics gives rise to the SiO_2–B_2O_3 borosilicate glass that can self-heal the imperfections, which are inevitably formed during high-temperature oxidation [11].

Hot pressing (HP) and spark plasma sintering (SPS), as well as the hybrid technologies (reaction HP and SPS), are the methods conventionally used to fabricate the consolidated boride-silicide ceramics [12, 13]. However, their application is associated with a number of technological challenges: for example, there is a significant difference in the melting points of the initial components, the powdered raw material has to be of high purity, etc. Self-propagating high-temperature synthesis (SHS) is an alternative technique [14, 15]. The technology of forced SHS compaction allows one to produce the boride-silicide ceramics with fine-grained structure and porosity of ~ 1.5–3.0% [16]. It is reasonable to use the combined method involving two stages—fabrication of a heterophase powder by SHS and its subsequent consolidation by HP or SPS—to manufacture nonporous ceramics with the retained disperse structure. It is promising to use magnesiothermic SHS, a rather energy-efficient technology, to produce heterophase powders based on refractory compounds [15]. This technology employs heat released during the course of chemical reactions in the combustion mode. This technology is economically sound as the cheaper oxide raw material can be used.

This study focuses on the potential of conducting magnesiothermic synthesis of heterophase $Me^{IV}B_2$–$(Me^{IV}, Me^{VI})Si_2$ ceramic powders (Me^{IV} = Hf, Zr and Me^{VI} = Mo) in a broad range of concentrations of the initial components followed by consolidation of the resulting powders by hot pressing.

2. Experimental

The $Me^{IV}B_2$–$(Me^{IV}, Me^{VI})Si_2$ ceramic powder was produced by magnesiothermic reduction of hafnium oxide HfO_2, silicon oxide SiO_2, molybdenum oxide MoO_3 (all these substances being of pure for analysis grade) and boron oxide B_2O_3; the content of the main substance in these powders was ≥ 98.0%. B_2O_3 powder with the grain size of 1–2 mm was ground in a planetary ball mill (PBM) at a 1:5 ratio until the fraction with grain size < 100 μm was obtained. Magnesium powder of MPF-3 grade (98.5–99.5%) with the average grain size of 100–150 μm was used as a reducing agent.

The initial components were pre-blended in the PBM at a 1:6 ratio between the hard-alloy grinding balls and powders. The resulting loose mixtures were subjected to SHS in an 8 L universal SHS-8 setup in argon environment under an excess pressure of 3 MPa. The combustion reaction was initiated using a tungsten filament coil and monitored using a gauge detecting pressure inside the reaction chamber. A general scheme of magnesiothermic synthesis of ceramic powders is described by formula (1):

$$(x+z)2SiO_2 + (x+y)(ZrO_2/HfO_2) + zMoO_3 + yB_2O_3 + wMg \rightarrow (x+z)2Si + (x+y)(Zr/Hf) + zMo + y2B + wMgO + Q \rightarrow x(ZrSi_2/HfSi_2) + y(ZrB_2/HfB_2) + zMoSi_2 + wMgO + Q \quad (1)$$

where x, y, z, and w are the coefficients depending on the given composition of the final product; Q is the quantity of heat released during the chemical reactions, J/mol.

Once synthesis had been completed, the resulting sinter cake was crushed and disintegrated until the grain size of the resulting powder was < 250 μm. The target product based on hafnium and molybdenum borides and silicides was separated from magnesium oxide and unreacted magnesium by chemical enrichment in an aqueous solution of hydrochloric acid of chemically pure grade. Acid-soluble magnesium salts were removed by filtering off the resulting precipitate of the target product, washing it with water, and drying at 100°C. The yield of the target product (η) was calculated using formula (2):

$$\eta = (m/M)\times 100\% \quad (2)$$

where m is the weight after the enrichment and M is the weight of the semi-finished product.

To ensure particle deagglomeration, the resulting powder was ground in a PBL and an MPP-1 planetary centrifugal mill (PCM).

The adiabatic flame temperature (T_f^{ad}) of the reaction mixtures was computed using the THERMO software (Merzhanov Institute of Structural Macrokinetics and Materials Science, Russia); the mass burning rate was calculated using formula (3):

$$U_b^m = G/\tau \text{ (g/s)} \quad (3)$$

where U_b^m is the mass burning rate (g/s); G is the weight of the reaction mixture loaded into a graphite container; τ is the combustion time that is defined as a section between the point where pressure in the reaction chamber starts to rise and the point where it starts to decrease.

The phase analysis of the ceramic and hot-pressed powders was carried out on a DRON-4 diffractometer (AO "Burevestnik Research and Production Enterprise", Russia) using Cu K_α monochromatic radiation. The XRD patterns were interpreted using the JCPDS database. The microstructure and morphology of the ceramic powders and hot-pressed samples were studied using an S-3400N scanning electron microscope (Hitachi High-Technology Corporation, Japan) equipped with a NORAN System 7 X-ray energy-dispersive spectrometer (Thermo Scientific, USA) to determine the elemental composition of the structural components. SEM images were recorded in two modes (secondary and backscattered electron imaging, magnifications of ×100–5000). The granulometric compositions of SHS powders were identified on an ANALYSETTE 22 MicroTec Plus laser particle sizer (Dr. Fritsch, Germany). Hot pressing of the ceramic powders was conducted using a Direct Hot Pressing-DSP-515 SA machine (Dr. Fritsch Sondermaschinen GmbH, Germany) in a graphite die mold 50 mm in diameter at the compaction temperature of 1200–1250°C and a pressure of 30 MPa.

3. Results

Thermodynamic calculations demonstrated that the average adiabatic flame temperature is ~ 2012 K (table 1). The experimentally determined values of mass burning rate are also rather high and fluctuate within the range of 8–10 g/s. The yield of the target product, the heterophase powder of boride-silicide ceramics, is 31–35% regardless of the composition of the initial mixture.

Table 1. Compositions of the initial mixtures, combustion parameters, and yield of the target product

Composition no.	Content of the initial components in the green mixture, wt %						T_f^{ad}, K	U_b^m, g/s	η, %
	SiO_2	ZrO_2	HfO_2	MoO_3	B_2O_3	Mg			
ZMSB-1	16.2	29.9	-	5.5	10.3	38.1	2120	9.4	34
ZMSB-2	23.0	30.5	-	6.4	2.7	37.4	2060	8.3	35
HMSB-1	16.2	-	29.9	5.5	10.3	38.1	2119	9.3	31
HMSB-2	23.0	-	30.5	6.4	2.7	37.4	1750	8.4	34

Table 2 shows the calculated and experimentally determined phase compositions of the synthesized powders. One can see from table 2 that the main difference in the calculated and experimentally determined phase compositions of ZMSB-1 and ZMSB-2 powders is that $ZrSi_2$ and ZrSi silicides are present instead of Zr_5Si_3; ZrO_2 oxide (~ 5%) is also present. The presence of the latter compound is attributed to heterogeneity of the initial mixtures, so its reduction during combustion is incomplete. The Zr_5Si_3 compound is metastable and exists in a narrow temperature range. It decomposes at 1745°C via the eutectoid reaction [17]:

$$Zr_5Si_3 \rightarrow Zr_3Si_2 + Zr_2Si \quad (4)$$

Therefore, the Zr_5Si_3 phase can be present among the synthesis products only if they are rapidly cooled down or at a certain Zr : Si ratio (3:2; 5:3; 2:1) [18]; i.e., when the initial mixtures are enriched in zirconium. Since the Gibbs free energies of the reactions of formation of ZrSi and $ZrSi_2$ at flame temperature are close (–133 and –136 kJ, respectively), these reactions can be considered to be parallel. It is quite possible that the Zr : Si ratio also has an effect on presence of the ZrSi and $ZrSi_2$

phases in the powder. Hence, according to [18], $ZrSi_2$ is the main phase at a 1:2 Zr:Si ratio, while the contents of these two phases are equal at the 1:1 ratio. However, one needs to bear in mind that this study focuses on a multicomponent system, where Zr and Si are involved in formation of zirconium boride and molybdenum silicide. The fact that the initial oxides have different affinities for oxygen and, therefore, the rates of their reduction to pure components also differ, should also be taken into account. All these factors make it difficult to find the exact ratios between the components in the combustion wave and eventually the exact composition of the final products. For HMSB-1 and HMSB-2 powders, the calculated phase composition fully coincides with the experimentally determined one. ZrB_2/HfB_2, $ZrSi_2$/$HfSi_2$, ZrSi, and $MoSi_2$ are the main structural components of the synthesized powders; the ratio between these phases varies depending on contents of the initial components in the reaction mixtures.

Table 2. Phase compositions of the SHS powders

Composition no	Calculated composition, wt.%						Experimental composition, wt. %					
	ZrB_2	$MoSi_2$	Zr_5Si_3	ZrSi	Si	MgO	ZrB_2	$ZrSi_2$	ZrSi	$MoSi_2$	Si	ZrO_2
ZMSB-1	16	6	23	-	-	55	47	21	12	15	-	5
ZMSB-2	4	7	-	39	7	43	10	37	32	16	-	5
	HfB_2	$MoSi_2$	$HfSi_2$	Mg	Si	MgO	HfB_2	$HfSi_2$	HfB	$MoSi_2$	Si	-
HMSB-1	14	4	-	-	11	71	64	-	-	17	19	-
HMSB-2	8	6	23	11	8	44	12	56	2	22	8	-

One can infer from the results summarized in table 3 that the powders have similar granulometric compositions. Whereas the maximum powder grain size before grinding was 33.7–78.2 μm and the average grain size was 10.3–15.0 μm, after grinding these values decreased to 23.3–34.1 and 6.1–8.1 μm, respectively.

Table 3. The integral granulometric composition of the synthesized powders

Composition no.	Grain size, μm							
	Before grinding				After grinding in a PCM			
	D_{10}	D_{50}	D_{99}	D_{av}	D_{10}	D_{50}	D_{99}	D_{av}
ZMSB-1	3.0	11.3	58.3	14.7	0.9	5.6	34.1	8.0
ZMSB-2	3.0	10.6	78.2	15.0	0.9	5.8	34.3	8.1
HMSB-1	4.0	10.3	37.4	12.3	0.7	4.3	26.5	6.1
HMSB-2	3.2	9.1	33.7	10.3	1.0	5.0	23.3	6.3

Figures 1a–1d show the typical microstructures and the granulometric distribution curves of the synthesized powders before and after grinding (for HMSB-1 powder as an example).

Prior to disintegration, the powders consisted of agglomerated particles > 25 μm in size that caused microstructural heterogeneity (Fig. 1a). The differential curves showed normal particle distribution with a single maximum (Fig. 1b). After grinding, the particle size decreased to 4–6 μm (Fig. 1c), which increased powder's homogeneity and shifted the maximum on the differential curve towards smaller particle sizes (Fig. 1d).

Table 4 summarizes the results of analyzing the phase composition of hot-pressed samples.

Identically to how it is in the synthesized powders (Table 2), the ZrB_2/HfB_2, $ZrSi_2$/$HfSi_2$ and $MoSi_2$ phases are the key components of the hot-pressed samples. The increased content of the $ZrSi_2$ phase in the hot-pressed ZMSB-1 and ZMSB-2 samples can be attributed to the fact that ZrSi becomes saturated with silicon during consolidation at a high temperature. Although the XRD data showed no unbound silicon within the structure of ZMSB-1 and ZMSB-2 powders, this element can exist in the amorphous state. The hot-pressed sample contained no HfB phase, which was present in the HMSB-2 powder. This phase is probably converted to HfB_2 during pressing as the content of HfB_2 in the consolidated sample increases. Zirconium silicate $ZrSiO_4$ in the ZMSB-1 powder can be formed via chemical interaction between zirconium oxide and silicon oxide. The latter compound can exist in the powder in an amorphous state due to its incomplete reduction during synthesis.

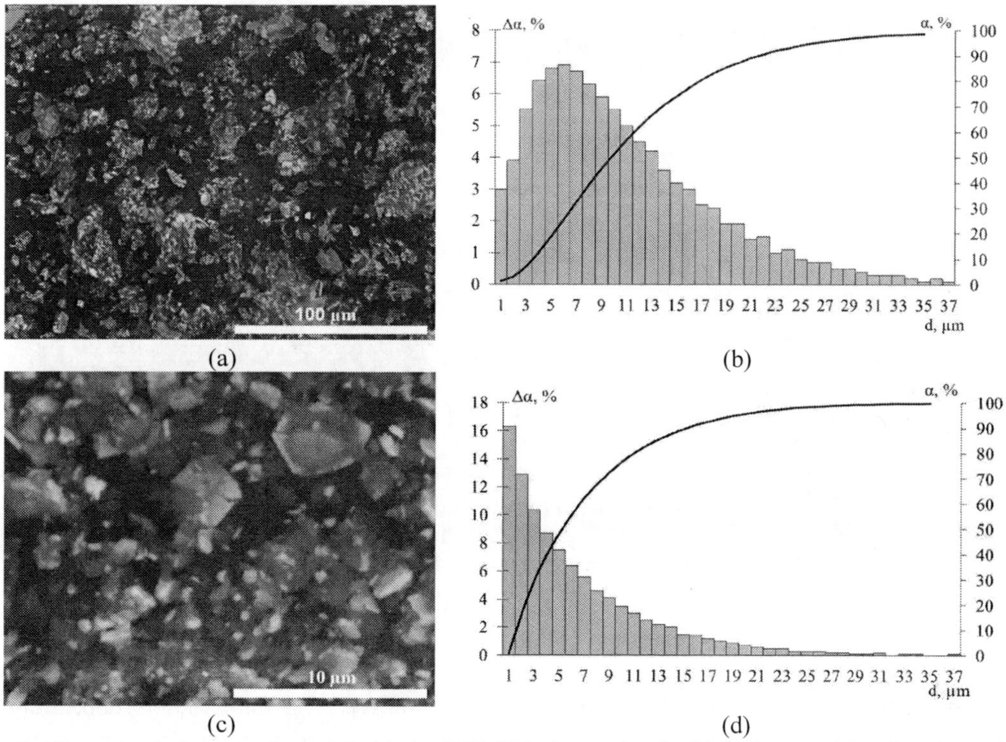

Figure 1. The morphology and particle size of HMSB-1 powder (a, b) before and (c, d) after grinding.

Table 4. Phase composition of hot-pressed samples

Composition no.	Phase content, wt.%				
	ZrB_2	$ZrSi_2$	$MoSi_2$	ZrO_2	$ZrSiO_4$
ZMSB-1	47	23	13	8	9
ZMSB-2	18	61	16	5	-
	HfB_2	$HfSi_2$	$MoSi_2$	HfB	Si
HMSB-1	61	-	23	-	16
HMSB-2	19	48	24	-	9

Figure 2 shows the microstructures of the hot-pressed samples prepared from the synthesized powders.

The scanning electron microscopy (SEM) data are in good agreement with the findings obtained by XRD. The main structural components of the compacted samples are rectangular ZrB_2/HfB_2 grains with longitudinal size of 3–10 μm and $ZrSi_2/HfSi_2$ and $MoSi_2$ inclusions in the form of irregularly shaped coarse grains. The ZMSB-1 and ZMSB-2 samples also contain ZrO_2 and $ZrSiO_4$ oxides, which are indiscernible within the structure. In the ZMSB-2 sample, the dispersed $MoSi_2$ segregates are the inclusions located inside the $ZrSi_2$ phase, being typical of the eutectic.

IOP Conf. Series: Materials Science and Engineering **558** (2019) 012033 doi:10.1088/1757-899X/558/1/012033

Figure 2. Microstructures of the hot-pressed samples prepared from SHS powders: (a) ZMSB-1; (b) ZMSB-2; (c) HMSB-1; and (d) HMSB-2.

After consolidation, the residual porosity of the samples is 1.1–7.4% and is dependent on the content of silicides in the synthesis products. The samples sintered from the powder with a higher content of silicides were characterized by lower porosity.

4. Conclusions

The heterophase ZrB_2–$ZrSi_2$–$MoSi_2$ and HfB_2–$HfSi_2$–$MoSi_2$ ceramics in the form of powders and materials consolidated to the residual porosity of 1.1–7.4% were produced by magnesiothermic synthesis followed by hot pressing. The ceramics have a composite structure with the average grain size of 6–8 μm. ZrB_2/HfB_2, $ZrSi_2$/$HfSi_2$, ZrSi and $MoSi_2$, as well as a certain amount of the oxide phase, are the main phase components. The phase composition of the pressure-sintered dense samples almost coincides with that of the SHS powders. The ZMSB-1 and ZMSB-2 compositions are an exception: in these powders, the ZrSi phase becomes saturated with silicon during hot pressing, eventually resulting in an increased content of the $ZrSi_2$ phase.

Acknowledgments

This work was supported by the Ministry of Science and Higher Education of the Russian Federation (Government Assignment project no. 11.1207.2017/PCh).

References

[1] Silvestroni L, Failla S, Neshpor I and Grigoriev O 2018 *J. Europ. Ceram. Soc.* **38** 2467–76
[2] Sonber J K, Murthy T S R C, Subramanian C, Kumar S, Fotedar R K and Suri A K 2011 *Int. J. Refract. Met. Hard Mater.* **29** 21–30
[3] Levine S R, Opila E J, Halbig M C, Kiser J D, Singh M and Salem J A 2002 *J. Europ. Ceram. Soc.***22** 2757–67

[4] Fahrenholtz W G, Wuchina E J, Lee W E and Zhou Y 2014 *Ultra-high temperature ceramics: materials for extreme environment applications* ed (Wiley & Sons Inc.) p 441
[5] Pastor H 1997 *Metallic borides: preparation of solid bodies – sintering methods and properties of solid bodies* ed V I Matkovich (New York: Springer-Verlag) p 457–493
[6] Parthasarathy T A, Rapp R A, Opeka M M and Kerans R J 2008 *Mater. Sci. Forum* **595–598** 823–832
[7] Wang Z, Niu Y, Hu C, Li H, Zeng Y, Zheng X, Ren M and Sun J 2015 *Ceram. Int.* **41** 14868–75
[8] Silvestroni L, Meriggi G and Sciti D 2014 *Corros. Sci.* **83** 281–291
[9] Fei X, Niu Y, Ji H, Huang L and Zheng X 2010 *J. Therm. Spray Tech.* **19** 1074–80
[10] Potanin A Yu, Pogozhev Yu S, Levashov E A, Novikov AV, Shvindina N V and Sviridova T A 2017 *Ceram. Int.* **43** 10478–86
[11] Feng T, Li H-J, Shi X-H, Yang X, Li Y-X and Yao X-Y 2012 *Corros. Sci.* **60** 4–9
[12] Licheri R, Orrù R, Musa C and Cao G 2008 *Mater. Lett.* **62** 432–35
[13] Hu P, Gui K, Hong W and Zhang X 2017 *Mater. Lett.* **200** 14–17
[14] Rogachev A S and Mukasyan A S 2015 *Combustion for Materials Synthesis* (New York: Taylor and Francis) p 398
[15] Borovinskaya I P, Gromov A A, Levashov E A, Maksimov Y M, Mukasyan A S and Rogachev A S 2017 *Concise encyclopedia of self-propagating high-temperature synthesis. History, theory, technology, and products* Elsevier p 466
[16] Pogozhev Yu S, Iatsyuk I V, Potanin A Yu, Levashov E A, Novikov A V, Kochetov N A and Kovalev D Yu 2016 *Ceram. Int.* **42**16758–65
[17] Lyakishev N P 2000 *State Diagrams of Binary Metal Systems* (Moscow: Mashinostroenie), p 448
[18] Bertolino N, Anselmi-Tamburini U, Maglia F, Spinolo G and Munir Z A 1999 *J. Alloys Comp.* **288** 238–248

IOP Conf. Series: Materials Science and Engineering **558** (2019) 012034 doi:10.1088/1757-899X/558/1/012034

Mechanical activation of metallic powders in planetary ball mills: multi-scale modeling and experimental observation

O Politano[1], A Fourmont[1], S Le Gallet[1], F Baras[1], A A Nepapushev[2], A S Sedegov[2], S G Vadchenko[3] and A S Rogachev[2,3]

[1]UMR 6303 CNRS-Université Bourgogne Franche-Comté, 9 Av. Alain Savary, BP 47870, 21078 Dijon Cedex, France
[2]National University of Science and Technology "MISIS", Leninsky prosp., 4, Moscow, 119049 Russia
[3]Merzhanov Institute of Structural Macrokinetics and Materials Science, Russian Academy of Sciences (ISMAN), Chernogolovka, Moscow Region, 142432 Russia

E-mail: rogachev@ism.ac.ru

Abstract. Ball motion and microstructure formation in the mixtures of Al with metals (Ti, Ni) during high-energy planetary ball milling are studied by means of high-speed video recording, scanning electron microscopy and computer modeling, including molecular dynamics simulation. Possibility of production reactive composite powders with rounded particles for using in selective laser melting techniques has been demonstrated.

1. Introduction

Despite the wide use of High Energy Ball Milling (HEBM) in labs [1], many open questions remain about the efficiency of the process as a function of grinding parameters. It was commonly admitted that the efficiency of HEBM results from high-energy hits with the balls. Recent experimental works contradict this assessment. Indeed, the dynamics of the ball motion during high energy ball milling was investigated in-situ using high speed video recording and the conclusion was that shear deformation due to friction does play a major role in the microstructural transformations [2, 3]. Another important aspect concerns the transformations induced by HEBM in reactive powder mixtures to produce new phase (i.e. mechanical alloying) or to enhance their reactivity. In order to investigate these aspects, we developed experimental technique and multi-scale modeling ranging from the ball motion in the vial to the effect of ball hit on the powder at the nanometric scale. In this work, we investigated the possibility of the production by HEBM of reactive composite powders in the Ti–Al system with various morphology especially with rounded particles with high flowability, which is important for use in selective laser melting (SLM) technology. The reactive composite powders allowed to initiate an exothermic reaction at the relatively low temperature of 650–700°C [4], which can be further decreased by using mechanical activation [5]. After reaction initiation, temperature increased sharply due to the heat release from chemical reaction, and the products, based on titanium aluminides, were formed. These products can be used as structural materials in aerospace engineering. Comparison of experimental observation with modeling demonstrates reasonable agreement.

2. Experimental and modeling

The in-situ images of milling (Activator-2S) were obtained using high-speed video camera MIRO-310, and the milling modes were described as a function of milling parameters (number of balls, size of the balls, vial and disk angular velocities). HEBM of the Ti+Al mixtures was performed in the planetary ball mill «Activator-2S» («Activator», Russia), with variation of the rotation speed of the planetary sundisc (ω_1 = 0–900 rpm) and jars (ω_2 = 0–2800 rpm). The ratio $K = \omega_2/\omega_1$ was varied from K = 1.0 to K = 2.0, which led to different regimes of balls movement in the jars. Initial powder mixture was loaded in the jars along with the milling balls. Volume of the jar was 250 ml, ball to mixture mass ratio 20:1, diameter of the steel milling balls was varied from 2 to 15 mm. HEBM was conducted in the argon atmosphere (4 atm). For the sake of video recording, the balls were placed in specially designed flat jars, without powders. In the modeling we focused on the case of HEBM with a small number of balls (1–5). The trajectory of grinding balls was calculated using discrete element modeling (DEM) in the conditions corresponding to experiments [6]. The behavior of a grinding ball was characterized and classified as stagnant, centrifugal or free-flying. In each case, the associated processes were evaluated. In the case of a free-flying ball, the number of collisions between the balls or between the balls and the walls was estimated. In the case of a stagnant or centrifugal ball, the sliding/rolling ratio was obtained. In the case of two stagnant balls, the specific rolling was investigated. In order to better understand the milling process at the miscroscopic scale, molecular dynamics simulations were carried out using the LAMMPS code [7].

3. Results and discussion

3.1. Experimental results

Figure 1 represents positions of the milling balls during HEBM at different values of K. One ball always locates near the farthest point, as compared to the axis of sun disc, due to centrifugal force. However, using of several balls (*4* in the example in figure 1) results in complex behavior of the balls due to their interaction (cooperative effect). Striking of the balls were observed at K=1.2–1.5, while friction between their surfaces prevailed at K = 1.0 and 2.0.

Initial Ti powder consisted of irregular particles of different sizes; Al particles had spherical shape and smaller size (figure 2). After HEBM, morphology of the powder particles significantly changes (figure 3). Rounded agglomerates appear after 60 min of HEBM, which become smoother after 120 min. Cross-sections of the agglomerates show that these particles are composed of deformed Ti particles in the Al matrix. Heating of the composite powders in Ar and vacuum showed that exothermic reaction was initiated at ~ 650°C resulting in formation of $TiAl_3$ and residual Ti; after initiation, temperature increases up to 1000–1100°C.

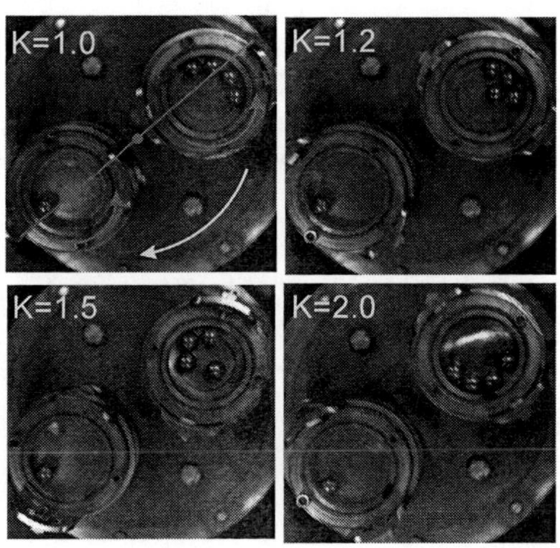

Figure 1. High-speed video frames of the balls motion during HEBM. Balls diameter is 15 mm.

IOP Conf. Series: Materials Science and Engineering **558** (2019) 012034 doi:10.1088/1757-899X/558/1/012034

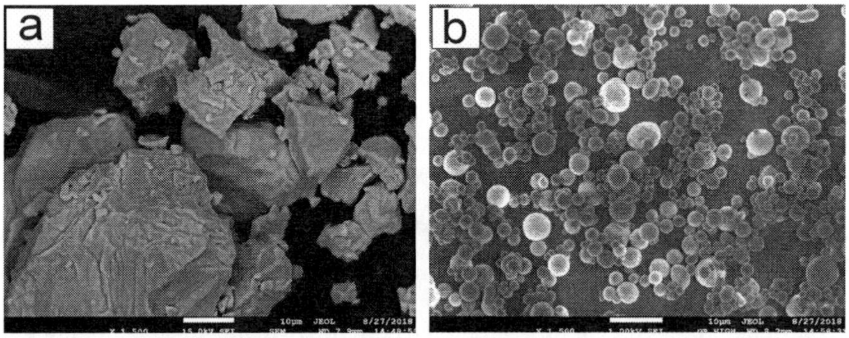

Figure 2. SEM of initial Ti (a) and Al (b) powders.

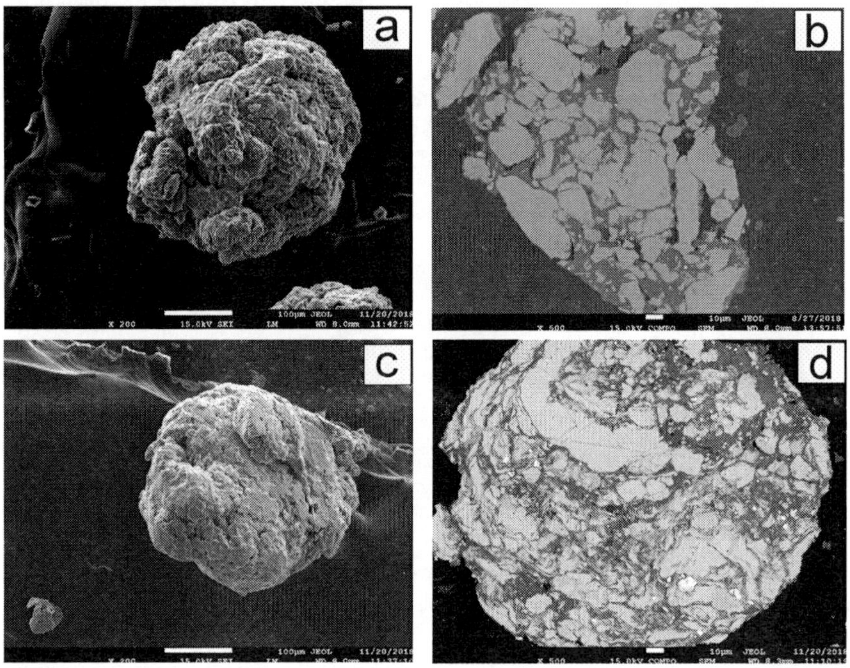

Figure 3. Reactive composite particles Ti/Al formed after HEBM during 60 min (a, b) and 120 min (c, d). SEM, external view (a, c) and cross-section (b, d); light phase Ti, dark grey Al (cross-section).

3.2. Results of modeling

An analysis based on one single ball and two balls (figure 4) demonstrates contradiction with the usual geometrical model [8] and good agreement with experimental observations (figure 1). It proves prevailed role of ball friction in HEBM and provides a guide for more efficient use of HEBM.

The effects due to the ball on a powder particles mixture were analyzed by means of a microscopic approach. The idea here is to develop molecular dynamics simulations in order to study the possible mechanisms (compaction, shear, plastic deformation, welding, fracturing, etc.) that can be observed during HBEM when we vary the ratio *K*. These elementary processes could be incorporated into a Monte-Carlo model to reach long time scale behavior [9]. We considered a set of spherical metal particles submitted to a rapid compression (the combination of Al and Ni was used for this modeling).

IOP Conf. Series: Materials Science and Engineering **558** (2019) 012034 doi:10.1088/1757-899X/558/1/012034

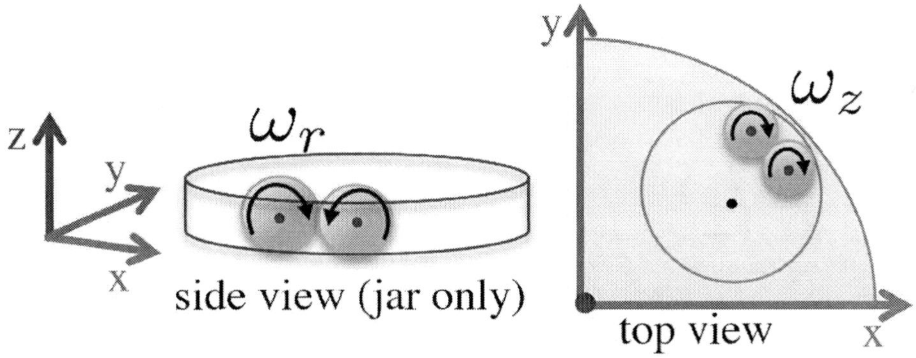

Figure 4. Schematic representation of two balls in the jar. Side view with rotation around the instantaneous radial direction of the ball inside the jar. Top view with rotation of balls around the z-axis. Only a quarter of the disk is represented.

As shown in figure 5, the compression produces a compaction of the powder by removing empty spaces between the particles. Once the particles are compacted, they start to undergo a plastic deformation. Indeed, the ductile Al mostly deforms whereas Ni remains more spherical. An increase of the number of atoms in mixing zones is observed. A detailed atomistic analysis allows us to follow the progressive amorphization, the induced chemical mixing and the possibility of recrystallization due to this first hit. Microstructure of the composite formed due to compression (figure 5) looks similar to the experimentally observed (figures 3b, 3d), despite the difference in scales and compositions.

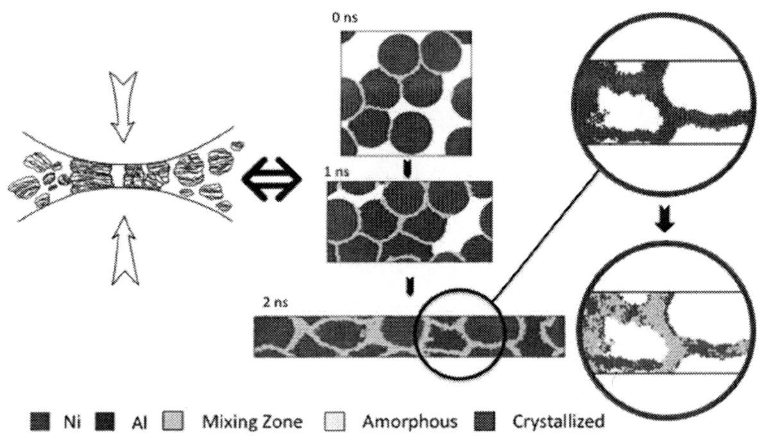

Figure 5. Schematic representation of the impact between two spherical grinding balls and powder (left). Snapshots of the simulated system (right). Initial system with randomly distributed particles in a cubic simulation box (0 ns). Final system after 75% of compression (2 ns). Red particles are Ni and blue ones are Al. Particle diameter is 9 nm. Yellow atoms are atoms that belong to a mixing zone. An atom is considered as "mixed" if one of its nearest neighbors was initially in another particle. For interpretation of thereferences to color in this figure legend, the reader is referred to the web version of this article.

4. Conclusions

Combination of experimental and modeling study of HEBM demonstrates the possibility for producing reactive composite powders (Ti/Al, Ni/Al, etc.) with rounded shape of the composite particles that can be recommended for using in SLM techniques.

Acknowledgments
This work was supported by the Ministry of Science and Higher Education of the Russian Federation, agreement no. 14.587.21.0051, project RFMEFI58718X0051, in the framework of a joint French-Russian PHC "Kolmogorov 2018" project between the laboratory ICB, University of Burgundy, Dijon, and the National University of Science and Technology MISiS, Moscow.

References

[1] Suryanarayana C 2001 *Progr. Mater. Sci.* **46** 1–184
[2] Rosenkranz S, Breitung-Faes S and Kwade W 2011 *Powder Tech.* **212** 224
[3] Rogachev A S, Moskovskikh D O, Nepapushev A A, Sviridova T A, Vadchenko S G, Rogachev S A and Mukasyan A S 2015 *Powder Tech.* **274** 44
[4] Merzhanov A G, Pis'menskaya E B, Ponomarev V I and Rogachev A S 1998 *Dokl. Phys. Chem.* **363**, 381
[5] Shkodich N F, Kochetov N A, Rogachev A S, Kovalev D Y and Sachkova N V 2006 *Russ. J. Non Ferr. Met.* **5** 44
[6] Kloss C, Goniva C, Hager A, Amberger S and Pirker S 2012 *Prog. Comput. Fluid Dynam. Int. J.* **12** 140
[7] Plimpton S, 1995 *J. Comp. Phys.* **117** 1
[8] Abdellaoui M and Gaffet E 1995 *Acta Met. Mater.* **43** 1087
[9] Pochet P, Tominez E, Chaffron L and Martin G 1995 *Phys. Rev.* B **52** 4006

IOP Conf. Series: Materials Science and Engineering **558** (2019) 012035 doi:10.1088/1757-899X/558/1/012035

Simulation of spatial ordering of powder mixture particles when packed in square matrices using an example of monodisperse balls

M A Ponomarev and V E Loryan

Merzhanov Institute of Structural Macrokinetics and Materials Science, Russian Academy of Sciences, Chernogolovka 142432, Russia

E-mail: map@ism.ac.ru; loryan@ism.ac.ru

Abstract. The peculiarities of self-ordering of large particles of powder mixture components with layer-by-layer pressing of long billets used in material production by the method of self-propagating high-temperature synthesis (SHS) are studied using the example of structure formation at compaction of thin layers of a model system of monodisperse calibrated solid balls (1 mm in diameter). In unilateral axial compaction, the case of packing of thin layers of balls (containing 1–5 monolayers) in a square matrix (with a ratio of the transverse dimension of the matrix to the diameter of the ball ~14) is considered at low pressures that do not cause significant deformation of the balls. In the disperse system under study, a structural ordering of the particles occurs with a characteristic stepped view of the curve of the compact height dependence on the number of balls in the initial charge of balls. With an increase in the number of balls in the charge, the density of thin layers changes periodically during the compaction. In the compacts, fragments of structures in which the particles are arranged orderly like regular packing of ideal balls appear. The square shape of the matrix provides the formation of fragments of rhombohedral and cubic packing. The influence of nonuniform distribution of the initial charge of balls inside the matrix on the structuring of thin layers during compaction is considered.

1. Introduction

The process of ordered arrangement of particles in the mold during the axial compaction of thin layers of powder multicomponent mixtures, including SHS ones, is still poorly known [1–3]. Investigation of the peculiarities of structural ordering in powder systems of different fractional compositions and control of this process are necessary to increase homogeneity of long compacted items and improve characteristics of synthesized materials as well as to study combustion in the model heterogeneous mixtures with ordered arrangement of components and reaction cells [4, 5]. At the beginning of the investigation, it is reasonable to reveal the compaction regularities in model systems of particle-balls in matrices of different cross sections. The aim of the work was to study experimentally the structuring features in a model system of monodisperse solid spheres when compacted in a square matrix. The influence of the cross-section shape of the matrix on the frequency of appearance of ordered structure fragments was studied; these structures consisted of the balls similar to ideal balls in regular packages [6].

IOP Conf. Series: Materials Science and Engineering **558** (2019) 012035 doi:10.1088/1757-899X/558/1/012035

2. Experimental

In the experiments, the samples consisted of calibrated steel balls of 1.000±0.005 mm in diameter were subjected to single-sided axial compaction in a square steel mold (square side L = 14.2 mm, $L/d \approx$ 14). The influence of uniformity of the initial distribution of the balls in the matrix on the process of compaction and ordering of spheres in a thin layer was studied (it is important for comparison with the compaction results of powder SHS mixture samples). Due to the fixation of the mold in a sloping position to the vertical axis (angles of slope α = 0°, 45°, 90°), the initial non-uniformity in the balls distribution by the matrix cross-section was achieved and compaction was carried out. At the chosen compaction pressures (P), the balls did not collapse or significantly deform – the density of the sample increased only due to the movement of balls during repackaging and their spatially ordered arrangement in the volume bounded by the punch planes and the matrix base and walls. The number of the balls in the sample (N) varied from 0 to 900 with a step of 5 particles for the range N = 0–570 and with a step of 10 for N = 570–900. The height (H) of the compacted sample was registered as a packing of a ball layer. The sample contained from 1 to 5 monolayers of balls in height. For N = 1–900 the condition of $H << L$ was satisfied. Relative density of the "thin layer" (ρ_o) was defined as $\rho_o = \rho/\rho_s$, (ρ_s is the density of the ball; ρ is the ratio of the balls mass to their full volume (V_c) in the "thin layer" ($\rho = \rho_s \cdot V_b \cdot N/V_c$, where, V_b is the volume of one ball)) and was calculated by the formula: $\rho_o = \pi \cdot d^3 \cdot N/(6H \cdot L^2)$. In the compacted "thin layers" various packages occurred; at certain values of N they could have a structure close to that of fragments of regular dense packages of ideal balls [6–9]. It was expected that the chosen shape of the matrix cross section (square) would stimulate the formation of fragments of the cubic package.

3. Results and discussion

Dependences of the height $H = f(N)$ and density $\rho_o = f(N)$ of the compacted "thin layer" on the number of calibrated balls in the sample were obtained (figures 1 and 2).

Most of the experimental points fall on the curves $H = f(N)$ with an increasing form for all the angles of slope. The characteristic feature of the experimental curves $H = f(N)$ is a discontinuous, periodic increase in the height of the thin layer with a constant interval; it occurs after each stage of monotonic growth of H. The compacted samples containing the same number of balls can have discretely different heights. In such samples, at the same values of N, various structures of balls are obtained – dense (with a minimum of voids) and non-dense (with vast voids and vacancies for balls), the interval of N values between the steps on the curves $H = f(N)$ is commonly kept constant.

In the square matrix, at the formation of thin layers, non-dense structures are observed more often, and the corresponding range of values of N is wider than that in the cylindrical matrices [9]. In the thin layer in the square matrix, up to four monolayers with dense structure can be contained, more than in the cylindrical matrices. In inclined matrices the non-uniformity of filling sample distribution is more pronounced, non-dense structures are more often realized. For α = 0°, voids are observed in the upper monolayers of the structure, near the punch; at α = 45°, 90° they are located at the lateral face of the matrix on the top.

Different sections of the stepped experimental curves $H = f(N)$ and horizontally located arrays of experimental points in figure 1 can correspond to the thin layers with fragments structurally similar to regular arrangement of ideal balls – cubic, tetrahedral (*1*), octahedral (*2*), rhombohedral (*3*) (the names are given in accordance with the type of the unit cell [6, 9]). The heights H_n of the regular packages consisting of n monolayers are shown in figure 1 with horizontal dashed lines (*1*–*3*). A certain correspondence of H in the horizontal arrays of experimental points to the values of H_n is observed.

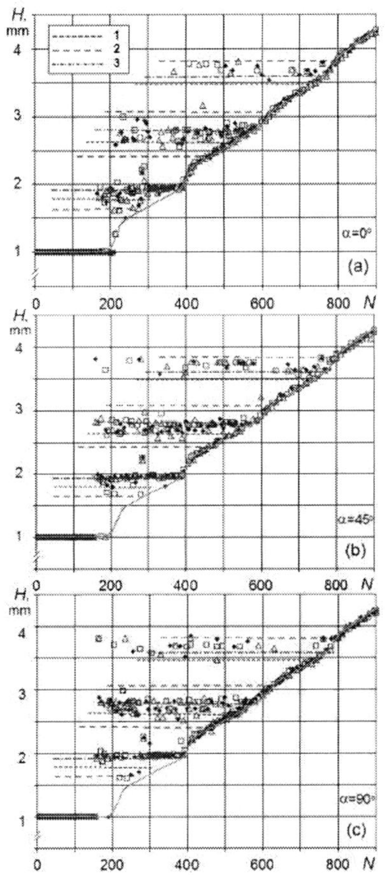

Figure 1. Dependence of balls layer height on their number: (a) α = 0°; (b) α = 45°; (c) α = 90°.

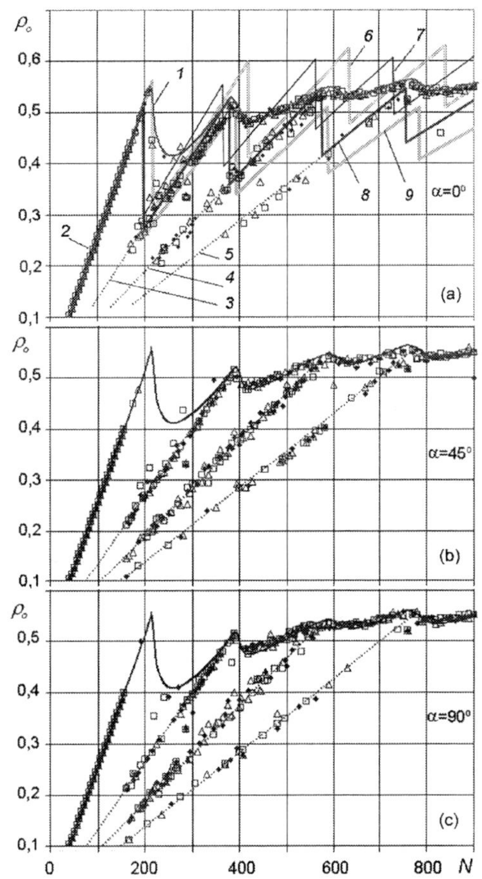

Figure 2. Dependence of relative density of thin layer on number of balls: *1* – experimental "envelope" curve (at α = 0°); *2–5* – lines corresponding to densities of loose structures of *1–4* monolayers; *6–9* – calculated densities of structures in square matrices ($L/d \approx 14$) at consecutive arrangements of balls into tetrahedral (*6*), octahedral (*7*), rhombohedral (*8*) or cubic (*9*) packages (α = 0°–90° (a–c)).

The frequency histograms in figure 3 have a pronounced discrete form. The values of N_c reach local maxima at certain values of H, which coincide or are close to vertical lines (dashed and dash-dotted lines) that correspond to the heights H_n for regular packages of balls of n = 1–5 monolayers (tetrahedral (*1*), octahedral (*2*), rhombohedral (*3*)). The location of N_c maxima near these lines indicates the possibility of formation of thin layers containing fragments of regular structures. When α changes from 0° to 90°, the characteristic shape of the frequency histogram, the number of N_c maxima and their location on the histogram, are basically preserved (figure 3). In the square matrix, when the balls are arranged along the faces near the vertex of the inner right angle in the matrix space, there are more favorable conditions for formation of fragments of such regular packages as rhombohedral, octahedral, and cubic ones. Due to the presence of right angles along the perimeter of the layer, the motion of the balls along the face, during movements and repackages, is limited to a perpendicular face, which increases the stability of the rhombohedral and cubic fragments in the corners of the matrix. In the volume of the thin layer removed from the corners and faces of the matrix, there are no obstacles for the appearance of fragments of stable and denser tetrahedral package.

IOP Conf. Series: Materials Science and Engineering **558** (2019) 012035 doi:10.1088/1757-899X/558/1/012035

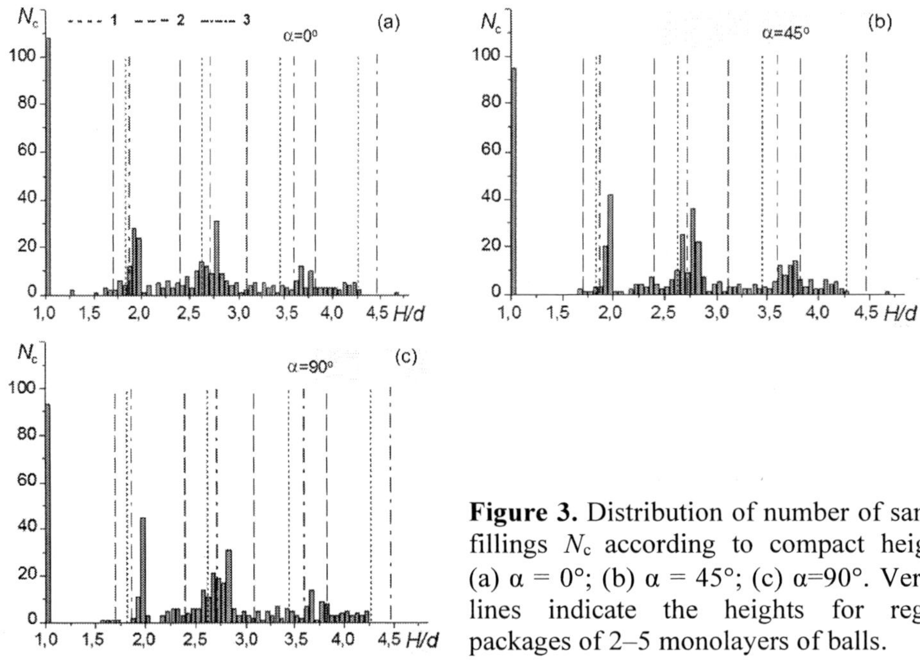

Figure 3. Distribution of number of sample fillings N_c according to compact heights: (a) α = 0°; (b) α = 45°; (c) α=90°. Vertical lines indicate the heights for regular packages of 2–5 monolayers of balls.

The relative density of thin layers is shown in figure 2. At all the values of α the dependences of the relative density on the number of balls $\rho_o = \rho_o(N)$ have a nonmonotonic form with periodical density maxima and minima.

The density extrema are similar to those for theoretical dependences for ideal balls. In figure 2, broad broken lines *6–9* correspond to the curves of the dependence of the density of thin layers of balls with regular structures (rhombohedral, octahedral, tetrahedral) on the basis of the square shape – for $L/d \approx 14$, which can be obtained by theoretical consideration of sequential package "ball-by-ball" of monolayers of a thin layer (only with minimal distortion of the structure) in the matrix. The period of the change in the amount of sample filling with which density maxima (ΔN_{max}) appear N_{max} = 180–205 balls.

Relative frequencies of appearance of denser structures in the volume of the matrix in fractions as a percentage of $W_i = (J_i/J)\cdot 100\%$ depending on the number of balls N in the sample filling are shown in figures 4 and 5. The data correspond to the "envelope" curve *1* in figure 2a, the value of J = 18 corresponds to the number of experimental points for the intervals $\pm\Delta N/2$ within the selected values of N with ΔN = 30 in the range of N = 0–570 and ΔN = 60 at N = 570–900. The period of appearance of W_i maxima is: $\Delta N \approx 190$ – for dense structure (curves *1–3*) and $\Delta N \approx 150$ – for loose structures (figure 5).

The peculiarities of structuring in thin layers of powders must be taken into account when producing long multilayer compacts from powder mixtures by sequential layer-by-layer compaction. The choice of a single sample filling with a value corresponding to one of the maxima of the thin layer density will contribute to the formation of a long compact containing significant fragments with ordered structure of coarse-grained component of the mixture. In this case, homogeneity of the multilayer compacts and constancy of the density along the length will increase. The obtained results can be used in developing recommendations for increasing density, homogeneity, and ordering of structure of compacts and products obtained by SHS or powder metallurgy when the initial mixtures contain spherical components, granular powder mixtures, clad or composite particles.

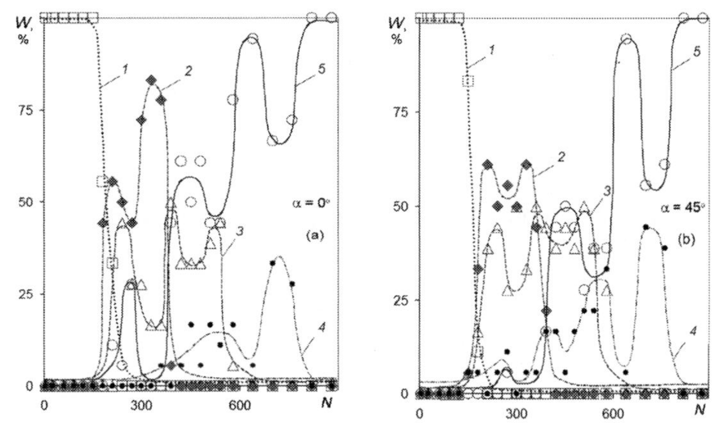

Figure 4. Frequency of realization of structures depending on *N*:
1–4 – loose structures of *1–4* monolayers; 5 – dense structures corresponding to the "envelope" curve; (a) α = 0°; (b) α = 45°.

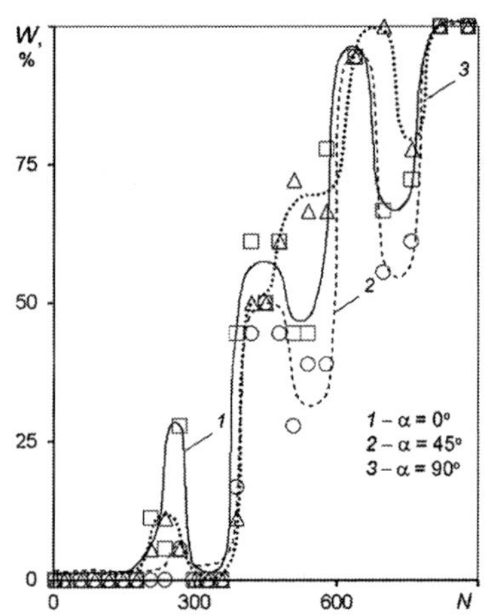

Figure 5. Dependence of frequency of denser structures corresponding to the "envelope" curve on the number of balls *N* in the sample filling – for different angles of matrix inclination: *1* – α = 0°; *2* – α = 45°; *3* – α = 90°.

4. Conclusion

In the matrix of square cross-section there is a structural ordering of particles in thin layers in a system of calibrated monodisperse balls, simulating powder SHS-mixture. Structuring manifests itself through the stepped course of the curve of the compact height dependence on the number of balls in the charge and is reflected in the periodic change in the density of the samples. The fragments with an ordered arrangement of balls in thin layers are similar in structure to regular packing of ideal balls. The square shape of the matrix section contributes to the formation of fragments of cubic and rhombohedral packing near the corners and walls of the form. The heterogeneity of the initial distribution of ball charges in the matrix provides the appearance of loose structures more often than that in cylindrical matrices [9].

The considered features of the structural ordering of particles in thin layers should be taken into account in preparation of long compacts by sequential layer-by-layer pressing of SHS-mixtures into molds with a square or rectangular cross-section.

References

[1] Galakhov A 2014 *Refract. Ind. Ceram.* **55** 209–17
[2] Maznoi A S and Kirdyashkin A I 2014 *Combust. Explos. Shock Waves* **50** 60–67
[3] Andriyanov D I, Amosov A P and Samboruk A R 2016 *Key Eng. Mater.* **685** 500–04

[4] Rogachev A S 2003 *Combust. Explos. Shock Waves* **39** 150–58
[5] Rogachev A S and Mukas'yan A S 2015 *Combust. Explos. Shock Waves* **51** 53–62
[6] Gol'dshtik M A 2005 *Transfer Processes in a Granular Layer* (Novosibirsk: Izd. Inst. Teplofiz.) p 358
[7] Ponomarev M A 2010 *Fizicheskaya mezomekhanika* **13** 103–10
[8] Ponomarev M A, Loryan V E and Merzhanov A G 2012 *Fizicheskaya mezomekhanika* **15** 85–90
[9] Ponomarev M A and Loryan V E 2016 *Int. J. Self-Propag. High-Temp. Synth.* **25** 43–49

IOP Conf. Series: Materials Science and Engineering **558** (2019) 012036 doi:10.1088/1757-899X/558/1/012036

Synthesis of materials with multiscale porosity from fine powders of boron, titanium, aluminum and coarse granules of VT6 alloy

M A Ponomarev and V E Loryan

Merzhanov Institute of Structural Macrokinetics and Materials Science, Russian Academy of Sciences, Chernogolovka 142432, Russia

E-mail: map@ism.ac.ru; loryan@ism.ac.ru

Abstract. The material with different-scale porosity was prepared from powder mixtures with significantly different particle sizes of components by the method of self-propagating high-temperature synthesis in the combustion mode. Model mixtures were formed from coarse particles of VT6 alloy previously coated with a layer of aluminum, and fine-fraction powders of titanium and boron. Synthesis in such a system took place in stages — the fine components were the first to enter into chemical interaction in the combustion wave, with the formation of a porous boride framework. A strongly exothermic reaction between titanium and boron provided the entry of coarse particles of the VT6 alloy with aluminum coating into the reaction. The melt of reacting components and formed intermetallic compounds impregnated the porous boride matrix. The stages of the synthesis are reflected in the thermograms of the combustion process. The EDS, SEM, and XRD data of the obtained material are consistent with the observed stages of the process. Up to four different scales in pore size are distinguished in the synthesis product.

1. Introduction

The aim of the study was to obtain a porous material by the method of self-propagating high-temperature synthesis (SHS) in one technological stage in the mixtures consisting of fine powders of titanium and boron with the addition of coarse spherical granules of titanium alloy VT6, previously clad with aluminum. In the model powder mixtures under study, the synthesis was possible without preheating, unlike the mixtures of Ti–Al and Ti–B systems with coarse titanium granules, which required for heating the compacts up to 1000°C [1]. The chemical reaction between the fine powders of titanium and boron, acting as a "chemical oven" [2, 3], allows to heat coarse clad granules of titanium alloy and their subsequent reaction. A wide range of particle dispersion of the metal components of the powder mixture, which melt with the combustion wave propagation, can provide a multiscale porosity of the resulting material: from macropores of several hundred microns to pores of submicron size. Stages in the course of the synthesis make it possible to obtain a product similar in structure to the composite material [3], since the fine pores of the boride matrix resulting from the chemical reaction between the fine components can be filled with a melt of the initial VT6 and aluminum as well as intermetallics resulted from the chemical reaction at the clad granule combustion.

2. Experimental

2.1. Preparation of model mixtures

Model mixtures from (Ti + 2B) and coarse clad granules of Ti(*s*) + Al alloy were prepared for the experiments. Black amorphous boron (99.8 wt %) with the particle size d_b = 0.2–6 μm was used as a nonmetal component in the mixtures. The metal components included Ti powder of the type PTS (average particle size $d_{d1} \approx$ 120 μm, chemical impurities content in titanium (wt %): N – 0.08, C – 0.05, H – 0.35, Fe + Ni – 0.4, Si – 0.1, Cl – 0.004) and coarse mono-dispersed spherical granules of Ti(*s*) alloy clad with aluminum layer. Ti (s) granules obtained by the method of plasma spraying of a "rotating electrode" had the following characteristics: the specific surface area was 0.64 m^2/g; the average diameter was d_1 = 290±30 μm. The chemical composition corresponded to VT6 alloy (major elements (wt %): Ti – 86.45–90.9, Al – from 5.3 to 68, V – from 3.5 to 5.3). To obtain clad particles, the powder mixture of Ti(s) + Al (36 wt % aluminum powder ASD–4 and 64 wt % VT6 alloy granules) was pressed into billets (up to a relative density of 0.5), heated to ~ 700°C in argon atmosphere, hold for 10 min and cooled. The billets were crushed into separate rounded forms of clad particles (Ti(s) + Al) with the coating thickness of ~ 20–30 μm. The average particle diameter d_2 = 320±30 μm. The composition, element distribution and structure of particles are shown in figure 1.

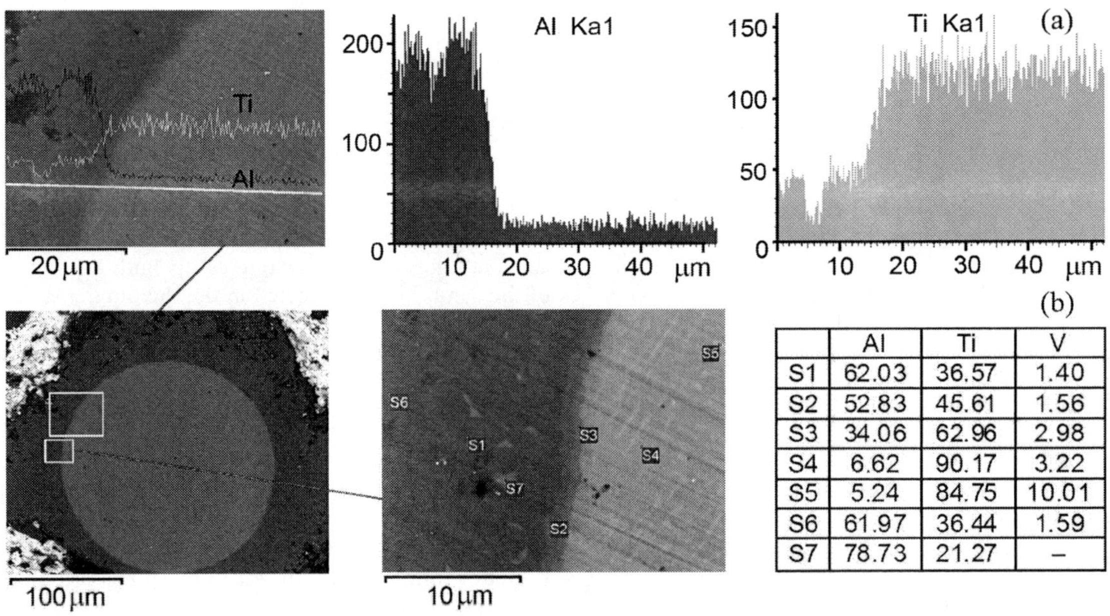

	Al	Ti	V
S1	62.03	36.57	1.40
S2	52.83	45.61	1.56
S3	34.06	62.96	2.98
S4	6.62	90.17	3.22
S5	5.24	84.75	10.01
S6	61.97	36.44	1.59
S7	78.73	21.27	–

Figure 1. Structure and composition (wt %) (polished particles in Wood alloy): (a) distribution of elements (Ti and Al) near the transition zone; (b) composition in the layer (S1, S6, S7), transition zone (S2, S3), and particles (S4, S5).

A mixture of Ti(*d*1) + 2B (69 wt % titanium and 31 wt % boron) was prepared from titanium Ti(*d*1) and boron powders. Mixtures of *a*(Ti(*s*) + Al) + (1–*a*)(Ti(*d*1) + 2B) were formed from the mixture Ti(*d*1) + 2B and clad particles (*a* is the mass fraction of the clad particles in the mixture).

2.2. Compaction of powder mixture

Cylindrical quartz glasses were filled with the green mixture by sequential portion packing of "thin" layers of the mixture [5]. The length of glasses were 43 mm, an inner diameter D = 4–4.4 mm and an outer diameter D_o = 7.8 mm ($D/d_2 > H$, $D/d_2 >> H$, D/d2 >> 1, $D/d_2 \sim 13$, where H is the thin layer height). Each layer of the mixture was packed due to the momentum received from the impactor with

IOP Conf. Series: Materials Science and Engineering **558** (2019) 012036 doi:10.1088/1757-899X/558/1/012036

m = 684 g falling from the height h = 43 mm. The sample contained 19–23 layers. The height of each layer was 1200–1400 µm. The size of the portions contributed to the formation of fragments with ordered arrangement of clad granules in the layers [5]. Packing within the "thin" layers was achieved due to the movements and stacking of clad particles into denser and more ordered fragments without particle destruction. Boron ($D\ d_b \sim 1000$, $d_2/d_b \sim 100$) and titanium particles Ti ($d1$) ($d_1/d_{d1} \sim 2.5$, $d_2/d_{d1} \sim 3$) served as an intermediate medium and "solid lubricant" for coarse-dispersed clad particles (Ti(s) + Al) and provided their arrangement.

2.3. Synthesis and measurements

The synthesis proceeded in argon atmosphere (at 1 atm) [3]. The quartz glass and the plugs at its ends prevented the change in the dimensions of the sample. The temperature in the synthesis wave was determined using tungsten–rhenium thermocouples (type A, the diameter of the junction with the protective layer of 0.25–0.35 mm), a QMBox 4050–8–1 temperature recorder and a PC ASUS–A52J. The combustion wave propagation was recorded using Sony HDR–CX130E. Electron microscopy was carried out using a field-emission SEM LEO 1450 VP Carl Zeiss.

3. Results and discussion

3.1. Combustion of pressed samples and stage-character of the synthesis

The synthesis in the samples from mixtures of a(Ti(s) + Al) + (1–a)(Ti($d1$) + 2B) proceeded in the combustion mode at $a \leq 0.6$ with the initial temperature T_0 = 20°C. At $a \geq 0.6$ the samples did not burn. The average combustion rate (u) decreased by 3 times from 5 cm/s at a = 0 to 1.5 cm/s at a = 0.5. With an increase in a, the maximum temperature T_b in the combustion wave decreased from 2260°C at a = 0 to 1515°C at a = 0.5 (figure 2).

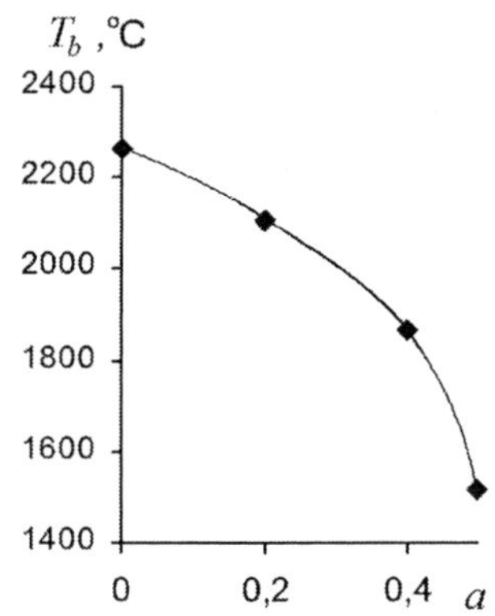

Figure 2. Dependence of maximum temperature in the combustion wave on a.

Peculiar jumps were observed on the thermograms $T_b = f\ (t)$, which are related to the stage-character of the synthesis process [3]. The thickness of interlayers of Ti($d1$) + 2B mixture between the titanium clad particles are 100–200 µm. In the heating zone, the titanium particles of the dendritic structure and, in particular, the finest particles (the particle fraction smaller than 40 µm in the Ti ($d1$) composition is ~ 30 mass %) undergo the fastest heating. In thin (up to ~ 150 µm) samples, in the form of a Ti + 2B mixture placed between the flat walls, the propagation of the combustion wave is possible, moreover, with high combustion rates [6]. This circumstance explains the observed combustion of Ti ($d1$) + 2B) mixture in the interlayers between the clad particles in a single transverse monolayer. The burning mixture of Ti($d1$) + 2B acts as a "chemical oven" and leads to heating of clad particles, melting of aluminum coating and Ti (s) granules and their mutual reaction. The resulting melt penetrates into the surrounding boride layer.

3.2. Macro, micro and porous structure of the product

The synthesis product has a developed porous structure (depending on *a*) and the total porosity of ~ 45–60%; the closed porosity being 1–5% (figure 3). With an increase in *a*, the total porosity decreases and the closed porosity increases.

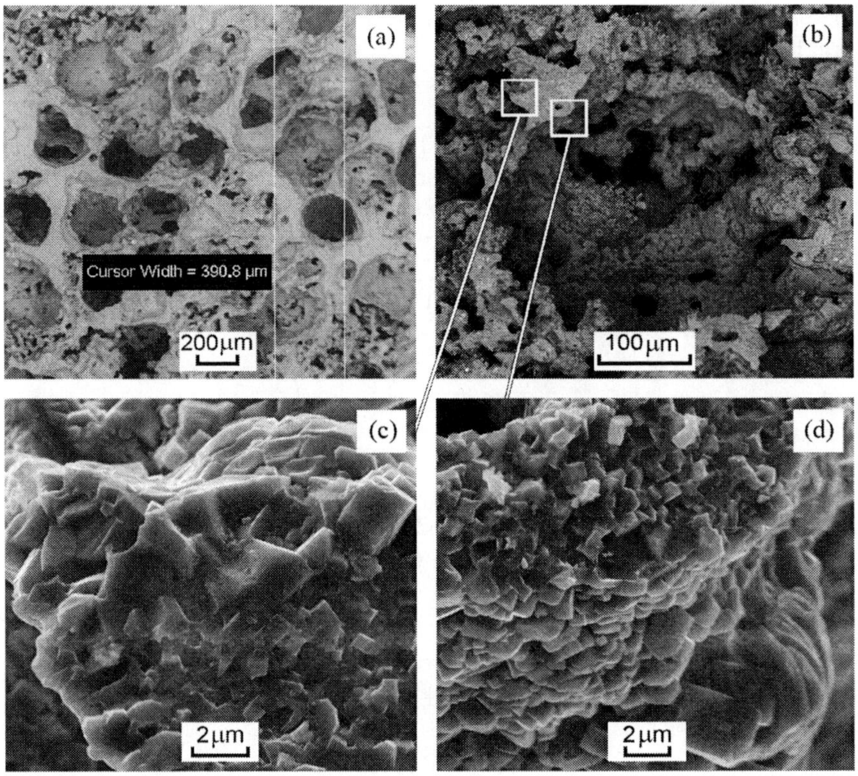

Figure 3. Structure consisting of (a) macropores (*a* = 0.4, polished section) and (b) macropore (*a* = 0.2, cleavage); cleavage structure near (c) medium size pores and (d) macropores.

Four types of the pore size are distinguished (figure 3). Coarse spherical macropores (~ 300–350 µm in diameter, figure 3a) connected with each other by capillaries; they appeared in place of clad particles as a result of capillary spreading of the melt from the granules to the pores of boride matrix; the spatial location of the macropores inherits the elements of order in the structure of the granules arrangement. Pores of irregular shape of medium size (~10–120 µm) appeared in place of Ti(*d*1) particles after melting and spreading of the melt into the surrounding boron at combustion of Ti(*d*1)+2B interlayers [7, 8]. Small round pores (~ 1–5 µm in diameter) appeared in the interaction stage of the finest titanium particles Ti(*d*1) with boron. When the melt spreads from the particles (Ti(*s*) + Al) into the boride matrix, the fine pores are partially filled with aluminum and titanium aluminides, VT6 alloy [9]. Pores of submicron size (< 1 µm) appeared during phase formation and located between the grains or were captured by the growing faces of crystalline boride grains.

The microstructure of the material around the macropores changes with an increase in the parameter *a* from fine-grained with *a* = 0.2 to coarse-grained with *a* > 0.4. At low values of *a*, the melt after melting of the clad particles completely flowed by the capillaries of the matrix, as a result, the granular structures of the matrix are clearly visible on the surface of the macropores and pores of the medium size (figures 3c and 3d). When *a* = 0.5, the specific fraction of the melt from the clad granules increases and, after capillary spreading, its excess remains inside the macropores. In this case, the structure of the matrix on the surface of the macropores becomes poorly distinguished under a thin layer of intermetallics. At $a = 0.5$, elongated crystallites are seen in the near-surface (~ 10 μm) layer in the structure of the boride matrix, while micron and submicron size crystallites (~ 200–1000 nm) are observed far from the macropores.

The phase composition of the synthesis product and the distribution of elements in different sections of the sample correspond to the staged nature of the process occurring in the synthesis wave (figures 4 and 5). Dark gray fragments correspond to an increased content of boron and titanium (points *S2* and *S3* in figure 4b and points *S1*–*S3* in figure 5), TiB_2 and TiB phase (according to XRD analysis). In light gray interlayers (0.3–0.6 μm) around the grains of the dark gray phase (0.5–1.5 μm) and in larger light gray fragments (1–15 μm), an increased content of aluminum, titanium and vanadium is determined, (points *S1* and *S4*, figure 4b). These fragments appeared as a result of the reaction in the clad granules and the penetration of the melt into the pores of the boride matrix. The composition in them is close to that of the original alloy VT6 and titanium aluminides (Ti_3Al and TiAl – according to XRD). The presence of aluminum in the absence of vanadium in the dark gray fragments (points *S2* and *S3*, figure 4b) is associated with filling part of the fine pores of the matrix only with aluminum melt from the cladding layer. The distribution of elements between macropores (figure 4a) indicates the spreading of the melt into the matrix at a distance of more than 70 μm.

The synthesis product contains TiB_2, TiAl, Ti_3Al phases and a relatively small amount of TiB.

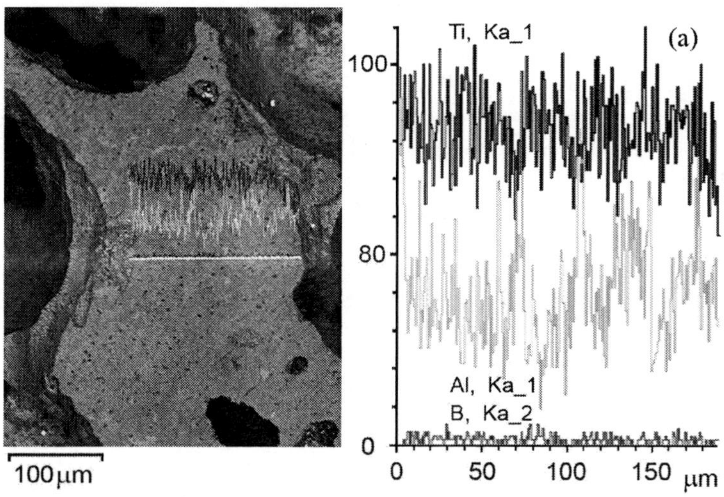

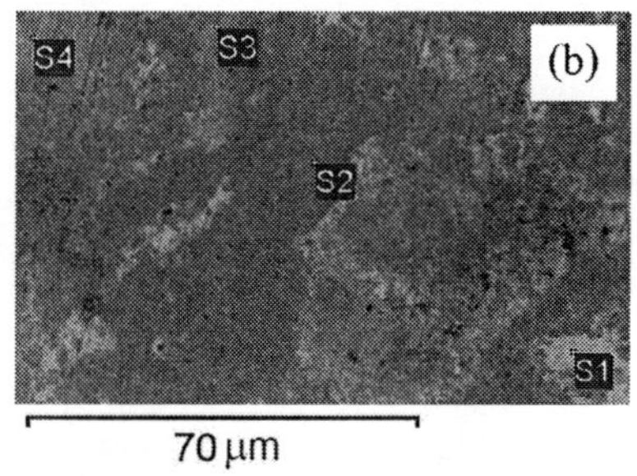

	B	Al	Ti	V
S1	-	24.58	72.38	3.04
S2	14.61	8.86	76.52	-
S3	7.45	13.83	78.72	-
S4	-	22.35	75.08	2.58

Figure 4. (a) Microstructure of product and element distribution along the line, and (b) composition in points (wt %), $a = 0.5$.

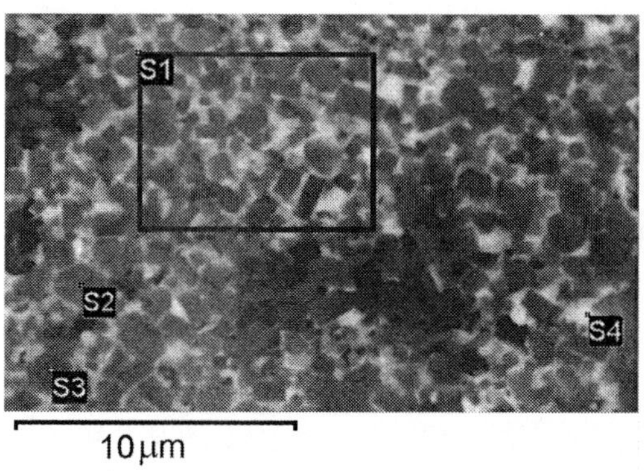

	B	Al	Ti
S1	10.38	4.13	85.49
S2	6.46	0.97	92.57
S3	7.34	1.34	91.33
S4	-	7.96	92.04

Figure 5. Microstructure and composition in points, $a = 0.2$ (wt %).

4. Conclusion

The purposeful selection of the components of the powder mixture according to the quality and dispersion composition, and sequential portion packing make it possible:

- to carry out the synthesis in the system with large components without preheating due to the stage nature of the process – one part of the mixture (of fine powders of boron and titanium) reacts in the combustion mode with a high temperature and high thermal effect, forms a porous refractory product – a boride matrix; at the same time, this mixture acts as a "chemical oven" for the inert and slightly exothermic part of the mixture – aluminum – clad granules of VT6 alloy, which react to form intermetallic compounds, filling the matrix pores with the melt;
- to obtain a porous product with developed multiscale porosity and characteristic fragmentarily ordered structure with coarse spherical pores.

The structure of the synthesized refractory product is a porous diboride matrix partially filled with titanium aluminides.

References

[1] Ponomarev M A, Loryan V E, Shchukin A S and Merzhanov A G 2013 *Int. J. Self-Propag. High-Temp. Synth.* **22** 202–09
[2] Merzhanov A G 2011 *Int. J. Self-Propag. High-Temp. Synth.* **20** 61–63
[3] Ponomarev M A and Loryan V E 2018 *Inorg. Mater.* **54** 772–78
[4] Lepakova O K 2000 *Combust. Explos. Shock Waves* **36** 575–81
[5] Ponomarev M A, Loryan V E and Merzhanov A G 2012 *Int. J. Self-Propag. High-Temp. Synth.* **21** 51–54
[6] Ponomarev M A, Shcherbakov V A and Shteinberg A S 1995 *Proceedings of the Russian Academy of Sciences (Doklady Chemical Technology)* **340** 642–45
[7] Azatyan T S, Mal'tsev V M, Merzhanov A G and Seleznev V A 1980 *Combust. Explos. Shock Waves* **16** 163–67
[8] Shkiro V M and Borovinskaya I P 1976 *Combust. Explos. Shock Waves* **12** 828–31
[9] Rosenband V and Gany A 2014 *Combust. Explos. Shock Waves* **50** 653–57

IOP Conf. Series: Materials Science and Engineering **558** (2019) 012037 doi:10.1088/1757-899X/558/1/012037

High-temperature reactive melt spreading: Experimental modeling of SHS reactions

A S Rogachev[1], S G Vadchenko[1], E V Illarionova[1], Yu B Scheck[1], S Rouvimov[2] and A S Mukasyan[2]

[1]Merzhanov Institute of Structural Macrokinetics and Materials Science Russian Academy of Sciences (ISMAN), Chernogolovka, Moscow Region, 142432 Russia
[2]Department of Chemical and Biomolecular Engineering, University of Notre Dame, Notre Dame, IN, 46556 USA

E-mail: rogachev@ism.ac.ru

Abstract. Initial stages of high-temperature spreading and inreaction of molten droplets with solid substrates were experimentally investigated in the systems Ti–C and Nb–B. Thin layer of solid product (carbide, boride) crystalline grains forms immediately in the spreading frontier, while the melt infiltrates through nano-scale gaps between grains. The formation of a continuous layer of oxidation- and wear-resistant solid refractory (~ 3000°C) product can be utilized for coating protection of refractory metals and carbon materials, as well as in high-temperature joining of carbon materials via a thin layer of titanium carbide.

1. Introduction

Reactive spreading of melts plays a key role in the mechanism of self-propagating high-temperature synthesis (SHS), especially in systems combining melting and non-melting reagents. The effect of capillary spreading was first observed in the Ti–C system [1] and then experimentally and theoretically explored in [2–4]. For different ratios of melting points of metal to non-metal, different scenarios of the spreading process can be expected to occur. For instance, in metal–carbon systems a metal melt spreads over the surface of solid carbon particles to yield carbide grains. Recently, we have found that melted Ti spreads over the carbon surface through the channels formed between TiC grains [5]. Less studied are the systems (such as Nb–B, Mo–B, Ti–Si) in which the melting point of non-metal is lower than that of metal, so that in this case non-metal melt has to react with solid particles of metallic reagent. Previously, some general features of the spreading process were explored in a particle–foil approximation for Nb–B [6] and Ti–Si [7] systems, although some subtle details remain unclear. Finally, there exist some systems (e.g. Ni–Al) in which both constituent components are low-melting and, as a result, SHS reactions take place between two melted reagents.

In this work, we report on comparative analysis of reactive spreading of melted metal and nonmetal reagents over the surface of solid reagent.

2. Experimental

The particles of low-melting reagent (B, Ti) were placed on thin substrate of high-melting reagent (Nb, C, respectively). The latter was heated (at 10^{-3} Pa) by a pulse of electric current (pulse duration between 0.5 and several seconds) to a temperature above the melting point of particles but below that of substrate.

The process was recorded with a high-speed video camera. Combustion products were characterized by SEM, STEM, and TEM.

3. Results and discussion

3.1. Nb–B System

Due to melting boron crystals on the surface of Nb foil, on the spread boron particles we observed the craters covered with finely dispersed reaction product (figure 1). According the results of EDS analysis, the product contains B, Nb and small admixtures of O, N, and C. Using focused ion beams, we managed to cut out thin plates oriented normally to the direction of spread propagation and used them in TEM studies (figure 2).

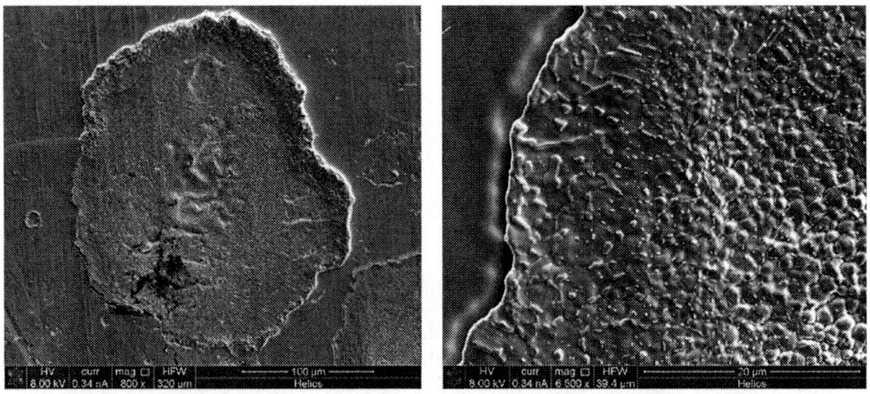

Figure 1. SEM image (in secondary electrons) of spreading reaction zone.

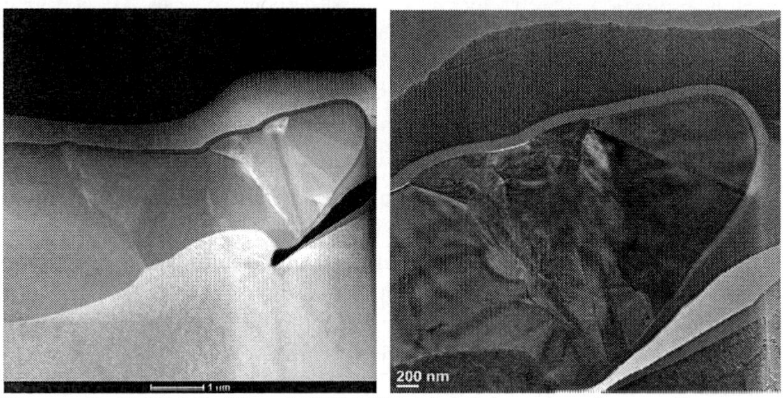

Figure 2. STEM (left) and TEM DF (right) images of spreading zone.

As can be inferred from figure 2, the Nb substrate gradually dissolved on the perimeter of spreading. The dissolution zone is about 5 μm in its width, while the depth of metal erosion makes a value of 3–4 μm (figure 2). The product layer was formed by 1-μm crystals of hexagonal NbB_2. No unreacted boron (as well as other boride phases) was detected in between the product grains. The absence of other borides can be explained by the fact that temperature of the melt was higher than crystallization temperatures of all the borides in the Nb–B system. Melted boron spread over the surface of NbB_2 layer and transformed to diboride on the perimeter of spreading.

3.2. Ti–C System

The spread of melted 100-μm Ti particle over the graphite surface was found to proceed in three stages. The first one is accomplished in a time period of about 50 ms (figure 3). It can be assumed that, at this stage, spreading rate V is controlled only by the rate of Ti melting. This is followed by a slow growth in diameter of the spreading spot (D) up to t = 500 ms (figure 3). Finally, the spreading area (spot) loses its rounded shape due to Ti spread along surface imperfections. The process of spreading decelerates and can last even for several seconds until all Ti transforms into TiC crystaline grains. Initially V has a value of 1.8 cm/s (in some experimental runs even 6–7 cm/s), which is close to burning velocity of powdered Ti–C mixtures. At the second stage, spreading rate decreases down to 1–2 mm/s.

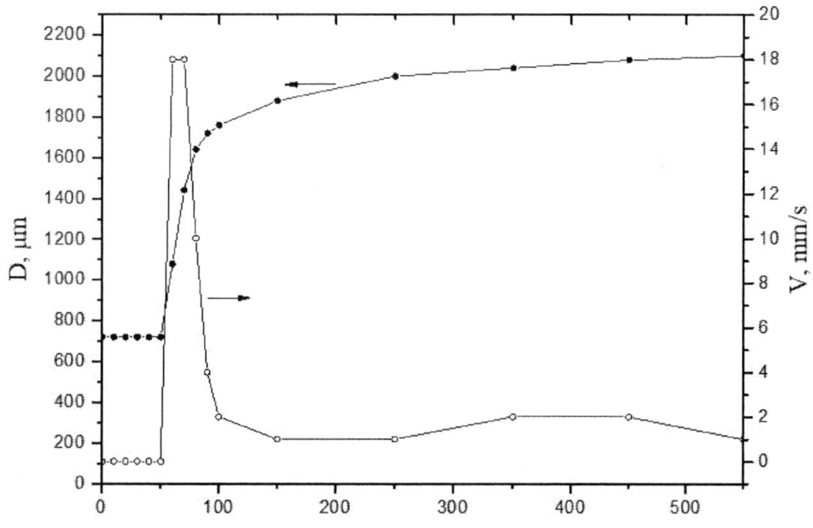

Figure 3. Diameter of spreading Ti spot (D) and its spreading rate (V) as a function of time t.

Figure 4 illustrates the fine structure of circular reaction zone. Platelet TiC grains 2–3 μm in size and ~ 0.5 μm thick are seen (figure 4a) to arise directly at the contact surface. There is a 100–150 nm gap between the grains through which Ti melt can infiltrate to the perimeter of spreading. The layer-structured grains are oriented along the substrate surface. The lowest TiC layer with a thickness of below 100 nm is extended and acts as a base for the growth of new carbide grains. This layer plays the role of 'leading' structural component in the process of Ti propagation over graphite: ahead of this layer, no Ti on the graphite surface was detected. Meanwhile, the results of atomic-force microscopy and profilometry have revealed the presence in the substrate, ahead of spreading perimeter, of some cave several μm wide and about 300 nm deep, which can be tentatively associated with dissolution of carbon in Ti melt.

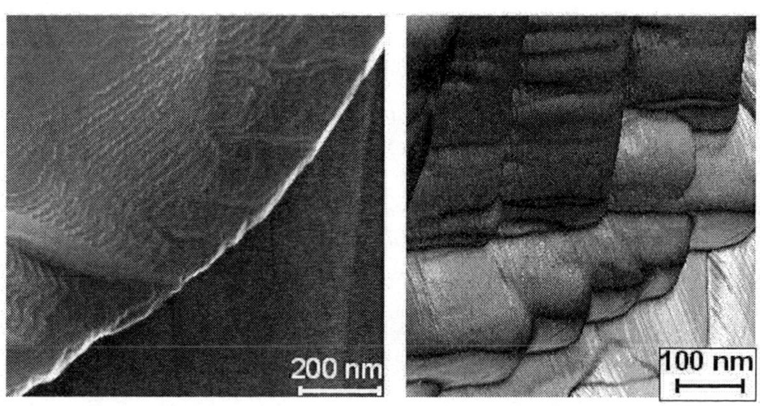

Figure 4. Fine structure of reaction zone (SAFM image).

The above results allow us to suggest the following mechanism for interaction between Ti and C in the conditions under consideration. The Ti melt formed at the area of contact between reagents is spread over the surface of solid carbon material and yields micron and sub-micron platelet TiC crystals. The occurrence of crystallization clearly indicates that the Ti melt saturated with carbon due to dissolution of carbon in melted titanium. In the contact area, the carbide grains form an island structure with a network of straits around 100 nm wide, through which melt spreading is effectuated. Initially, melt is spreading over the framework on newly formed TiC grains with open porosity. Despite the formation of solid product, direct contact between melted titanium and solid carbon still remains effective, which ensures a high rate of the reaction. Apparently, such a mechanism plays a key role in SHS of refractory compounds and materials.

When the Ti particle was placed on the surface of laminated carbon (pyrographite), it was found that interaction of Ti droplet with a substrate of pyrographite lead to the splitting of the substrate into thin layers (figure 5) that were drawn into the Ti drop to form TiC grains. At this, one of these layers is seen to crawl over the drop surface. This phenomenon can be used for creating textured microstructures and multilayer graphene sheets.

Figure 5. Disintegration of pyrographite and carbidization of nanolayers.

4. Conclusions

The formation of a continuous layer of oxidation- and wear-resistant solid refractory (~3000°C) product can be utilized in deposition of protective layers onto the surface of refractory metals and carbon materials, as well as in high-temperature joining of carbon materials via a thin layer of titanium carbide.

Acknowledgements

This work was financially supported by the Russian Foundation for Basic Research, grant RFBR 16-03-00592a.

References

[1] Shkiro V M and Borovinskaya I P 1976 *Combust. Explos. Shock Waves* **12** 828
[2] Nekrasov E A, Maksimov Yu M, Ziatdinov M Kh and Shteinberg A S 1978 *Combust. Explos. Shock Waves* **14** 575
[3] Seplyarskii B S 2004 *Dokl. Akad. Nauk* **396** 640
[4] Rogachev A S and Mukasyan A S 1987 *Dokl. Akad. Nauk* **297** 1425
[5] Rogachev A S, Vadchenko S G, Sachkova N V, Illarionova E V, Bagin V P, Kostikov V I and Merzhanov A G 2018 *Dokl. Akad. Nauk* **478** 425
[6] Shugaev V A, Rogachev A S and Ponomarev V I 1992 *Dokl. Akad. Nauk* **324** 1240
[7] Rogachev A S, Varma A, Shugaev V A, Khomenko I O and Kachelmeyer C R 1995 *Combust. Sci. Tech.* **109** 53

Applicability of molecular dynamics method to the prediction of the melting point of refractory metals and compounds

S A Rogachev

Merzhanov Institute of Structural Macrokinetics and Material Science, Russian Academy of Sciences, Chernogolovka, Moscow, 142432 Russia

E-mail: RogachevSA@gmail.com

Abstract. Various methods for measuring the melting point using embedded atom model (EAM) and from first principles modeling are considered. The features of each of them are shown. The results of the work of these methods are compared with the experimental values on the example of systems consisting of pure metals such as Fe, Pd, Ti, Pt Zr, Mo, Ta, W. Good agreement of the estimated melting points with experimental values is shown.

1. Introduction
Many modern industries, such as the aerospace industry, energetics, transport, need inexpensive refractory materials capable to work in extreme conditions. As known, the most refractory pure metal is tungsten (3695 K), and the most refractory of non-metals is carbon (slightly less than 4000 K). Currently the most refractory compound is Ta_4HfC_5, with a melting point of 4215 K [1].

However, empirical methods for researching such materials are complicated by the too large variability of components and the complexity of the experimental measurement of high temperatures. With the increasing computing power in the last few decades, molecular dynamics simulation (MDS) has established itself as a great tool for reasonably accurate prediction of various material properties [1, 2]. Thus, MDS is becoming more and more promising to estimate the melting temperature of a material before the start of an experimental study.

In the literature, there are various meanings of the concept of refractoriness, we will understand by this those materials whose melting point is higher than the melting point of iron (1812 K), which is the basis of many alloys.

In this paper, we compared various methods for estimating the melting temperature for pure metals such as Fe, Pd, Ti, Pt, Zr, Mo, Ta, W. The melting temperatures were measured using two different approaches: using the EAM and from first principles (ab-initio). Some simple compounds, such as TaW, TiC and TaC, were also considered.

2. Methods
One of the distinguishing features of almost any simulation is the idealization of the modeled situation. For example, in molecular dynamics (MD) if we take a box with periodic boundary conditions (in molecular dynamics, this means that if an atom at some integration step flies out one of the edges of the calculation area, then it will be returned to the system from the opposite edge) consisting of tungsten in bcc lattice structure and gradually begin to raise the temperature in it in the iso-enthalpic-isobaric (NPH) ensemble, then it will begin to melt at a temperature significantly higher than it should. This

IOP Conf. Series: Materials Science and Engineering **558** (2019) 012038 doi:10.1088/1757-899X/558/1/012038

phenomenon has a physical meaning. The fact is that melting requires some center of nucleation. In real life irregularities in the structure or boundary of the sample act as such centers. In this regard, various procedures are used in MD to measure the melting point.

2.1 Methods for EAM modeling

For EAM modeling, the LAMMPS software package was used [3]. Interaction potentials were taken from [4]. In all EAM calculations, an integration step of 1 femtosecond was used.

The first approach is to gradually increase the temperature in spheres. As a model system, spheres of different radii (2 nm, 3 nm, 4 nm, 5 nm, 6 nm,) consisting of crystal lattices corresponding to each metal at 300 K were considered. The temperature of the spheres increased at a rate of 10^{12} K/s, before becoming a completely liquid drop, as shown in figure 1 for a TaW particle. Figure 1a shows the initial structure of a TaW particle at 300 K. Figure 1b shows the process of melting a particle (at a temperature of 3574 K). It is clearly seen that melting goes from the boundary of the particle to the center. Figure 1c shows a fully melted drop of TaW.

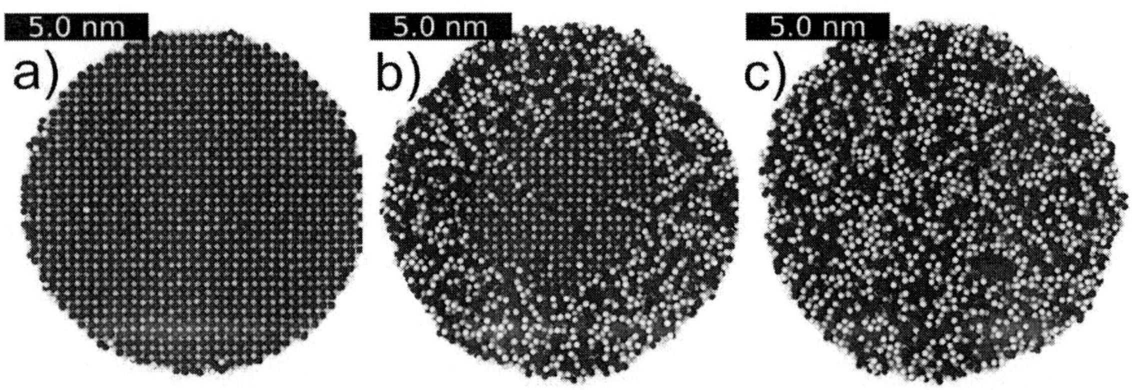

Figure 1. Spherical particle of TaW with a radius equal to 6 nm. The blue spheres correspond to tantalum atoms. The yellow spheres correspond to tungsten atoms.

The melting point for each particle was calculated as the temperature at which the enthalpy jump occurred in the system. As is known, particles with a small radius will have a lower melting point. To calculate the melting point of the bulk metal, linear approximation of the values obtained for spheres of different radii to a sphere of infinite radius was carried out (1 / (Radius of the sphere) values were taken along the X axis and enthalpy jump temperatures measured along the Y axis). For example, for TaW particles with radii of 2 nm, 4 nm, 6 nm, we have 3000 K, 3450 K, 3574 K, respectively. Approximation to a sphere of infinite radius gives the melting temperature for TaW equal to 3872 K, which is 300 K higher than the experimental value. The melting temperature T_m^{eam} for pure metals obtained by this method is shown in table 1.

The second approach is known as the solid-liquid coexistence method. The idea of the method consists in the equilibrium coexistence of the solid and liquid phases of a substance in one simulated system at a certain temperature. For this, a simulated box of 50 nm in x direction, 10 nm in y direction and 5 nm in z direction was created. Periodic boundary conditions along axes y and z were used. Initially, the lattice corresponding to the compound was set in the simulated system. Then the system was divided into the left and right equally large half. The left half of the sample was heated to a temperature above the melting point in the isothermal-isobaric ensemble (NPT) until completely liquid phase is reached, while the right half remained untouched in the iso-enthalpic-isobaric (NPH) ensemble at room temperature. A typical model system is shown in figure 2. Then we need to guess the approximate melting point for a metal or compound, let's call this value T_i.

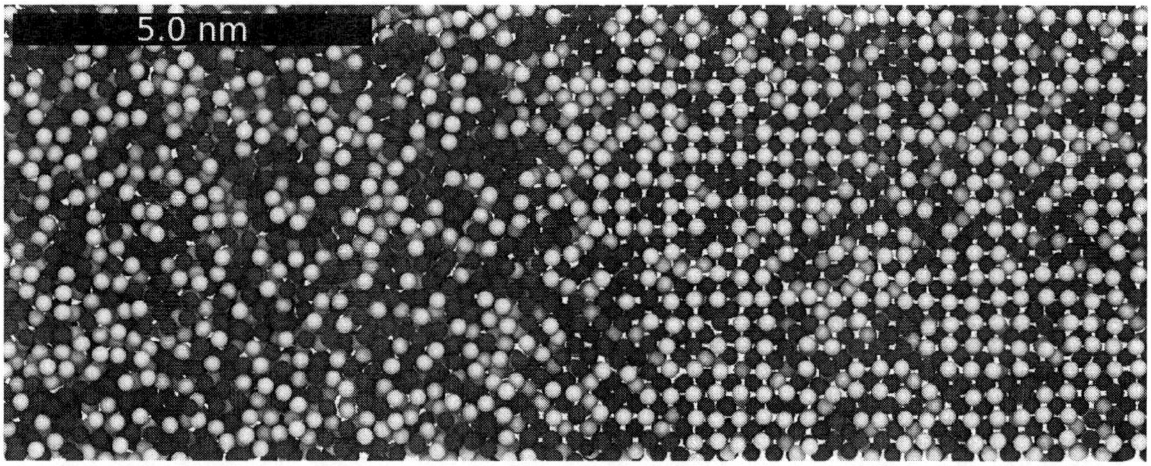

Figure 2. The structure of the coexistence of solid and liquid phases on the example of TaW system. The blue spheres correspond to tantalum atoms. The yellow spheres correspond to tungsten.

In each of the halves, we set the temperature T_i in the NPT ensemble. Then the simulation is again carried out in the whole system in the NPH ensemble. Depending on how closely we guessed the T_i, there may be different behavior in the evolution of such a system. If the phase boundary remains stationary, the melting point is equals to T_i. If the phase boundary shifts towards the crystalline component, but at the same time there remains at least some crystalline phase in the system, the melting point will be equal to the new steady-state temperature in the system. Similarly, if part of the liquid crystallizes at the phase boundary, then the melting point will be equal to the new steady-state temperature in the system. It is also possible that the whole system become completely liquid or a completely crystallized. In these cases, we can only say that the melting point is lower or higher, respectively, than the chosen initial temperature. And the simulation needs to be done again with the modified value of T_i. With this approach, the larger the size of the simulated system, the less accurately you need to be to guess the initial temperature T_i. The values obtained in this way agree fairly accurately with the values T_m^{eam}.

2.2 Methods for ab-initio modeling

For ab-initio modeling, the VASP software package was used. Model systems contained from 100 to 200 atoms. PAW (projector augmented wave) pseudopotentials were used for the simulation. The integration step was equal to 1 femtosecond.

The methods used for EAM modeling are only partially suitable for first-principle modeling. This is due to the fact that computational complexity in ab-initio modeling grows as the third power of the number of valence electrons in the system. Therefore, long-time system modeling with more then 500 atoms using first-principle methods is still considered challenging task even for powerful computing systems.

The first approach to measure the melting point is to gradually heat the system in canonical ensemble (NVT) until reaching a completely liquid state. For this for each metal, a crystal lattice corresponding to it was set at 300 K and simulation was performed for 0.1 picoseconds and then the temperature increased abruptly by 50 K and calculations were again performed for 0.1 picoseconds and so continued until a fully molten state was reached. The fact that the volume remains constant makes the melting process easier, however, the melting point obtained in this way will correspond to a rather strongly increased external pressure. Such an approach can only serve as a rough estimate of the melting point.

The second approach for a more accurate calculation of the melting point is a variation of the solid-liquid coexistence method. As well as during EAM modeling, a system consisting of a certain amount of a substance in a liquid and crystalline states was created. Due to computational limitations, the number of atoms in the system was less than 200. For such a small system, reaching an equilibrium state in which both the crystalline and liquid phases would remain is almost impossible. Thus for each compound, a series of experiments was carried out in some of which, the system went into a completely liquid state, and the others became completely crystalline. Temperature equilibrium of a liquid and a solid with such an approach will be that temperature at which the transition to both completely liquid and fully crystalline state is statistically equally likely. The melting points obtained by this method for pure metals $T_m^{ab\text{-}in}$ are listed in the table 1. It should be taken into account that the error of this approach is about 50 K, but can be reduced by increasing the statistical base, i.e., carrying out a larger number of computational experiments.

Whether the structure is liquid or crystalline was determined by the method of CNA (common neighbor analysis) [5], which allows to determine the belonging of atoms to a cubic face-centered lattice, a cubic body-centered lattice and a hexagonal close-packed lattice (and all the pure metals we consider have one of the above structures). The system was considered liquid when the number of atoms defined as crystalline by CNA method became less than 5%.

3. Results and discussion

The experimental melting points T_m^* are taken from [6]. As can be seen, the use of EAM potentials gives rather far melting points from real values. The difference between T_m^{eam} and T_m^* is from 74 K to 443 K. As can be seen, the difference between the temperature $T_m^{ab\text{-}in}$ and the experimental values is about 50 K, which fits into the error of the calculation method.

Table 1. Melting points calculated by various methods and the experimental values T_m^*

	Fe	Pd	Ti	Pt	Zr	Mo	Ta	W
T_m^{eam},K	2073	1704	1678	1598	1840	3263	3364	4035
$T_m^{ab\text{-}in}$,K	1850	1850	1950	2050	2150	2900	3300	3700
T_m^* , K	1812	1828	1943	2041	2128	2896	3290	3695

The EAM modelling requires potentials of cross-pair interaction that can not always be found in the literature. For example, the calculated melting point for titanium carbide using the potential from work [7] was equal to 3400 K, which is consistent with the experimental data. And the melting point of tantalum carbide calculated by the potential developed by a method similar to the method in work [7] was equal to 3800 K, which is about 350 degrees lower than the experimental value. Ab-initio modeling using the method described above gives a melting point 3450 K for TiC and 4150 for TaC, which agrees with experimental values, with an error of 50 degrees.

4. Conclusions

EAM modeling is widely used to study the processes occurring in some systems, including it can be useful for studying the processes of synthesis and consolidation of materials. This is due to the fact that it is possible to model systems consisting of millions of atoms in a reasonable time. But for accurate measurement of the melting point, like many other physical quantities, EAM modeling is not the best choice. Ab-initio simulation gives melting points quite close to experimental data. The main disadvantage is the smallness of the simulated system, due to the increasing complexity of the calculation. Another advantage of ab-initio modeling is the presence of pseudopotentials for all elements of the periodic system.

Acknowledgments
This work was supported by the Russian Foundation for Basic Research, project no. 18-33-00641.

References

[1] Hong Q van de Walle A 2015 *Phys. Rev.* B **92** 020104(1–6).
[2] Wang L G van de Walle A 2012 *Phys. Chem. Chem. Phys.* **14** 1529–34
[3] Plimpton S J 1995 *Comput. Phys.* **117** 1–19

[4] Zhou X W, Johnson R A and Wadley H N G 2004 *Phys. Rev.* B **69** 1–10
[5] Faken D and Jonsson H 1994 *Comput. Mater. Sci.* **2** 279–286
[6] Gale W F and Totemeier T C 2003 *Smithells Metals Reference Book* pp 2080
[7] Kim H K, Jung W S and Lee B J 2009 *Acta Mater.* **57** 3140–47

IOP Conf. Series: Materials Science and Engineering **558** (2019) 012039 doi:10.1088/1757-899X/558/1/012039

Experimental studies of the fundamental mechanism for phase formation in reactive solutions toward creation of the functional materials

S I Roslyakov[1], D O Moskovskikh[1], A A Nepapushev[1], K V Kuskov[1], G V Trusov[1], A S Sedegov[1], A S Mukasyan[2,1] and K V Manukyan[3]

[1]National University of Science and Technology "MISIS", Leninskii 4, Moscow, 119049 Russia
[2]University of Notre Dame, Department of Chemical and Biomolecular Engineering, Notre Dame, Indiana 46556, United States
[3]University of Notre Dame, Department of Physics, Notre Dame, Indiana 46556, United States

E-mail: roslyakov.si@misis.ru

Abstract. In this work, we examined several modifications of solution combustion synthesis (SCS) method to produce new nanoscale non-oxide phases (nitrides, intermetallics, and alloys). In particularly two approaches were outlined: (i) so-called, catalytic SCS, where existence of one metal promotes reduction during combustion; (ii) synthesis by using media, which contains both reactive solution and solid metals, i.e. combination of solution and heterogeneous combustion.

1. Introduction

Solution combustion synthesis (SCS) has attracted considerable attention of scientists from around the world for versatile synthesis of a variety of nanoscale materials [1–7]. SCS is based on generic idea of using exothermic self-sustained (combustion) reactions, which can be initiated in solutions of oxidizer (typically metal nitrate) and organic fuel (glycine, urea, citric acid etc.). One of the unique advantages of SCS is the ability to prepare high-quality multi-element compounds with complex crystal structures, such as garnets, perovskites, spinels, silicates, phosphates, and others [4, 6, 8, 9]. The mixing of reactants at the molecular level enable efficient doping of materials, even with a trace amount of elements. Proper selection of fuel and optimization of the process allows for avoidance of secondary calcination step and prepare materials in one macroscopic stage. Finally, the process can be performed in continuous schemes providing high throughput synthesis of nanoscale materials [3].

A limitation of SCS is related to the relatively low level of control over the morphological uniformity of the fabricated materials. The rapid increase in temperature leads to spontaneous nucleation, growth, and agglomeration of the products' particles [1, 5]. The most critical role in determining the morphology of the final products is played by maximum temperature and gas release. Another limitation of synthesis related to chemical nature of the mixture is preparation of predominantly oxide materials. It is well-known the high temperature decomposition of metal nitrates leads to the formation of corresponding metal oxides [10, 11]. However, recently it was demonstrated that this method also allows synthesis of pure metals (Ni, Cu, Co, Pt etc.) and alloys (NiCu, NiCo etc.). It was found that under certain conditions (typically excess of fuel) the intermediate gaseous combustion products have a reduction (typically hydrogen-based) nature and thus allow reduction of metal oxide to form metals or alloys during SCS [12].

These promising results suggest that by understanding the fundamentals of combustion in reactive solutions we will able to synthesize other compounds including intermetallics, carbides, nitrides and others, thus significantly widen specter of SCS materials. It is worth noting that since reactants are mixed on molecular level in the solutions it is typically leads to formation of high surface area, nano-structured powders with small crystallites, which is not easy to produce by other techniques. Such materials already found applications in different fields, including catalysis, microelectronics, are used for further fabrication of nanostructured ceramics and alloys [1, 13–17]. In the above context, it is difficult to overestimate the importance of the research targeting routes for spreading the application of SCS to produce wide variety of non-oxide based compounds.

In this work, we examined several modifications of SCS method to produce new non-oxide phases (nitrides, intermetallics, and alloys). In particularly two approaches should be outlined: (i) so-called, catalytic SCS, where existence of one metal promotes reduction during combustion; (ii) synthesis by using media, which contains both reactive solution and solid metals, i.e. combination of solution and heterogeneous combustion.

2. Results and Discussion

2.1. Iron-based systems: SCS of Iron nitride

Reactive solutions with different fuel (hexamethylenetetramine, $C_6H_{12}N_4$) to oxidizer (iron nitrate nonahydrate, $Fe(NO_3)_3{\cdot}9H_2O$) ratio (φ) were investigated. In all cases after drying of the solutions reactions were carried out in inert (argon) atmosphere and were locally initiated by "hot" tungsten wire. The typical temperature-time profiles of the combustion wave for different solutions are shown in figure 1. As it is expected an increase of φ value leads to the decrease of the maximum combustion temperature. Indeed in inert atmosphere the highest combustion temperature corresponds to the stoichiometric ration $\varphi = 1$. It can be also seen that maximum combustion temperature is in the range 900–1600 K and total duration of the reaction (calculated from the rapid increase of the temperature to its decrease) is in the range 2–3 s.

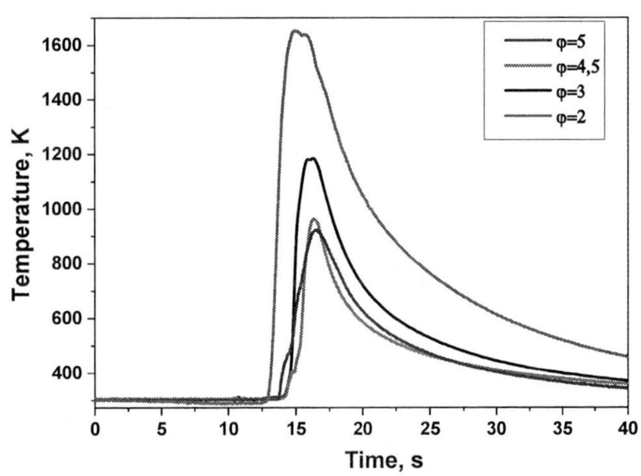

Figure 1. Time-temperature profiles of the combustion wave for solutions with different fuel to oxidizer ratio (φ)

The XRD data for the combustion products obtained for different reactive solutions are presented in figure 2. It can be seen that at $\varphi = 2$ the main products are Fe_2O_3 and FeO oxides, while essentially pure FeO was obtained for solutions with $\varphi = 3$. However, further increase of amount of

IOP Conf. Series: Materials Science and Engineering **558** (2019) 012039 doi:10.1088/1757-899X/558/1/012039

hexamethylenetetramine (φ = 4.5; 5) leads to the formation of iron nitride (Fe_3N). Thus we were able to obtain essentially oxygen free combustion product but it required large amount of carbon/nitrogen containing fuel ($C_6H_{12}N_4$). These elements reacted with iron forming nitrides and carbonitrides.

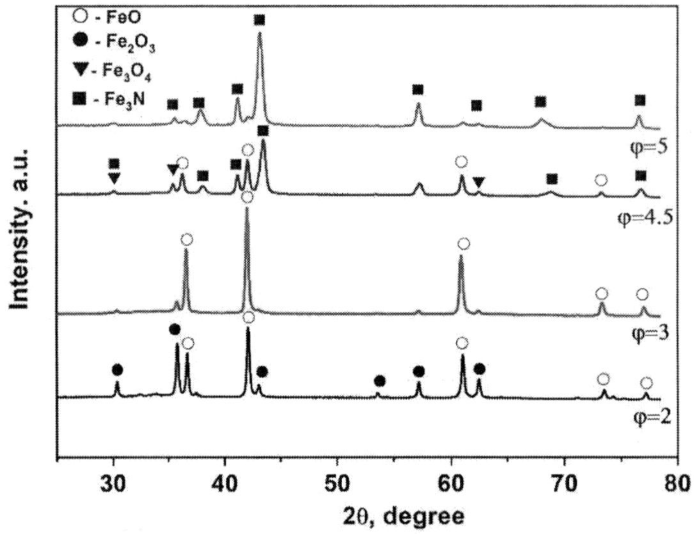

Figure 2. XRD patterns for combustion products after SCS with different fuel to oxidizer ratio (φ).

Microstructure for obtained powders are shown in figure 3. It can be seen that increase φ, while leads to decrease of combustion temperature (figure 1), results in increase of the size of the obtained powder agglomerates (figure 3a), as well as size of the individual grains (figure 3b). The morphology of the individual grains also changes: from less than 100 nm spherical type particles (FeO) for φ =3, to platelet 500 nm particles (Fe_3N).

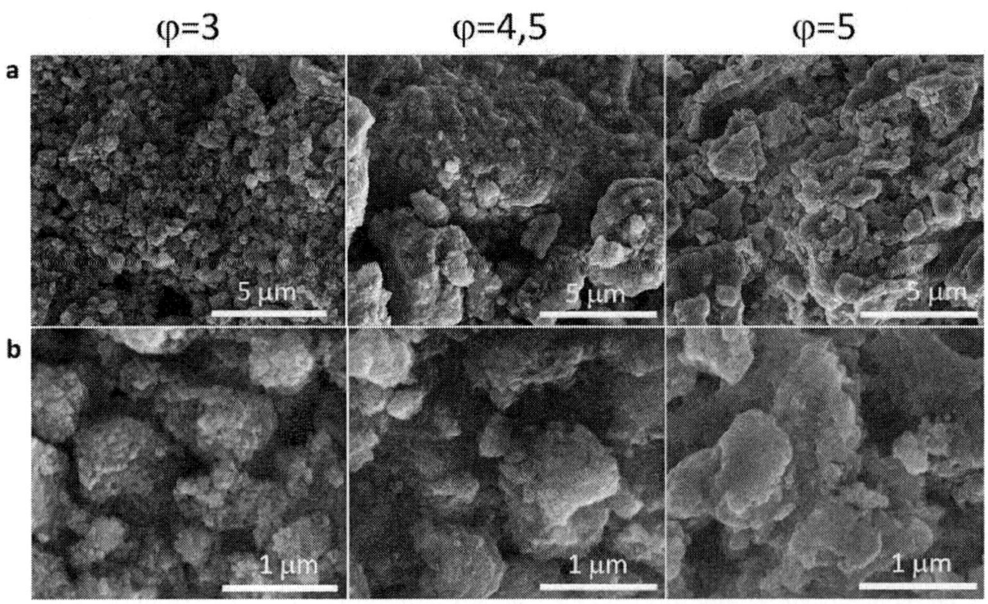

Figure 3. Microstructure for combustion products after SCS with different fuel to oxidizer ratio (φ).

2.2. Iron–Nickel based systems: SCS of NiFe Alloy

Again reaction was organized in so-called self-propagating mode when the reactive solution was locally preheated by "hot" tungsten wire and combustion front propagates along the reactive media in self-sustained manner. Typical time-temperature profile of the combustion wave shows that maximum reaction temperature is about 1100 K, which is well above boiling point of all used precursors. The duration of the reaction, which can be estimated from the time of rapid increase of the temperature to the beginning of the cooling stage, does not exceed 3 s. The XRD patterns of the combustion product is presented in figure 4. Only one sets of peaks was detected, which corresponds to FeNi alloy (figure 4a).

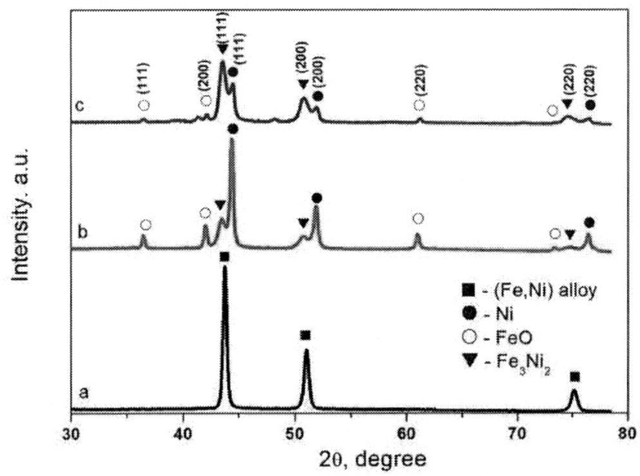

Figure 4. XRD patterns for combustion products in: (a) iron nitrate + nickel nitrate + hexamethylenetetramine system; (b, c) nickel powder + iron nitrate + hexamethylenetetramine system.

SEM images of the microstructure of the obtained powders with corresponding results of EDS analysis are shown in figure 5. It can be seen that as-synthesized alloys has morphology of porous agglomerates. It is important to note that the amount of oxygen in the alloys is very small.

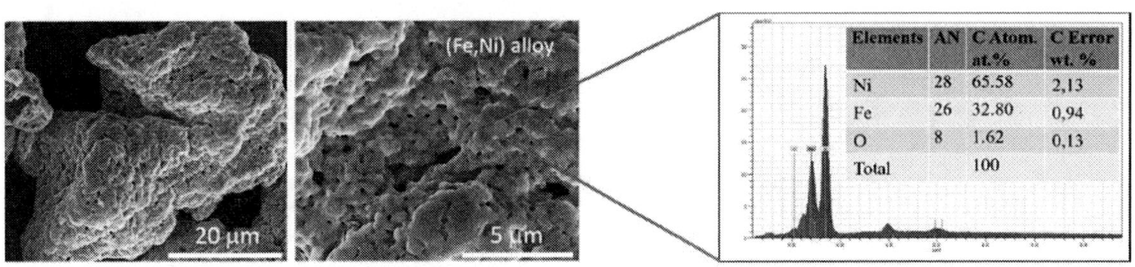

Figure 5. Microstructure and EDS analysis of obtained NiFe alloy.

Thus, while we could not completely reduce iron oxide to iron in iron nitrate - hexamethylenetetramine system, we were able to obtain essentially oxygen free NiFe alloy in complex metal nitrates (iron nitrate and nickel nitrate) - hexamethylenetetramine system. It is interesting to figure out the role Ni in the reduction process of iron oxide.

To light a shade on this question the following two experiments were conducted: different amount (0,7 g and 0,2 g) of Ni powders (particle size 4–5 µm) were poured into the 5 ml of aqueous solution of iron nitrate and hexamethylenetetramine ($\varphi = 3$). As it was shown above, the combustion of such solution without addition of Ni resulted in formation of FeO phase (figure 2). Initiation of reactions in Ni-powder/solution system in inert atmosphere were accomplished by local preheating of solution. The XRD pattern for thus obtained combustion product is shown in Figure 4 b,c. It can be seen that the main

IOP Conf. Series: Materials Science and Engineering **558** (2019) 012039 doi:10.1088/1757-899X/558/1/012039

newly produced phase is Fe_3Ni_2 alloy. Taking into account that the only highly exothermic reaction in such system is the reaction between Fe-nitrate and $C_6H_{12}N_4$, which without presence of Ni leads to the formation of FeO, we may suggest that Ni catalyzes the reduction reaction. Indeed, it is know that in presence of Ni iron oxide reduced faster by both carbon and hydrogen [18]. From the standpoint of the combustion synthesis method, such approach, of mixing of reactive solutions with fine reactive or catalytic solid powders, open a new venue for fabrication of the variety of materials, e.g. intermetallics.

2.3. Aluminum–Nickel based systems: SCS of NiAl intermetallics

Two chemical routes have been investigated to synthesized Ni_xAl_y intermetallics in the solution combustion wave. First, we try to optimize reaction conditions, similarly to that for Fe–Ni system, in the solution of metal nitrates (aluminum nitrate and nickel nitrate) and the oxygen free fuel ($C_6H_{12}N_4$) in the wide range of fuel to oxidizer ratio (φ = 1.5–3). Typical XRD patterns of the combustion products in such solution demonstrate that pure well crystalline Ni and amorphous Al_2O_3 phases were formed.

The second route consider idea of impregnated active layer [10]. The following experiments were conducted. The aqueous solution of nickel nitrate and fuel (glycine) was prepared and then impregnated into specific amount of solid Al powder. Reaction in such heterogeneous media was initiated in the volume combustion mode (uniform preheating) or by local preheating to accomplish the self-propagating combustion mode.

The XRD patterns for combustion products obtained after SCS in the volume combustion mode are shown in figure 6. It can be seen that one of the main parameters that control phase composition of the synthesized materials is an amount of water used. If the initial amount of water is relatively large, e.g. 100:1 wt.% for H_2O-Al ratio, then even applying intensive stirring we could not reach sufficient contact between the initially solid Al and formed in combustion wave Ni particles and the product consists of pure Ni and Al phases. Decrease amount of water allows us initially form essentially solid media, with solution impregnated into the porous Al skeleton. In this case, aqueous solution spread along the Al particles forming thin reactive layer. Volume combustion in such media is extremely vigorous and results in formation of intermetallic (Al_3Ni_2) phase. By changing Al to Ni ratio as well as fuel to oxidizer ratio one may control the phase composition of the obtained materials.

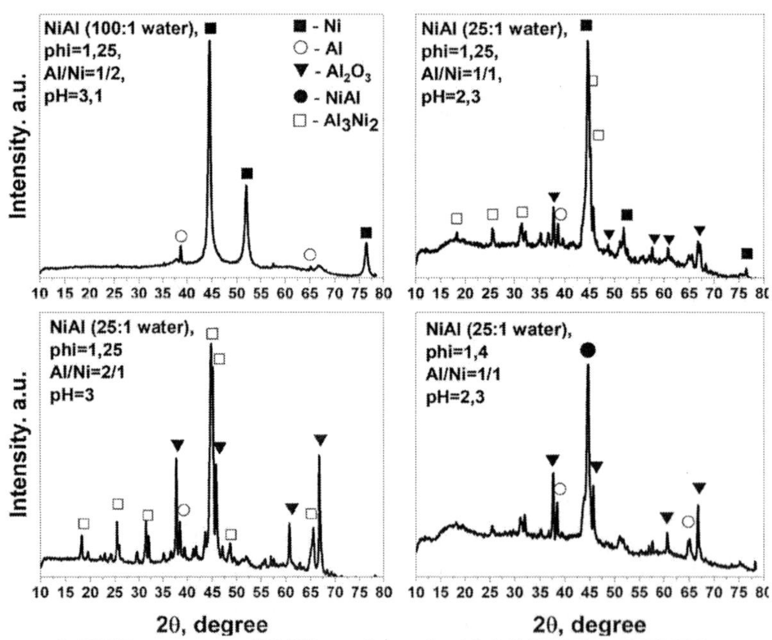

Figure 6. XRD patterns of SCS products in Al–$Ni(NO_3)_2$–$C_2H_5NO_2$ system; Volume combustion synthesis mode.

Based on the above findings additional experiments were conducted in self-propagating mode. In this case, the reactive solution of Ni nitrate and glycine (φ =1,75) was impregnated with Al:Ni ratio 3:2 into porous skeleton formed by Al powder. Thus prepared media was dried in the oven, under the following conditions (vacuum, T = 450 K; t = 2 h). Obtained solid samples were uploaded to the reactor and burned in self-propagating mode under Ar atmosphere. The XRD data on the obtained product demonstrates that in this case the main formed intermetallic phase is NiAl and essentially zero counts for alumina phase was detected.

3. Conclusions

To conclude, it was demonstrated that based on the knowledge of the combustion mechanism and using of the inert environment, oxygen free fuel and combination of reactive solutions and solid precursors, one may significantly widen the range of synthesized compounds by using SCS. Not only pure metals and metal alloys, but also intermetallics and metal nitrides can be thus fabricated.

Acknowledgments
The reported study was funded by RFBR according to the research project no. 18-03-00306.

References

[1] Varma A, Mukasyan A S, Rogachev A S and Manukyan K V. 2016 *Chem. Rev.* **116** 14493–14586
[2] Yu X, Smith J, Zhou N, Zeng L, Guo P, Xia Y, Alvarez A, Aghion S, Lin H, Yu J, Chang R P H, Bedzyk M J, Ferragut R, Marks T J and Facchetti A 2015 *Proc. Natl. Acad. Sci.* **112** 3217–22
[3] Patil K C, Aruna S T and Mimani T 2002 *Curr. Opin. Solid State Mater. Sci.* **6** 507–12
[4] Wen W, Yao J-C, Jiang C-C and Wu J-M 2017 *Int. J. Self-Propag. High-Temp. Synth.* **26** 187–98
[5] Nersisyan H H, Lee J H, Ding J R, Kim K S, Manukyan K V and Mukasyan A S 2017 *Prog. Energy Comb. Sci.* **63** 79–118
[6] Pendashteh A, Rahmanifar M S, Kaner R B and Mousavi M F 2014 *Chem. Commun.* **50** 1972–5
[7] Specchia S, Ercolino G, Karimi S, Italiano C and Vita A 2017 *Int. J. Self-Propag. High-Temp. Synth.* **26** 166–86
[8] Suresh K, Kumar N R S and Patil K C 1991 *Adv. Mater.* **3** 148–50
[9] Pandozzi F, Vetrone F, Boyer J C, Naccache R, Capobianco J A, Speghini A and Bettinelli M 2005 *J. Phys. Chem. B* **109** 17400–5
[10] Mukasyan A S, Epstein P and Dinka P 2007 *Proc. Comb. Inst.* **31** 1789–95
[11] Deshpande K, Mukasyan A and Varma A 2004 *Chem. Mater.* **16** 4896–904
[12] Podbolotov K B, Khort A A, Tarasov A B, Trusov G V, Roslyakov S I and Mukasyan A S 2017 *Comb. Sci. Tech.* **189** 1878–90
[13] Rajeshwar K and De Tacconi N R 2009 *Chem. Soc. Rev.* **38** 1984–98
[14] Aruna S T and Mukasyan A S 2008 *Curr. Opin. Solid State Mater. Sci.* **12** 44–50
[15] Schuyten S, Dinka P, Mukasyan A S and Wolf E 2008 *Catal. Lett.* **121** 189–98
[16] Cross A, Kumar A, Wolf E E and Mukasyan A S 2012 *Ind. Eng. Chem. Res.* **51** 12004–8
[17] Nagaveni K, Sivalingam G, Hegde M S and Madras G 2004 *Appl. Catal. B Environ.* **48** 83–93
[18] Il'chenko N I 1972 *Russ. Chem. Rev.* **41** 47–53

IOP Conf. Series: Materials Science and Engineering **558** (2019) 012040 doi:10.1088/1757-899X/558/1/012040

Electroconductivity of Al_2O_3/graphene nanocomposite processed by SPS technique

A Sanchez-Bolinchez[1], E Klyatskina[1], F Segovia-Lopez[1], A G Zholnin[2] and V V Stolyarov[2,3]

[1]Department of Mechanical and Materials Engineering, Universitat Politècnica de Valencia, Spain
[2]National Research Nuclear University MEPhI, Moscow, Russia
[3]Mechanical Engineering Research Institute, RAS, Moscow, Russia

E-mail: vlstol@mail.ru

Abstract. Electrical conductivity (σ), relative dielectric permittivity (ε_r) and dissipation factor (D) measured in graphene-alumina composites. Samples obtained by plasma spark sintering (SPS) from a mixture of raw powders: δ-alumina (36 nm average particle size) and graphene flakes (3 nm thickness and 2–3 μm length). Graphene content in samples was 0, 1 and 2% by weight. The study carried out for frequencies from 50 Hz to 100 kHz. Both σ and ε_r were higher for Al_2O_3–2% graphene: up to 90 μS/m and 19 respectively; while alumina with 1% graphene showed similar values to the pure alumina samples: 50 μS/m to electrical conductivity and 16 to relative permittivity. The dissipation factor was similar in the three materials tested. D increased with the frequency, reaching high values (0.7) at 100 kHz. Composites with 1 and 2% graphene content showed a dissimilar dielectric behavior with the frequency. Alumina reflected a classical behavior of the permittivity dependence with the frequency. Graphene composites also show the same behavior at frequencies above 100 Hz. Below this frequency, the presence of graphene increases the relative permittivity to exceed that from pure alumina. The graphene content leads to rise of relative permittivity, which means easier polarizability.

1. Introduction

The study of the properties of ceramic composite materials is one of the priority trends in the development of modern materials science. Of particular interest is the introduction of a two-dimensional structure in the graphene composite. This chemically inert material has unique properties: very high electronic conductivity, thermal conductivity above 5000 W/mK and an elastic modulus close to 1 TPa, due to the large specific surface area of graphene plates [1–3].

The method of plasma spark-sintering (SPS) is one of the most used for the production of composite ceramic materials with a nanostructure, since it significantly reduces the sintering time. This makes it possible to avoid thermally activated structure-phase transformations in the nanopowder and to obtain a non-porous material [4, 5].

Al_2O_3 ceramics is the most studied material for a matrix of composites with graphene due to its wide use as a structural material in the automotive, aerospace and biomedical fields.

The present study of Al_2O_3/graphene composites is associated with the problem of increasing the electrical conductivity of aluminum oxide. It is assumed that the mechanism of the occurrence of conductivity can be associated with overlapping of the surfaces of graphene plates, which form a more efficient network than graphene nanotubes, in which the contact is point like and leads to a higher resistance. When a sufficiently high electrical conductivity of 1 S/m is reached, this material can be used for the manufacture of complex ceramic parts by the methods of electric spark machining [6]. The authors of this paper investigated the dependence of the electrical conductivity measured on alternating current on the content of graphene and showed that when graphene is added above 8 wt %, the required electrical conductivity is achieved. With a graphene content below the percolation limit the electrical conductivity did not exceed 10^{-9} S/m. The authors of another work carried out similar studies of the electrical conductivity at a direct current of a composite with 2 wt % graphene and showed that the electrical conductivity was 10^{-2} S/m [7]. Such a large difference in the results may be due to different methods of measuring electrical conductivity. The goal of this paper is to investigate the effect of graphene additives in the range from 0 to 2 wt % on the electrical properties of the Al_2O_3/graphene nanocomposite in the alternating current measurement process.

2. Experimental method and materials

Nanopowders δ-Al_2O_3 and graphene were used in the work. The starting powder δ-Al_2O_3 was obtained by oxidizing disperse aluminum in air plasma (IMET RAS). The average particle size was 36 nm, the surface area was 35–40 m^2/g. The particles of the original powders are rounded, but most of the particles are collected in agglomerates. Graphene was produced by the method of ultrasonic exfoliation by the firm Graphene-tech (Spain) and was not oxidized flakes whose thickness and length were 3 nm and 2–3 μm, respectively. A mixture of the components of the composite was ultrasonically dispersed in dimethyl formamide using a Sonicator Q500 ultrasound disperser at a power of 250 W. The optimum dispersion time was determined experimentally and was 40 min. The resulting suspension was air-dried by heating and constant stirring for 1 h, and then triturated in a mortar.

SPS powder mixtures were conducted in a vacuum on a Labox-625 SinterLand in a graphite matrix with an internal diameter of 15 mm. To reduce heat losses due to radiation and to reduce the temperature difference between the matrix surface and the sintered object, the matrix was enveloped by a graphite felt 6 mm thick with a hole for the pyrometer. Thus, the pyrometer measured the temperature of the lateral outer surface of the matrix at the level of location of the sintered object. The sintering regime was chosen empirically for pure alumina from the condition of obtaining a high homogeneity of compacts with a maximum average microhardness of their surface [8]. The heating rate was 100°C/min; the maximum heating temperature measured by the pyrometer on the surface of the matrix was 1550°C; holding at maximum temperature was 10 min. To the punches a pressure of 50 MPa was applied. To prevent welding to the mold, the sintered powder is isolated from it by graphite paper. The thickness of the sample disk after sintering was 2 mm. The same sintering regime was also used in the preparation of composites with a content of graphene by mass of 0, 1, and 2%. The electrical properties, the relative permittivity (ε_r) and the loss tangent (D), in the thickness direction of sintered tablets, were measured in an alternating electric field from 50 Hz to 100 kHz using precision Hameg 8118 and Agilent 4294A devices with an accuracy of 0.05%.

3. Experimental results

Figure 1 shows the frequency dependences of the electrical properties of ε_r and D of a sintered composite with various graphene contents. In the absence of graphene (0%), these dependences reflect the classical behavior of the dielectric - a decrease in the dielectric constant (figure 1a) due to the disconnection of one of the polarization mechanisms (electronic or ionic) and an increase in dielectric losses (figure 1b). The samples with graphene show the same dielectric behavior for frequencies above 100 Hz.

The increase in the graphene content is manifested in two ways. At low frequencies, up to 100 Hz (figure 1a), the presence of graphene leads to dielectric constant values less than one of the pure alumina which indicates the appearance of conductivity. For higher frequencies the increasing in graphene content leads to the rise of the relative permittivity which means an easy sample polarizability.

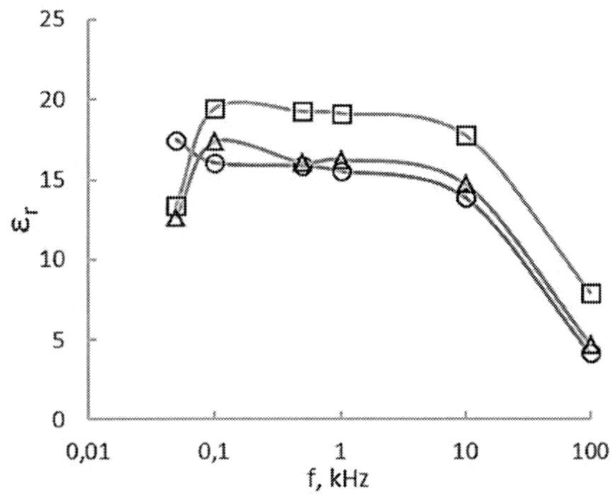

(a)

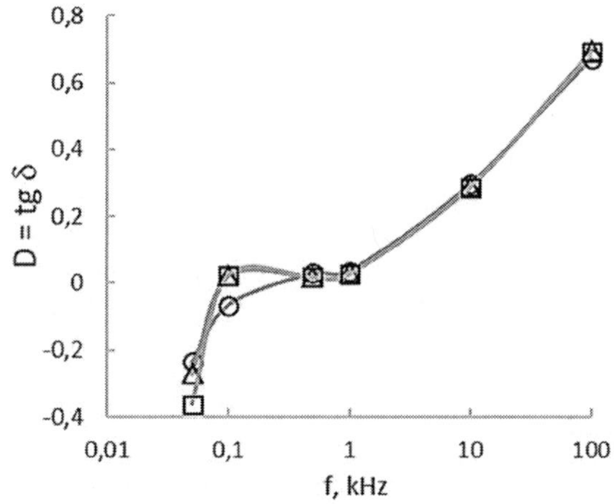

(b)

Figure 1. Dependence of the (a) relative permittivity and (b) dielectric loss of the composite on the field frequency and graphene content. ◯ 0%; △ 1%; □ 2%

The curves for *D* (figure 1b) reflect an almost monotonic increase with frequency, and are very similar for more than 700 Hz in all samples. The coincidence of all three curves for frequencies above 700 Hz and their consistency with the curves of the relative permittivity indicates the prevalence of dielectric losses in aluminum oxide. The curves from the samples with graphene indicate the existence of ohmic conductivity below 700 Hz.

The composites conductivity (σ) was calculated as the real part of their admittance: $\sigma = (2\pi f \varepsilon_r \varepsilon_0)/D$. Its dependence on the frequency and graphene content (figure 2) show that from the 0% of graphene the value of σ is quite high. It can be explained by the presence of traces of carbon due to the details of the SPS method. The growths of the conductance with the frequency is a quite general trend for the composite dielectrics [9].

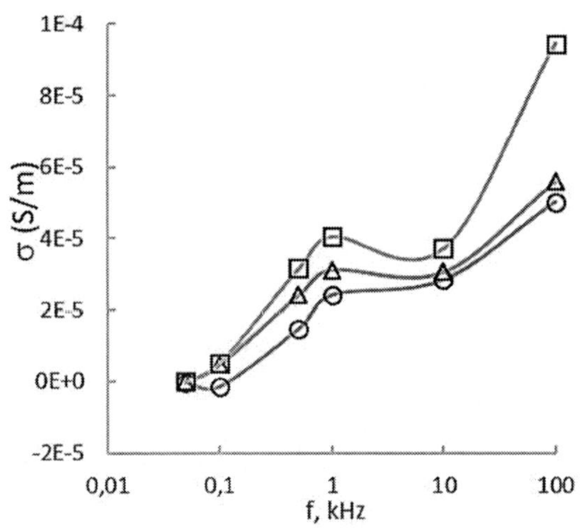

Figure 2. Dependence of the composite relative conductivity on the field frequency and graphene content. ◯ 0%; △ 1%; □ 2%.

4. Conclusions

The effect of the additive graphene and the frequency of alternating current on the electrical properties of SPS composites is established. It is shown that at frequencies up to 70 Hz, the addition of graphene up to 2% to the nanopowder of aluminum oxide can increase the electrical conductivity of the sintered composite in comparison with the value for pure corundum. This is manifested in a decrease in the dielectric constant and an increase in the dielectric loss. An increase in the frequency of the current in the region up to 100 kHz promotes a stronger increase in electrical conductivity, with the effect of graphene weakening or even absent.

Acknowledgments

Authors wishing to acknowledge the financial support of the Ministry of Education and Science of the Russian Federation, project no. 11.1957.2017/4.6 (composite compaction processing); RNF (properties investigation), grant no. 16-19-10213

References

[1] Markandan K, Chin J and Tan M 2017 *J. Mater. Res.* **32** 84–106
[2] Nieto A, Bisht A, Lahiri D, Zhang C and Agarwal A 2017 *Int. Mater. Rev.* **62** 241–302
[3] Miranzo P, Belmonte M and Osendi M 2017 *J. Eur. Ceram. Soc.* **37** 3649–72
[4] Tubío C, Rama A and Gomez M 2018 *Ceram. Int.* **44** 5760–67
[5] Kostecki M, Grybczuk M, Klimczyk P, Cygan T, Woźniak J, Wejrzanowski T, Jaworska L, Morgiel J and Olszynaet A 2016 *J. Eur. Ceram. Soc* **36** 4171–79
[6] Malek O, González J, Vleugels J, Vanderauwera W, Lauwers B and Belmonte M 2011 *Mater. Today* **14** 496–501
[7] Zholnin A, Klyatskina E, Grigoriev E, Salvador M, Misochenko A, Dobrokhotov P, Isaenkova M, Sinaysky M and Stolyarov V 2018 *Inorg. Mater: Appl. Res.* **9** 498–503
[8] Zholnin A, Kovaleva I, Rytenko V, Pahilo-Daryal I, Litvinova I, Goltsov V and Grigoriev E 2016 *Phys. Chem. Treat. Mater.* **2** 71–77
[9] Almond D, Bowen C and Rees D 2006 *J. Phys. D: Appl. Phys.* **39** 1295–1304

Mill scale recycling by SHS metallurgy for production of cast ferrosilicon and ferrosilicoaluminium

V N Sanin, D M Ikornikov, D E Andreev, N V Sachkova and V I Yukhvid

Merzhanov Institute of Structural Macrokinetics and Materials Science, Russian Academy of Sciences, Chernogolovka, 142432 Russia.

E-mail: svn@ism.ac.ru

Abstract. The possibility of utilizing the industrial wastes of metallurgical production and secondary raw materials (aluminum) by organizing self-propagating high-temperature synthesis (SHS) and obtaining cast ferroalloys (Fe–Si; Fe–Si–Al, and Fe–Si–Al(Cr, Mn)) was investigated. The main production operations of the preliminary preparation of the starting raw materials were worked out and cast ferroalloys providing high concentration Si, Al and B were revealed. The possibility of processing the industrial wastes of metallurgical production and obtaining cast ferroalloys was shown.

1. Introduction.

The metallurgical industry is one of the most massive material forming industries. High amount of by-products in the form of mill scale sludge is generated every year in each section of steel making plants due to hot rolled metal processing at elevated temperatures. Mill scale, often called as scale, is the flaky surface consisting of the iron oxides iron (II) oxide (FeO), iron (III) oxide (Fe_2O_3), and iron (II,III) oxide (Fe_3O_4, magnetite). It can be considered as a valuable metallurgical raw material for iron and steel making industry [1, 2]. The strict environmental regulations as well as the valuable metallic content cause the ability to recycle mill scales become an important issue. Currently, many metallurgical plants have accumulated high amounts of mill scale in the dumps containing oils (4–12%) and moisture (4–18%). After the removal of oil and water, this material contains up to 70% iron with a minimum amount of sulfur compounds and can be effectively used in both blast and steel smelting operations. The solution of this task is becoming more effective due to the world deficit and the constant growth of the cost of iron ore, as it allows to reduce raw materials dependence and improve the ecology. So, there has been a big tendency to recover the valuable iron content in the production process.

Accordingly, to recycle the iron content wastes, briquettes are smelted using blast furnace or electric arc furnace (EAF) [3]. These processes allow better utilization of iron oxide, but are not cost-effective. For this reason, energy efficient technologies should be used. It is also required to expand the nomenclature of products obtained from the recycled scale.

In this regard, the present study was aimed at the evaluation of the possibility, efficiency, and consequences of the reduction of mill scale by combustion. A promissory route to production of cast ferroalloys is the SHS technology of high temperature melts or SHS metallurgy [4, 5]. This is an energy-saving technique due to the use of internal energy released in high-caloric combustion reactions. Previously, we first showed the possibility of obtaining cast ceramic coatings of the inside of metallic pipes by means of reprocessing the mill scale [6].

IOP Conf. Series: Materials Science and Engineering **558** (2019) 012041 doi:10.1088/1757-899X/558/1/012041

2. Experimental

2.1. Experimental Procedure of Investigations

This study involved three stages:

(i) the preliminary investigation of starting reagents (industrial mill scale) and the selection of optimal modes of their heat treatment for purification;

(ii) the study of correlation between the green mixture composition and final composition of ferroalloys (ferrosilicon, ferrosilicoaluminum and ferroboron);

(iii) the investigation of phase and chemical composition of synthesized ferroalloys and the study of the effect of experimental conditions on formation of the final composition.

In our experiments, iron oxide mill scale (Fe_xO_y) generated from steel making plants, recycled aluminum (Al), and alloying additives (AD) were used as starting reagents for preparation of cast ferroalloys. The average particle size of scale and aluminum (brand PA-4) was 50–100 μm and less than 140 μm, respectively. The weight of the green mixture was constant (1200 g). Combustion was carried out in graphite crucible 80 mm in diameter. The internal surface of graphite crucibles was laminated with Al oxide (Al_2O_3) to avoid reactive interaction between the crucible and metal melt.

The overall scheme of reaction can be represented by the following scheme:

$$\nu_1 Fe_xO_y + \nu_2 Al + \nu_3 AD \rightarrow \nu 4 Fe_xSi(B)_y + \nu 5 Al_2O_3$$

where AD is the alloying additive such as Si, SiO_2, B, and B_2O_3.

The initial compositions of mill scales contained admixtures, such as iron, oil, paraffin, etc. In order to remove the organic additives, we calcined the scales in shallow trays. It was found that such heat treatment should be carried out for 2–3 h at the temperature of 300–350°C. After calcining, the scales were refined to a particle size below 400 μm.

Figure 1 schematically illustrates the SHS process for fabrication of cast materials. After initiation with an igniting coil, the reaction frontally propagates downward. Because the attained temperatures (up to 3000°C) exceed the melting points of reaction products, the melt represents a mixture of mutually insoluble metallic (alloy) and oxide (Al_2O_3) phases. Due to strongly different specific weights, these phases undergo gravity-assisted phase separation and subsequent crystallization. As the result, the lower layer is target material (ferroalloys); the upper one, Al_2O_3.

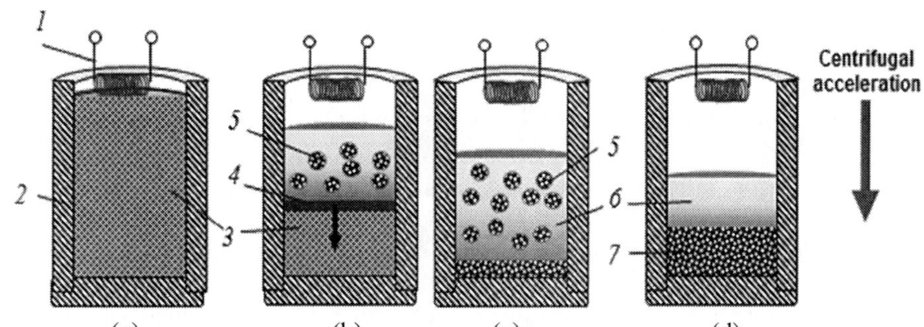

Figure 1. Schematic of the SHS process for fabrication of cast materials: (a) charge preparation, (b) combustion, (c) phase separation, (d) pattern formation. Here, *1* igniting coil; *2* refractory crucible; *3* green mixture; *4* combustion front; *5* MMC globules in the metal melt; *6* oxide phase (Al_2O_3); and *7* cast alloy (ingot).

2.2. Experimental installation

Our previous studies have demonstrated that the SHS process carried out under high gravity conditions affords the best separation of the target product (ingot) from the slag (Al_2O_3) and convective mixing of all alloy components, which becomes especially important for alloys with an increased number and concentration of components. Therefore, the synthesis of the as cast alloy under study was carried out using a centrifugal SHS setup (figure 2) with the possibility of forming the necessary synthesis conditions in a broad range of specified overloads. The main features of the SHS installation include systems for initiating the reaction mixture, controlling the rotation velocity of samples, and measuring the sample temperature after synthesis (at the cooling stage). The centrifugal installation has a working rotation range from 50 to 3000 rpm, which makes it possible to implement the overloads from 10 to 1200 g.

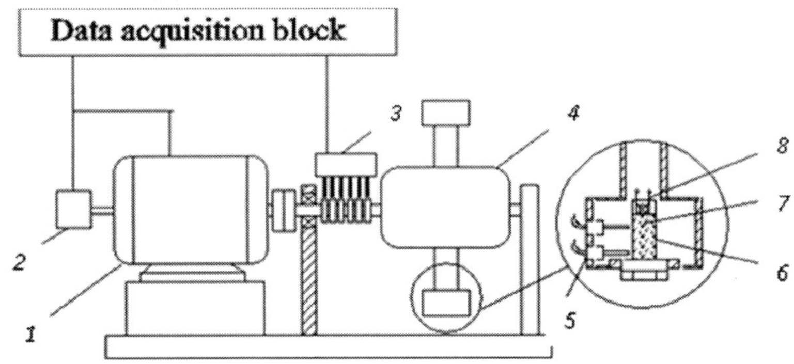

Figure 2. Experimental setup: *1* electric motor; *2* tachometer; *3* collector; *4* rotor; *5* photodiodes; *6* graphite or silica container; *7* green mixture; and *8* igniting coil.

3. Results and Discussion

3.1. Analysis of Starting Components

To calculate the ratio of the starting components, we should know the mutual concentration relation of iron and oxygen in the scale (x and y values) exactly. The analysis of the chemical and phase composition of two studied scale (table 1), which were supplied from different plants, revealed the difference in the mutual relationship of iron oxides.

Table 1. Main characteristics of the scale types.

Sample no.	$FeO:Fe_2O_3$, wt %	Particle size, mm	Impurities, wt %
1	46:48	1–5	6
2	68:28	0.5–10	4

The first party contained a larger amount of Fe_2O_3 (48 wt %), while the second one preferentially consisted of FeO. The size of scale particles varied from 0.5 to 10 mm. Both compositions had impurities including inclusions of iron, silicon oxide, alumina, and organic compounds (oil, paraffin, etc.). The presence of the latter can substantially affect the SHS process by intense gas liberation (gasifying the impurities) and can lead to the complete or partial emission of reaction products in the course of combustion. Therefore, it became necessary to carry out preliminary the calcination of scale. The use of various heat treatment modes and the subsequent analysis of the composition of the processed scale allowed us to reveal the optimal temperature and time range of calcination: $T = 300–350°C$ for 2–3 h (in the plane bottom plates). Calcination at $T < 300°C$ led to an essential increase in the calcination time (up to 10 h). At $T < 200°C$, it led only to the partial gasifying of organic impurities. The annealing at $T > 400°C$ is undesirable because the selfignition of the evaporating gases occurs in some cases.

After calcination and cooling, the scale was subjected to mechanical milling. It was found experimentally that the optimal particle size should be smaller than 400 μm. It should be noted that the most expensive component of the starting mixture is the aluminum powder. According to our analysis, it was established that replacing chemically pure aluminum (brand PA-4) by secondary one (the worked scrap) made it possible to decrease the cost of the starting mixture by more than 25%. However, the processed aluminum contains mainly impurities, the basic ones being iron (up to 5 wt %) and silicon (up to 1 wt %). It should be noted that these impurities cannot be attributed to "harmful" from the viewpoint of implementation of the SHS process. It even seems likely that their presence is desirable, since makes it possible to somewhat reduce the synthesis temperature, while the presence of silicon promotes the formation of target product (ferrosilicon and ferrosilicoaluminum).

3.2. The phase and chemical composition of synthesized ferroalloys

It is well known that complete melting of system components is a prerequisite for good sensitivity of the process to gravity forces [5, 6]. The investigation on III stage (the effect of experimental conditions on the formation of final compositions) revealed that the burning velocity (U) increases with increasing a/g (overload). The effect is most pronounced within the range of a = 10–200 g.

Strong influence of mass forces on the b urning velocity was associated wi th intensification of convection and deformation within the reaction zone, which improves the completeness of combustion reaction. An increase in a/g (where a is the artificial gravity, g is the earth gravity) is also accompanied by a decrease in material loss (η_{sput}) and an increase in the yield of target product (η_{ingot}). More than 50(±5) g, the η_{ingot} values get close to theoretical ones.

As is seen in figure 3, synthesized ferrosilicoaluminum alloys can be obtained within a wide range of parameters α Si (the mass fraction of Si in final ingot). Thus, prepared ferrosilicon samples exhibited no residual porosity that is typical of cast materials and a small contraction cavity at the center. An increase in αSi was found to slightly affect the U: it decreased from 4 to 3 cm/s.

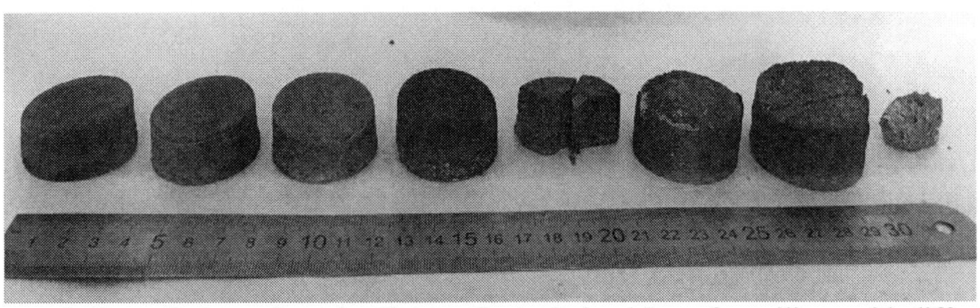

Figure 3. Overall view of combustion products derived from ferrosilicon with different Si content (Si varies in range from 10 to 45% mass).

XRD data presented in figure 4 show that cast ferrosilicon contains three (Fe_2Si, FeSi, and $FeSi_2$) phases. The increase in αSi in the initial mixture favors the formation of high concentrated ferrosilicon phase ($FeSi_2$).

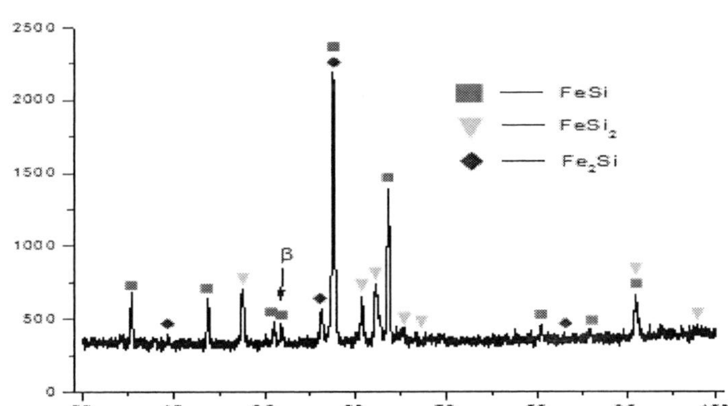

Figure 4. X-ray diffraction pattern of synthesized Fe–40Si cast ferrosilicon by SHS.

Figures 5 and 6 show the microstructure and composition of SHS produced as-cast ferroalloys by reprocessing of mill scale wastes.

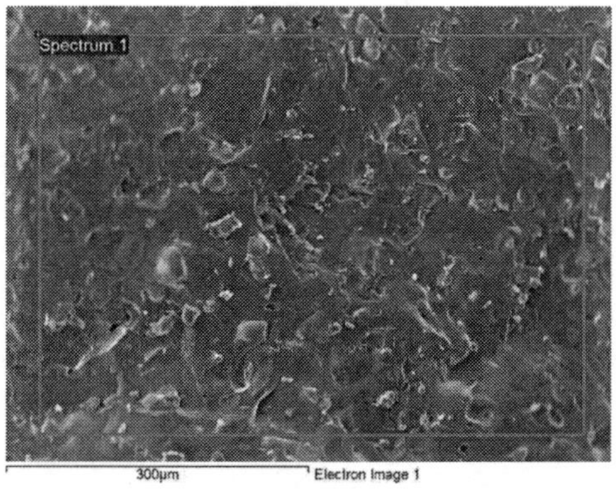

	Al	Si	Mn	Fe	Total
Sp.1	4.3	23.2	0.6	71.9	100

Figure 5. SEM image and EDS data of synthesized cast ferrosilicoaluminium Fe–Si–Al.

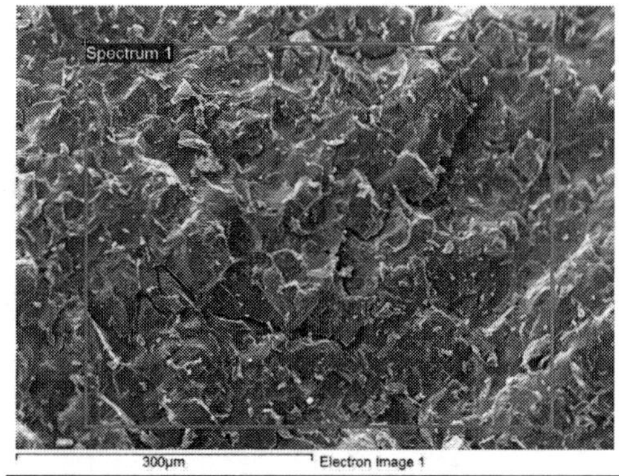

	Al	Si	Cr	Mn	Fe	Total
Sp.1	4.0	20.0	3.4	4.8	67.8	100.0

Figure 6. SEM image and EDS data of synthesized additionally alloyed cast ferrosilicoaluminium Fe–Si–Al(Cr, Mn).

4. Conclusions

The analysis of the obtained data allows drawing a conclusion about the prospects of materials under investigation and the method of their production for the formation of bulk materials for steel making industry. The production of complex alloyed ferroalloys can be realized in combustion mode (SHS) for powder mixtures based on scale. The process of obtaining high-alloy ferroalloys is completely energy independent, which makes it attractive for practical realization.

Acknowledgments

The research was supported by the Program of Fundamental Research of the Presidium of RAS, no. 39.

References

[1] Shatokha V I, Gogenko O O and Kripak S M 2011 *Res. Cons. Recyc.* **55** 435–440
[2] Cartwright D and Clayton J 2000 *Steel Times Int.* **24** 42–43
[3] Yang Q, Holmberg N and Bjorkman B 2009 *Steel Times Int.* **80** 422–428
[4] Sanin V N, Ikornikov D M, Andreev D E and Yukhvid V I 2014 *Russ. J. Non Ferr. Met.* **55** 613–619 doi 10.3103/S1067821214060212
[5] Sanin V N, Yukhvid V I, Ikornikov D M, Andreev D E, Sachkova N V and Alymov M I 2016 *Dokl. Phys. Chem*. **470** 145–149 doi:10.1134/S001250161610002X
[6] Sanin V N, Andreev D E and Yukhvid V I 2013 *Russ. J. Non Ferr. Met.* **54** 274–279

IOP Conf. Series: Materials Science and Engineering **558** (2019) 012042 doi:10.1088/1757-899X/558/1/012042

Self-propagating high-temperature synthesis of Fe_2TiSn based Heusler alloys with following spark plasma sintering

A S Sedegov, Zh S Yermekova, A I Taranova, A P Novitskii, A I Voronin, A A Nepapushev and D O Moskovskikh

National University of Science and Technology MISIS, Leninskii 4, Moscow, 119049 Russia

E-mail: mos@misis.ru

Abstract. The Fe_2TiSn based Heusler alloys were synthesized and studied. The samples were fabricated by a combination of self-propagating high-temperature synthesis (SHS) followed by spark plasma sintering (SPS). An optimal mechanical activation (MA) regime, favoring for successful SHS process was experimentally defined for the stoichiometric mixture of elements powder. The phase evolution of the $Fe_2TiSn_{1-x}Si_x$ ($x = 0$, 0.10, 0.15, 1) compound and the morphology of a bulk specimens were studied. It was shown that with the presence of an impurity side phase Fe_2Ti the electrical resistivity of material decreased by an order of magnitude, while the Seebeck coefficient remains unchanged in the range from 3 to 8 μV K^{-1}.

1. Introduction

With the human population constantly increasing the issue of the energy efficiency becomes more and more important. Involving recycling process into the any technology is a necessary step and it also concerns recycling of the wasted heat. There is a class of materials which can convert the heat energy directly into electricity and vice versa, which is called thermoelectric materials. Thermoelectric (TE) devices can have variety of applications such as for solar heat conversion [1], waste heat recovery [2, 3], local cooling of electrical devices [4] and air conditioning [5]. However, relatively low conversion efficiency of the commercially available TE materials hinders their wider application.

Conversion efficiency mainly defined by the temperature difference between cold and hot side of thermoelectric module and the dimensionless figure of merit $zT = \alpha 2\sigma T/\kappa$, where α is the Seebeck coefficient, σ is the electrical conductivity, T is the absolute temperature and κ the total thermal conductivity [6]. For high efficiency, high power factor (α2σ) coupled with low thermal conductivity (κ) is essential. Semiconducting intermetallic compounds with general formula Fe_2YZ (so called full-Heusler alloys), where Y is basically 3d- or 4d- transition metal, and Z is a metalloid have attracted considerable attention as thermoelectric materials [7]. It has been reported that theoretically Fe_2TiSn based compounds can achieve the Seebeck coefficient as high as –300 μV K^{-1} [8]. It was proposed that substitution of titanium or tin may improve the TE characteristics of the alloy.

Voronin with others [9] investigated the $Fe_2TiSn_{1-x}Si_x$ Heusler alloys fabricated by the conventional melting procedure followed by annealing and reported that it is hart to obtain the single-phase material by the traditional approach, more over the impurity phases decrease expected thermoelectric performance. There are several ways to enhance the figure of merit, (a) carrier concentration optimization, (b) band structure engineering, (c) nanostructuring [10–12] and (d) defects engineering

[13, 14]. All of the listed parameters can be controlled in a fabrication process which leaves a lot of space for the better materials development.

There are few traditional synthesis approaches which include multi-step techniques with melting and annealing as a main step and some additional pre- and post-treatment procedures [15, 16]. Such methods are hard to adopt for large-scale production of TE materials especially of nanostructures ones. Among the relatively new techniques the self-propagating high-temperature synthesis (SHS) could be a most promising approach for the direct single-phase TE material fabrication at a minimal cost and on the timescale of seconds [17–24].

In the present work we adapt the spark plasma sintering method for the thermoelectric bulk material fabrication. $Fe_2TiSn_{1-x}Si_x$ ($x = 0$, 0.10, 0.15, 1) single-phase compound were synthesized by the SHS with the mechanical activation of the precursor powders. Product structural characteristics and thermoelectric properties was investigated.

2. Materials and Methods

The following reagents were used: commercially available powders of iron (99.7%, 50–200 μm) titanium (99.1%, 50 μm), tin (99.1%, 45–80 μm) and silicon (99.9% less than 5 μm) were used as starting materials.

The batch mechanical activation (MA) was conducted in an Activator-2S planetary ball mill from 5 to 20 min. The stoichiometric mixture of the pure element powders was ball milled under argon atmosphere using steel jars with steel grinding balls 6 mm in diameter at the powder-to-balls weight ratio of 20:1. MA was performed under centrifugal acceleration of 250 m/s^2 for different times. After mechanical activation the powder was pressed into the pellets with a diameter, *d,* of10 mm, a height, *h,* of 15 mm, and a relative density of 40–50%. After compression the pellet were placed inside the SHS reactor and the sample was ignited by tungsten coil connected to the external current lead. Each time before the synthesis the reactor was evacuated (~ 300 Pa) and refilled with the argon (2 atm). The resultant specimens were grinded to powder and then were densified using the spark plasma sintering (SPS) machine (Labox 650, SinterLand, Japan) according to the scheme shown in Figure 1.

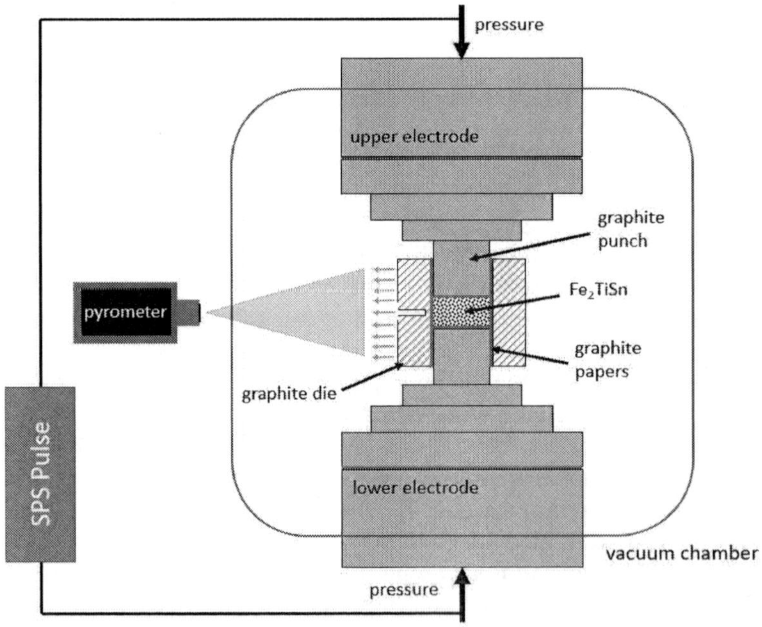

Figure 1. Schematic of spark plasma sintering and direct hot pressing (without SPS puls) process.

The powders were put into a cylindrical graphite die which was placed in the SPS chamber evacuated to 10 Pa. Uniaxial pressure was then applied through top and bottom plungers. The sintering temperature was ranged from 700–900°C with the heating rate of 50–100°C/min. Pressure of 50 MPa and a dwelling time of 10 min were kept constant for all the samples, respectively. The compacted disc specimens had a diameter of 12.7 mm and height of 2–3 mm. In case of Fe_2TiSi the direct hot-pressing DSP-515 SA (Dr. Fritsch, Sondermaschinen GmbH, Germany) were used. The sintering temperature was ranged from 800 to 1000°C while other parameters were the same as for the case of $Fe_2TiSn_{1-x}Si_x$ ($x = 0, 0.10, 0.15$).

The phase analysis of the samples was performed at room temperature by X-ray diffraction (XRD) with a Difray 401 diffractometer (Scientific Instruments, Russia) using CrK_α ($\lambda = 2.2911$ Å) radiation. Scanning electron microscopy in conjunction with energy-dispersive X-ray spectroscopy (SEM, EDX; Vega 3 SB, Tescan, Czech Republic) was employed to determine the chemical composition and to observe microstructure of as-synthesized specimens. Temperature dependencies of the electrical resistivity and the Seebeck coefficient were simultaneously measured in He atmosphere from 298 to 473 K by the four-probe and differential methods, respectively. The uncertainty of the Seebeck coefficient and the electrical conductivity measurements was within 8%.

3. Results and discussion

Preliminary thermodynamic calculations of the adiabatic combustion temperature (T_{ad}) and other thermodynamic parameters were performed for all the compositions of $Fe_2TiSn_{1-x}Si_x$ ($x = 0, 0.10, 0.15, 1$) Estimated values of the adiabatic combustion temperature depending on the composition are presented in table 1 (for Fe_2TiSn and $Fe_2TiSn_{0.9}Si_{0.1}$). The calculations revealed that values of the adiabatic temperature of combustion increase with respect to silicon doping level.

Table 1. Results of the adiabatic temperature calculations for the $Fe_2TiSn_{1-x}Si_x$ system ($x = 0$ and 0.1)

Parameter (units)	Composition			
	Fe_2Ti_1 (C)	1.0000	Fe_1Si_1 (C)	0.0769
	Sn_1 (L)	1.0000	Fe_2Ti_1 (C)	0.9615
			Si_3Ti_5 (C)	0.0077
			Sn_1 (L)	0.9000
Volume of gas products (l)	0.0000		0.0000	
Pressure of gas products (atm)	1.0000		1.0000	
Temperature (K)	1002.1988		1068.1669	
Gas product amount (mol)	$4.28\cdot10^{-15}$		$8.69\cdot10^{-14}$	
Products heat capacity (J/K)	121.4135		122.8608	
Products entropy (J/K)	275.6490		279.3739	
Products enthalpy (KJ)	0.0736		0.0790	

It is known that during the high energy mechanical treatment not only the material mixing occurs, but the morphology and inner energy of material can be changed as well. Chemical reaction between mixture components is also possible. Defect accumulation and powder grinding scientifically affect the following SHS process and the products which can be formed. Experimentally we have found an optimal regime for the mechanical activation (MA) of the initial powders which effectively promotes the SHS process. It appears that 10-min mechanical pretreatment of the pure element's mixture leads to the easier ignition of the pallet. Carrying out the synthesis in an argon atmosphere leads to the formation of a phase Fe_2TiSn, but there are also minor peaks of titanium oxide (figure 2). Figure 3 shows phase change in a mixture of Fe/Ti/Sn powders. After MA of stoichiometric components ratio some intermediate phases such as Fe_3Sn, FeSn, FeTi and SHS, Heusler Fe_2SnTi were formed. However, during the following SHS process almost all secondary phases disappeared and two-phase sample Fe_2TiSn–FeSn were synthesized. Such phase evolution is an in good agreement with the Fe–Ti–Sn system phase equilibria [25].

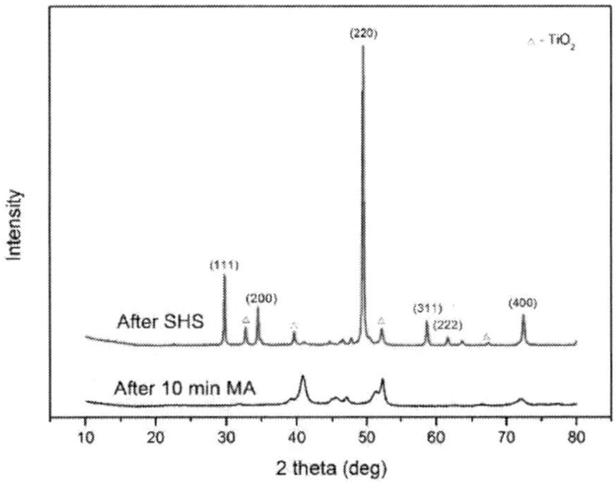

Figure 2. XRD patterns for batch mixture after 10 min MA (black line), Fe_2TiSi after SHS in argon atmosphere (red line).

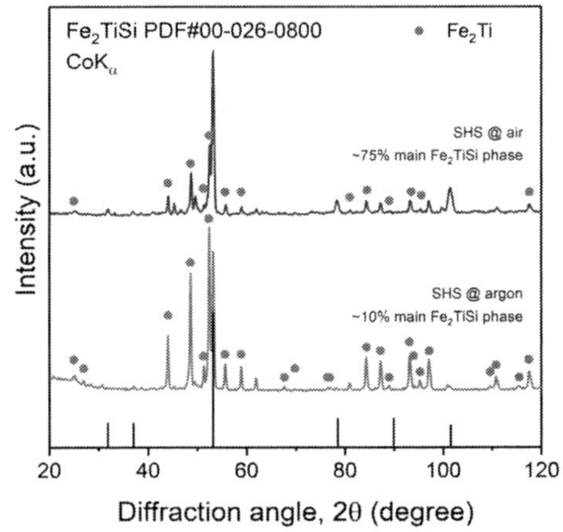

Figure 3. XRD patterns for Fe_2TiSi after SHS in argon flow (red line) and air (blue line) atmospheres.

The same phase evolution was observed for the samples with different amount of silicon: $x = 0.10$, 0.15, respectively. Additional $TiSi_2$ impurity phase were formed after MA for Si doped samples.

During the synthesis of the silicon containing material such as Fe_2TiSi another phase equilibrium of the Fe–Ti–Si system should be considered [26]. Under the same parameters of MA and SHS as for the composition without the silicon, only ~ 10% of the desirable Fe_2TiSi phase was obtained. Leading the SHS process in air atmosphere is one of the ways to increase the temperature of synthesis to the level where the Fe_2TiSi formation could be promoted. Indeed, only Fe_2Ti impurity phase was observed while the amount of the main Fe_2TiSi phase increased from 10 % to 75% for SHS performed in air atmosphere (figure 3). It should be noted, that EDX revealed a small amount of TiO_2 at the grain boundaries for both of the samples. Formation of TiO_2 can be caused by an air atmosphere during SHS.

For the SPS process the initial material were hot pressed under the 800 and 1000°C. As it was expected, for the specimen compacted at higher temperature the amount of the main phase is higher (see figure 4a). The electrical resistivity of the Fe_2TiSi sample hot pressed at 1000°C was higher by an order of magnitude comparing to the Fe_2TiSi sample pressed at 800°C (figure 4b). The Seebeck coefficient was ranged from 3 to 11 $\mu V\ K^{-1}$ in the whole temperature range, which is in an good agreement with other experimental results for Fe_2TiZ (Z = Sn, Si, Sb) based systems.

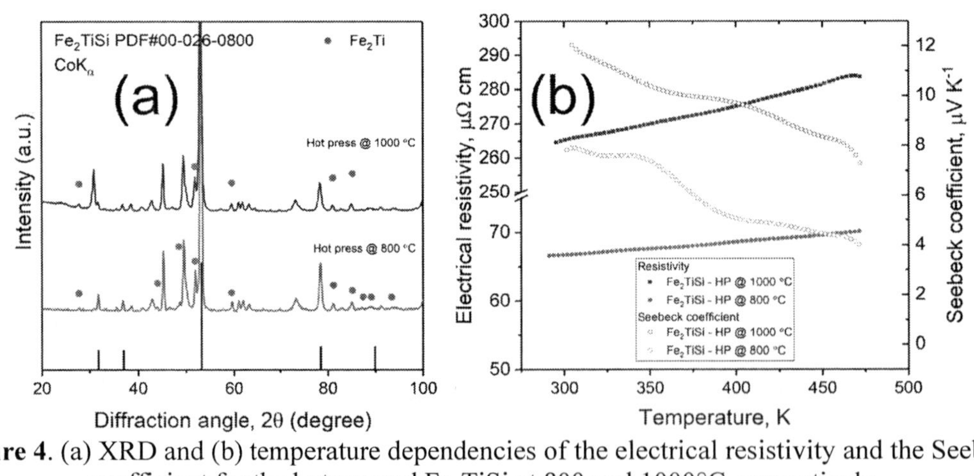

Figure 4. (a) XRD and (b) temperature dependencies of the electrical resistivity and the Seebeck coefficient for the hot pressed Fe_2TiSi at 800 and 1000°C, respectively.

Heating rate significantly affects the microstructure of the $Fe_2TiSn_{1-x}Si_x$ samples (x =0, 0.10, 0.15). For the bulk samples compacted with heating rate of 100°C/min the microstructure looked like for non-reacted powder with the matrix of Fe_2TiSn containing grains of impurity phases (see figure 5). Completely another microstructure was observed for the samples SPSed at 50°C/min heating rate (see figure 6). It can be suggested, that for this Fe–Ti–Si–Sn system more time required for elements diffusion and formation of the targeting Fe_2TiSn (Si) phase. However, even for these specimens some TiO_2 was observed at the grain boundaries (figure 6).

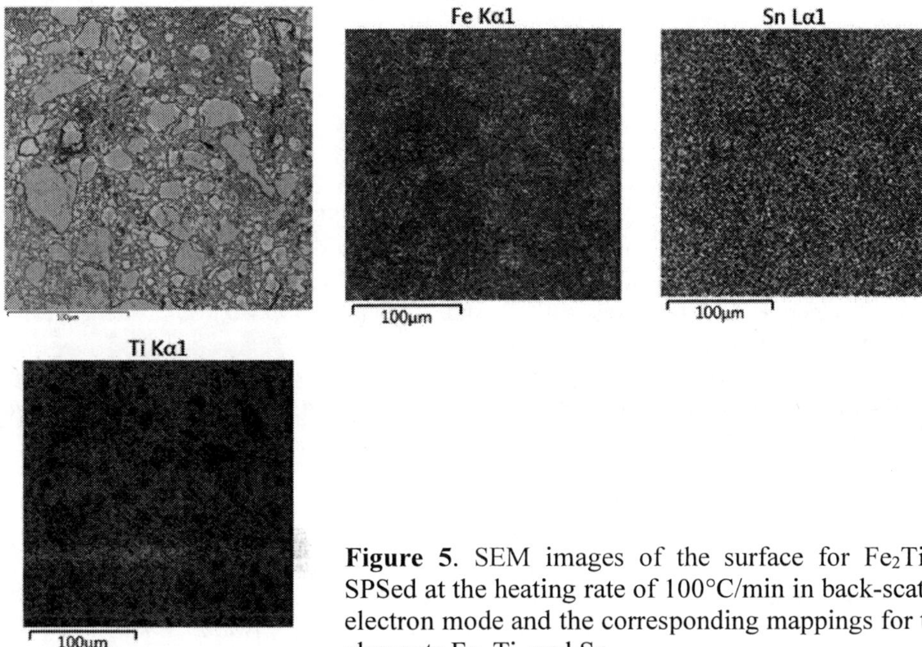

Figure 5. SEM images of the surface for Fe_2TiSn SPSed at the heating rate of 100°C/min in back-scatter electron mode and the corresponding mappings for the elements Fe, Ti, and Sn.

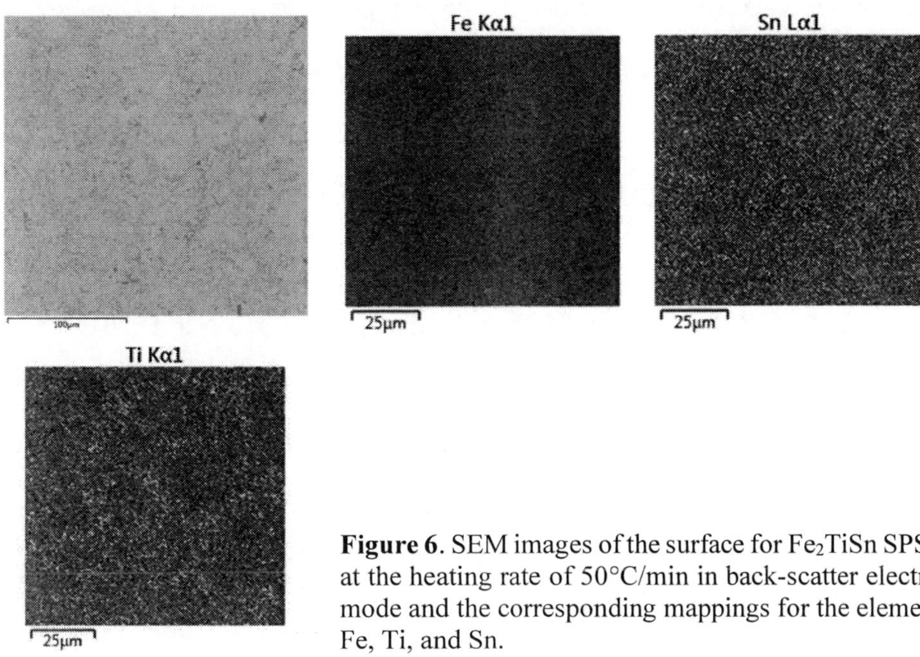

Figure 6. SEM images of the surface for Fe_2TiSn SPSed at the heating rate of 50°C/min in back-scatter electron mode and the corresponding mappings for the elements Fe, Ti, and Sn.

4. Conclusion

The theoretical prediction of possible increase of thermoelectric figure of merit for the Fe_2TiSn was contradicted with experimental data. Partial substitution of Sn with Si did not improve the material thermoelectrical properties, because the formation of a multiphase states in $Fe_2TiSn_{1-x}Si_x$ (x = 0; 0.1; 0.15) alloys. In general, impurity phases lead to a decrease in the electrical resistance and Seebeck coefficient and an increase in the thermal conductivity. Based on data obtained in this study, it can be assumed that the Seebeck coefficient in Fe2TiSi in the vicinity of room temperature is much lower than in Fe_2TiSn, while the electrical resistance and thermal conductivity of these compounds are close.

Acknowledgments

The reported study was funded by RFBR according to the research project no. 18-38-20170.

References

[1] Tritt T M, Böttner H and Chen L 2008 *MRS Bull.* **33** 366–368 doi:10.1557/mrs2008.73
[2] Bell LE 2008 *Sci.* **321** 1457–61 doi:10.1126/science.1158899
[3] Snyder G J and Toberer E S 2008 *Nat. Mater.* **7** 105–114 doi:10.1038/nmat20 90
[4] Tritt T M 1999 *Sci.* **283** 804–805 doi:10.1126/science.283.5403.804
[5] Mei V C, Chen F C, Mathiprakasam B and Heenan P 1993 *J. Sol. Energy Eng.* **115** 200 doi:10.1115/1.2930050
[6] Ioffe A F 1957 *Semiconductor thermoelements, and Thermoelectric cooling* (Infosearch, London)
[7] Khovaylo V V, Voronin A I, Zueva V Y, Seredina M A and Chatterdjee R 2017 *Semiconduc.* **51** 718–721 doi:10.1134/S1063782617060136
[8] Yabuuchi S, Okamoto M, Nishide A, Kurosaki Y and Hayakawa J 2013 *Appl. Phys. Express.* **6** 025504 doi:10.7567/APEX.6.025504
[9] Voronin A I, Zueva V Y, Karpenkov D Y, Moskovskikh D O, Novitskii A P, Miki H and Khovaylo V V 2017 *Semiconduc.* **51** 891–893 doi:10.1134/S1063782617070363
[10] Alam H and Ramakrishna S 2013 *Nano Energy* **2** 190–212 doi: 10.1016/j.nanoen.2012.10.005
[11] Liu W S, Yan X, Chen G and Ren Z F 2012 *Nano Energy* **1** 42–56 doi: 10.1016/j.nanoen.2011.10.001
[12] Xie W, Weidenkaff A, Tang X, Zhang Q, Poon J. and Tritt T. 2012 *Nanomater.* **2** 379–412 doi:10.3390/nano2040379
[13] Yang L, Chen Z G, Dargusch M S and Zou J 2017 *Adv. Energy Mater.* doi:10.1002/aenm.201701797
[14] Li J F, Liu W S, Zhao L D and Zhou M 2010 *NPG Asia Mater.* **2** 152–158 doi:10.1038/asiamat.2010.138
[15] Mori T 2017 *Small.* **13** 1702013 doi:10.1002/smll.201702013
[16] Poudel B, Hao Q, Ma Y, Lan Y, Minnich A, Yu B, Yan X, Wang D, Muto A, Vashaee D, Chen X, Liu J, Dresselhaus M S, Chen G and Ren Z 2008, *Sci.* **320** 634–638 doi:10.1126/science.1156446
[17] Se B, Liu R, Tan X, Ren G, Liu Y, Zhou Z, Liu C, Lin Y and Nan C 2017 *Crystals* **7** 257 doi:10.3390/cryst7090257
[18] Su X L, Fu F, Yan Y G, Zheng G, Liang T, Zhang Q Q J, Cheng X, Yang D W, Chi H, Tang X F, Zhang Q Q J and Uher C 2014 *Nat. Commun.* **5** 1–7 doi:10.1038/ncomms5908
[19] Zhang Q, Su X, Yan Y, Xie H, Liang T, You Y, Tang X and Uher C 2016 *ACS Appl. Mater. Interfaces* **8** 3268–76 doi:10.1021/acsami.5b11063
[20] Yang D, Su X, Yan Y, Hu T, Xie H, He J, Uher C, Kanatzidis M G and Tang X 2016 *Chem. Mater.* **28** 4628–40 doi: 10.1021/acs.chemmater.6b01291
[21] Liang T, Su X, Yan Y, Zheng G, Zhang Q, Chi H, Tang X, Uher C 2014 *J. Mater. Chem.* A. **2** 17914–18 doi:10.1039/c4ta02780a

[22] Lan J-L, Deng C, Ma W, Ren G K, Lin Y-H and Yang X 2017 *J. Alloys Comp.* **708** 955–960 doi: 10.1016/j.jallcom.2017.02.290
[23] Ren G-K, Lan J-L, Butt S, Ventura K J, Lin Y-H and Nan C-W 2015 *RSC Adv.* **5** 69878–85 doi:10.1039/c5ra13191j
[24] Xing Y, Liu R, Sun Y-Y, Chen F, Zhao K, Zhu T, Bai S and Chen L 2018 *J. Mater. Chem.* A **6** 19470–78 doi:10.1039/C8TA07411A
[25] Cai Y, Wu Y, Xie Z Y, Iv K L, Liu H S and Jin Z P 2015 *Calphad* **49** 110–119 doi: 10.1016/j.calphad.2015.04.005.
[26] Marker M C J, Duarte L I, Leinenbach C and Richter K W 2013 *Intermet.* **39** 38–49 doi: 10.1016/2013.03.007.

IOP Conf. Series: Materials Science and Engineering **558** (2019) 012043 doi:10.1088/1757-899X/558/1/012043

Synthesis and study of high-entropy ceramics based on the carbides of refractory metals

A Sedegov, S Vorotilo, V Tsybulin, K Kuskov and D Moscovskikh

National University of Science and Technology MISIS, Leninskii 4, Moscow, 119049 Russia

E-mail: a.sedegoff@gmail.com

Abstract. Production of single-phase carbide high entropy ceramic ($Hf_{0.2}Ta_{0.2}Ti_{0.2}Nb_{0.2}Mo_{0.2}$ or $Zr_{0.2}$,)C by self-propagating high-temperature synthesis (SHS) from a mechanically activated (MA) reaction mixtures and mechanical synthesized (MS) materials is discussed. The influence of mechanical treatment time at ball mill, on the formation of the phase of a high-entropy material is investigated. Spark plasma sintering and their structure, phase composition, obtained bulk samples and mechanical properties were studied. A self-propagating high-temperature synthesis of high-entropic carbide based on refractory elements was also carried out. The microstructure and mechanical properties of high-entropy ceramics obtained by the methods of mechanical synthesis and SHS were compared.

1. Introduction

Conventionally, the majority of alloys are based on one main constituent element. For the increase of properties and performance, they are alloyed by alloying elements which enhance the alloy's strength, corrosion resistance and other properties. Thus, the families of alloys are made. However, the number of elements is limited; therefore, the number of alloys which could be developed on the base of one or two main constituent elements is finite. However, what if one is to go outside the classical definition and increase the number of constituent elements? This conception was first suggested in 1995 [1], and these materials were later branded as high-entropy alloys [2].

High-entropy materials by definition contain five or more main constituent elements with the concentration between 5% and 35% [3]. The currently available data in the areas of physical metallurgy and binary and ternary diagram suggests that such multicomponent alloys might contain dozens of phases and intermetallics, which are usually brittle and have limited applicability. However, in contrary to these expectations that the increased entropy in these alloys leads to the formation of solid solutions with a relatively simple structure, thus diminishing the content of unwanted phases or eliminating them. These peculiar features, related to high entropy, are of the utmost importance for the development and applications of the high-entropy alloys.

During the last years of investigation of high-entropy materials, more than 300 compositions with wide functionality were developed on the basis on non-ferrous and refractory metals. Currently, the hottest topic is the development of high-entropy alloys (HEA) for high-temperature applications, which require high phase stability upon heating up to 1800°C, oxidation resistance and high-temperature strength.

Due to their unique multi-component compositions, HEAs might possess a complex of outstanding properties, including high strength and hardness, unmatched high-temperature strength, good structural stability, and corrosion resistance. The close interest in the development of these materials is driven by the demand for the materials with increased operating temperature from the aerospace industry.

Equimolar composition HfTaTiZrNbMo was investigated in the article [4] and demonstrated the high mechanical properties at room temperature and 1200 C. Xueliang Yan et al. in their recent article [5] synthesized a high-entropy with the composition $(Hf_{0.2}Zr_{0.2}Ta_{0.2}Nb_{0.2}Ti_{0.2})C$ from the monocarbides, which were mixed in a mill and consolidated by spark plasma sintering at 2000 C. The produced material is characterized by high thermal conductivity, thermal diffusivity, and the coefficient of thermal expansion, which was compared to the initial monocarbides. In various articles dedicated to the investigation of HEAs, the samples are usually produced by arc melting. However, this production route often leads to the formation of a dendritic structure, leading to the degradation of mechanical properties. To retain the fine-grain structure, which is most beneficial for mechanical properties, the powder metallurgy route is most perspective, since it ensures the homogeneity of material. Also, as a new class of high-entropy, Jian Luo et al. [6] described the ultra-high-temperature diboride ceramic based on the Hf–Ta–Ti–Nb–Mo–Zr system. In our work, a high-entropy alloy Hf–Ta–Ti–Nb–Mo–Zr and the related high-entropy carbide were produced by high-energy ball milling and spark plasma sintering.

2. Materials and Methods

The majority of high-entropy materials are produced by vacuum arc melting, which requires high energy expenditures. In this work, high-entropy materials were produced by various routes, which involve the mixing of metal powders and carbon, subsequent synthesis, and sintering.

High-purity metallic powders of Hf, Ta, Ti, Nb, Zr, Mo (> 99%) and high-purity graphite powders with the particles size of 7–10 μm (> 99.9 %).

For the production of high-entropy ceramics (HEC-1 corresponds to the composition $(Hf_{0.2}Ta_{0.2}Ti_{0.2}Nb_{0.2}Mo_{0.2})C$; HEC-2 corresponds to the composition $(Hf_{0.2}Ta_{0.2}Ti_{0.2}Zr_{0.2}Nb_{0.2}Zr_{0.2})C$), the mixture of powders were treated in the planetary ball mill Aktivator-2S (Russia). High-energy ball milling (HEBM) was organized according to two schemes: 1) milling of the mixture of metals and graphite during 120 min (mechanochemical synthesis); 2) milling of the mixture of metals during 90 min (+ 5 min with hexane to ensure the unloading the powder from the surface of jars and balls), then milling of pre-milled metal powders with graphite for 5 min (mechanical activation of reaction mixtures).

After the treatment according to scheme 2 (mechanical activation), combustion synthesis of the reaction mixtures was conducted in the lab reactor developed in the framework of the U.M.N.I.K. project. Combustion temperature and rate were measured during the synthesis.

Sintering of the powders produced by mechanochemical synthesis (120 min of co-milling of metals and graphite) and by combustion synthesis (90 min milling of metals, 5 min milling of mixture of metals and graphite and subsequent combustion synthesis in lab reactor) was conducted on the hot pressing installation Dr. Fritsch DSP 515 (Germany) and spark plasma sintering installation SPS Labox 650 (Japan) in a graphite matrix in vacuum at temperatures 1500, 1800 and 1900°C, pressure 30 and 50 MPa, dwelling time 5 min. Heating and cooling rates were equal to 100 and 50°C min^{-1}, correspondingly. The structure of powders and sintered materials were analyzed by means of scanning electron microscopy (SEM) on the Vega 3 (TESCAN, Czech Republic) and Nanolab Helios 600 (FEI, United States) systems. The last one, exhibiting the ion and electron beams, allows for the acquisition of a cross section of the particle in any direction, while the electron beam enables one to analyze its microstructure. The X-ray diffraction analysis was conducted on a DRON-3 diffractometer (Burevestnik, Russia). Material's hardness was measured on the Struers Durascan 70 tester with a load of 3 kN.

IOP Conf. Series: Materials Science and Engineering **558** (2019) 012043 doi:10.1088/1757-899X/558/1/012043

3. Results and Discussion

Long HEBM treatment (120 min) allowed the homogeneous distribution of graphite in the mixture and eliminated agglomerates. SEM of the cross-cut polished powders is presented in figure 1, featuring the uniform structure and absence of phase segregation in the powders and suggesting the completion of mechanochemical synthesis during the HEBM.

As the SEM of $Hf_{0.2}Ta_{0.2}Ti_{0.2}Nb_{0.2}Mo_{0.2}C$ ceramic demonstrates (figure 2a) the material has relatively high residual porosity due to insufficient pressure and dwelling time. At higher resolution, bright inclusions with the size ≤ 1 μm can be seen. EDS analysis suggests that these inclusions are comprised of hafnia (figures 2b, 2c). Unlike the $Hf_{0.2}Ta_{0.2}Ti_{0.2}Nb_{0.2}Mo_{0.2}C$, the microstructural investigation of HEC with the composition $Hf_{0.2}Ta_{0.2}Ti_{0.2}Nb_{0.2}Zr_{0.2}C$ (figures 2d–2f) demonstrated the absence of oxides or secondary phases. The material is porous as well, but its structure is uniform. Usually, the uniform structure provides higher mechanical properties, including hardness and high-temperature strength.

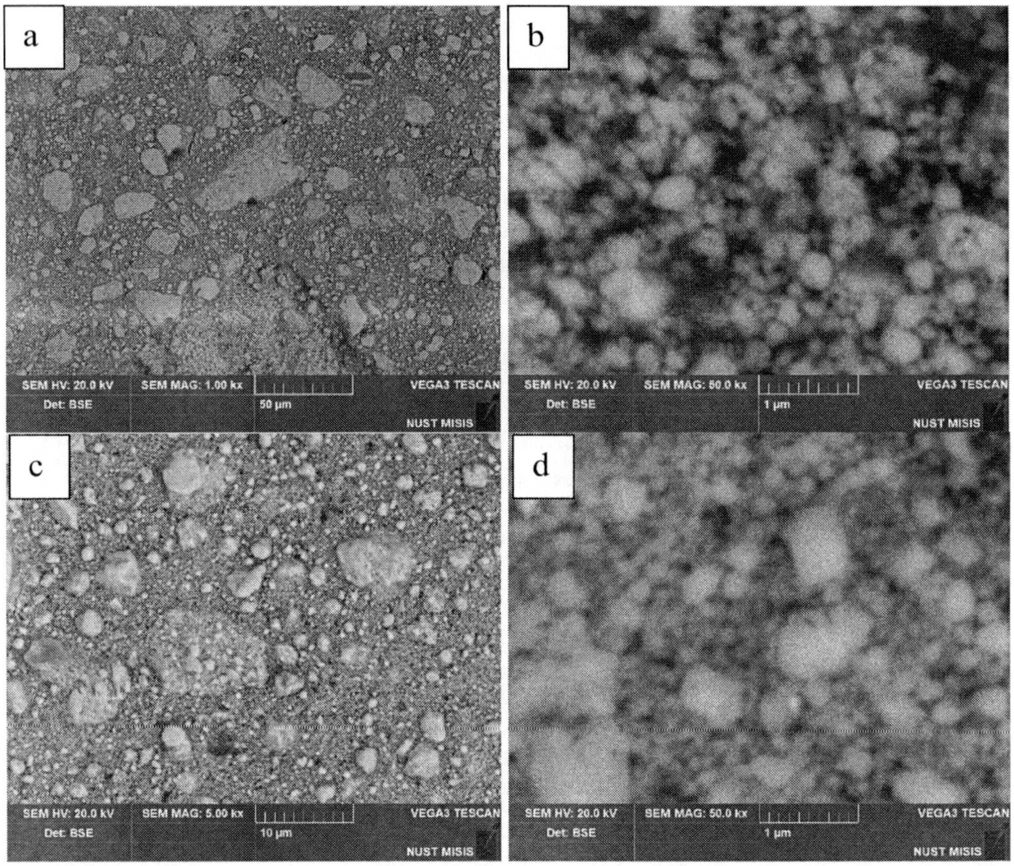

Figure 1. SEM of the cross-cut powders with the composition $Hf_{0.2}Ta_{0.2}Ti_{0.2}Nb_{0.2}Mo_{0.2}C$ (a, b) and $Hf_{0.2}Ta_{0.2}Ti_{0.2}Nb_{0.2}Zr_{0.2}C$ (c, d) after 120 min HEBM.

Sintering at 1800°C of the powders produced by SHS yielded more densified material, but SEM (figure 4) reveals the presence of minor phase decomposition and oxide contamination. Grain structure is somewhat coarser as compared to the materials sintered at 1500°C.

IOP Conf. Series: Materials Science and Engineering **558** (2019) 012043 doi:10.1088/1757-899X/558/1/012043

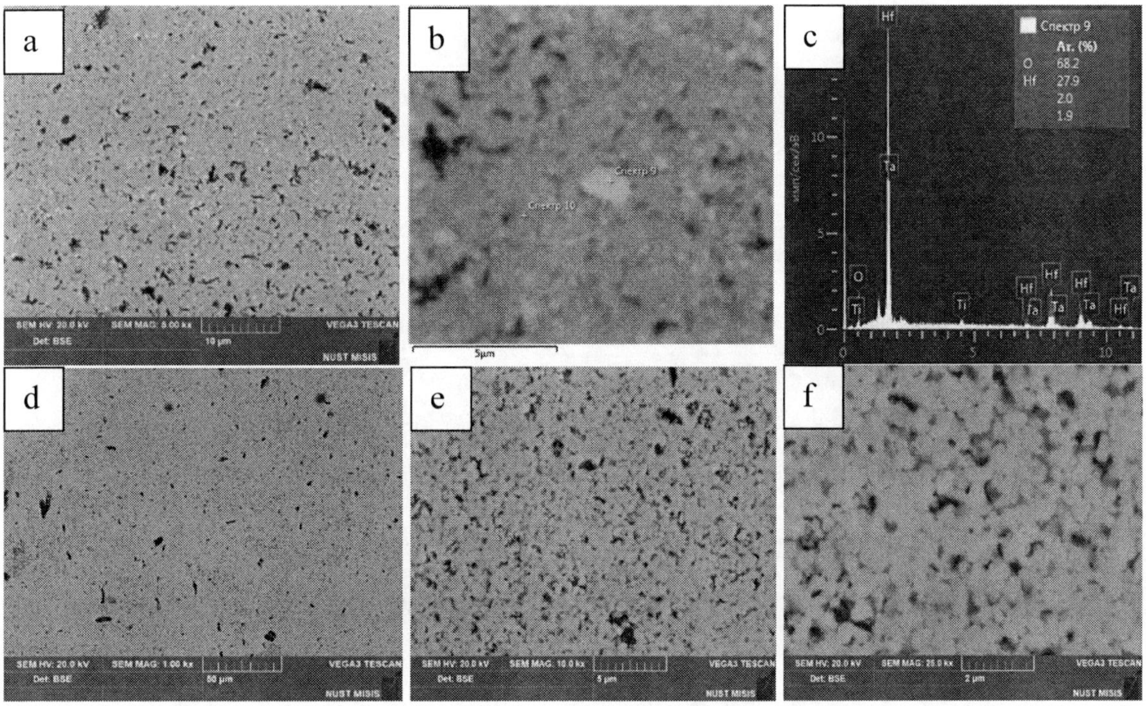

Figure 2. SEM of ceramic $Hf_{0.2}Ta_{0.2}Ti_{0.2}Nb_{0.2}Mo_{0.2}C$ (a) and $Hf_{0.2}Ta_{0.2}Ti_{0.2}Nb_{0.2}Zr_{0.2}C$ (d–f); EDS results (b, c).

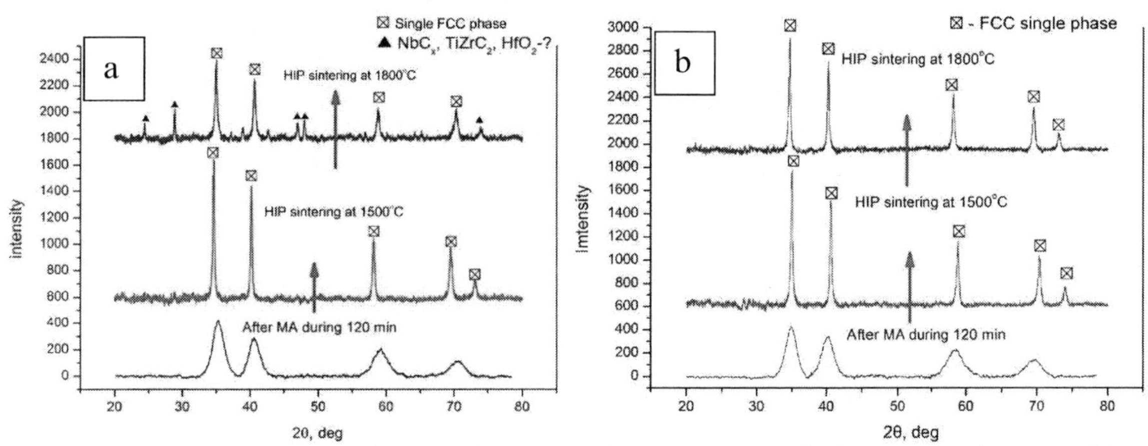

Figure 3. XRD patterns of ceramics $Hf_{0.2}Ta_{0.2}Ti_{0.2}Nb_{0.2}Mo_{0.2}C$ (a) and $Hf_{0.2}Ta_{0.2}Ti_{0.2}Nb_{0.2}Zr_{0.2}C$ (b) produced by hot pressing of powders after 120 min HEBM.

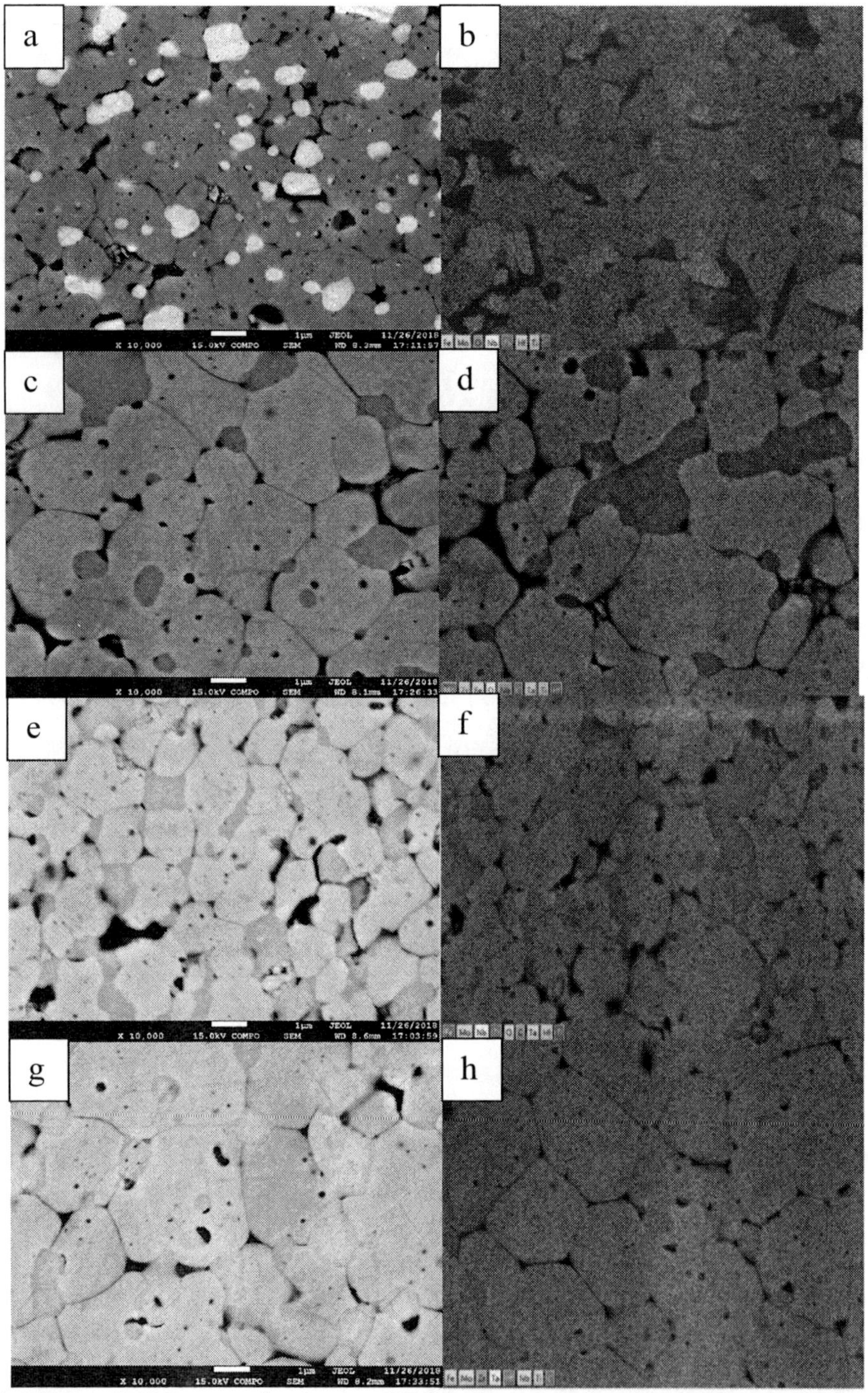

Figure 4. SEM and EDS maps of ceramics: HEC-1, hot-pressed at 1800°C (a, b); HEC-1 SPSed at 1800°C (c, d); HEC-2 hot-pressed at 1800°C (e, f); HEC-2 SPSed at 1800°C (g, h).

The material produced by SHS and SPS demonstrated higher hardness as compared to the specimens produced by mechanochemical synthesis and hot pressing (table 1).

XRD data (figure 3) demonstrates the formation of single-phase solid solution with FCC lattice during the HEBM and sintering. Sintering at higher temperature leads to the formation of impurities, probably due to oxygen contamination during the unloading of jars.

The second approach includes preparation of metallic HEA, mixing it with carbon and combustion synthesis. First-stage HEBM produced amorphous powders, which did not react with the carbon during the second stage of HEBM. Then the carbide was synthesized in the lab reactor in an argon atmosphere. Combustion temperature and rate were measured during the synthesis. Combustion rate was 4.17 mm/s and 7.69 mm/s for HEC-1 and HEC-2 respectively. XRD data (figure 5) reveals the formation of single-phase with FCC lattice and absence of oxides or monocarbides in the combustion products.

Table 1. Hardness test result

Material	Method of synthesis	Method of sintering	Hardness, GPa
HEC-1	MA 120 min	Hot pressing 1500°C	14.33
HEC-1	MA 120 min	Hot pressing 1800°C	15.95
HEC-1	SHS	SPS 1900°C	22.62
HEC-2	MA 120 min	Hot pressing 1500°C	16.75
HEC-2	MA 120 min	Hot pressing 1800°C	18.58
HEC-2	SHS	SPS 1900°C	20.53

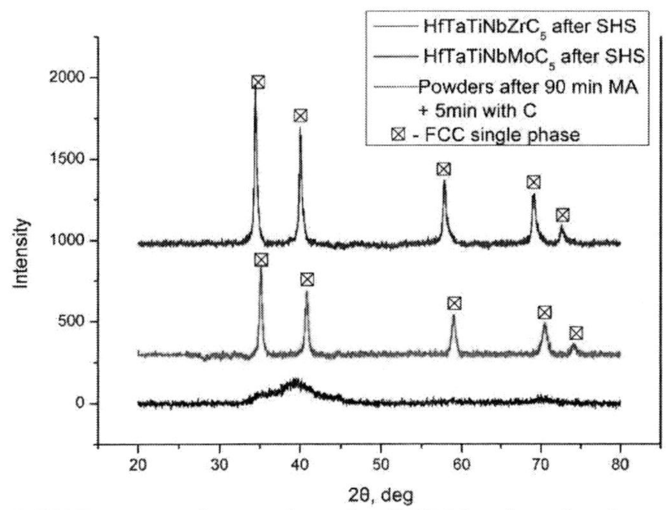

Figure 5. XRD patterns for powders after HEBM and combustion synthesis.

4. Conclusions

The following conclusions could be made:

1. High-entropy ceramics can be produced by the mechanochemical synthesis and combustion synthesis
2. Increase if the sintering temperature allows the production of more dense material with high hardness and fracture toughness.
3. These methods were applied for the first time for the production of high-entropy carbide ceramics by SHS.

Acknowledgments
This work was conducted with the finantial support of the Russian Science Foundation (grant no. 18-79-10215). A. Sedegov acknowledges the support of Fund for Assistance to Small Innovative Enterprises in Science and Technology (the Foundation for Innovation) (grant no. 11018GU/2016 from 13.02.2017) (creation of a laboratory reactor for SHS).

References
[1] Huang K H and Yeh J W 1996 *A study on multicomponent alloy systems containing equal-mole elements* M.S. thesis (Hsinchu: National Tsing Hua University)
[2] Yeh J W, Chen S K, Lin S J, Gan J Y, Chin T S, Shun T T, Tsau C H and Chang S Y 2004 *Adv. Eng. Mater.* **6** 299–303
[3] Yong Zhang, Ting Ting Zuo, Zhi Tang et al 2014 *Prog. Mater. Sci.* **61** 1–93
[4] Juan C C, Tsai M H, Tsai C W and Lin C M 2015 *Intermet.* **62** 7683
[5] Xueliang Yan, Loic Constantin, Jean-Francois Silvain et al doi: 10.1111/jace.15779
[6] Jian Luo, Joshua Gild, Tyler Harrington et al 2017 High-entropy metal diborides: a new class of ultra-high temperature ceramics *Ultra-High Temperature Ceramics: Materials for Extreme Environment Applications IV* (The University of Birmingham, Edgbaston, United Kingdom Bill Lee, Imperial College, London, United Kingdom Eds, ECI Symposium Series) http://dc.engconfintl.org/uhtc_iv/29

IOP Conf. Series: Materials Science and Engineering **558** (2019) 012044 doi:10.1088/1757-899X/558/1/012044

Thermally coupled processes in a composite granular mixture (Ni + Al)–(Ti + C)

B S Seplyarskii*, R A Kochetkov, T G Lisina and N I Abzalov

Merzhanov Institute of Structural Macrokinetics and Materials Science, Russian Academy of Sciences, Chernogolovka, 142432 Russia

*E-mail: seplb1@mail.ru

Abstract. For the first time combustion of a thermally conjugated granulated mixture consisting of the mixture of Ni + Al and Ti + C granules, differing in burning velocity and temperature has been investigated experimentally in co-current flows of argon and nitrogen. It has been established that in the whole investigated gas flow rates the ignition of Ni + Al granules occurs in the combustion front. The high efficiency of Ti + C donor granules energy usage has been shown, since the time of heat exchange between the granules does not depend on the sample size, but is determined only by the characteristics of internal heat transfer. A strong increase in the burning rate of the composite mixture at low nitrogen flow rates has been explained via interaction of aluminum with nitrogen and effective reduction of heat loss of the donor. The possibility of isolating the target synthesis product was shown even if its melting point is below the adiabatic temperature of combustion of the composite mixture.

1. Introduction

The use of the thermal conjugation method for synthesis in layered powder mixtures, proposed in [1, 2], is complicated when the process is scaled. The problems are concerned, first of all, with ensuring the uniformity of ignition and heating of the charge and removal of impurity gases [3] and with the separation of the products of combustion of different layers. A promising way to solve the problem of scaling can be the use of granulated mixtures instead of powder layers [4–9]. The purpose of this work is to study the regularities and mechanisms of combustion of composite granular mixtures, where a fast burning Ti + C mixture is used as a donor, and a slowly burning Ni + Al mixture is used as an acceptor.

2. Experimental results

The original substances and their characteristics are given in table 1. The method of preparation of the granules, the experimental setup scheme and the experimental technique are described in detail in [4].

First, we determined the burning rates U of each of the granular mixtures of Ti + C and Ni + Al separately and the composite mixture (Ni + Al) + (Ti + C) in the absence of a gas flow (see table 2). The adiabatic temperatures of combustion T_{ad} are calculated using software THERMO (http://www.ism.ac.ru/thermo/).The experiments have shown that the maximum content of the Ti + C donor mixture, which ensures synthesis in the steady-burning mode, was 46 wt %. The data of table 2 show the average burning velocity of the composite charge (12.5 mm/s) to be much higher than that of the acceptor mixture Ni + Al. Therefore, due to the low burning rate of the Ni + Al granules, we assumed they behave as inert in the reaction front. To test this assumption, we replaced them in charge with inert Al_2O_3 granules of the same size and weight ratio: 54% Al_2O_3 and 46% (Ti + C). Our experiments have shown that such a mixture is incapable of burning, which means that in the studied mixture 54% (Ni + Al)/46% (Ti + C) the ignition of Ni + Al granules occurs in the combustion front.

Table 1. Materials and reagents

Substance	Grade	Particle size	
		to 50 w t%, μm	to 90 wt %, μm
Titanium	PTM-1	< 105	< 169
Soot	P-803	< 2.5	< 4
Aluminum	АСД-4	< 11	< 16
Nickel	ПНЭ-1	< 43	< 51
Argon	High purity (99.987%)		
Nitrogen	High purity (99.987%)		
Ethanol	95%		
Polyvinylbutiral	SD-4		

Table 2. Experimental U and theoretical U_f burning velocities of the granular mixtures

Ni + Al T_{ad} = 1912 K			Ti + C T_{ad} = 3290 K			54% (Ni + Al)/46% (Ti + C) T_{ad} = 2518 K		
Q, l/h	U, mm/s	U_f, mm/s	Q, l/h	U, mm/s	U_f, mm/s	Q, l/h	U, mm/s	U_f, mm/s
0	3	3	0	22	22	0	12.5	12.5
630 (argon)	4,5	3.8	750 (argon)	25	24	720 (argon)	16,8	15,7
220 (nitrogen)	3	3.3	200 (nitrogen)	23	22.8	200 (nitrogen)	15,5	13
700 (nitrogen)	6,2	4.2	650 (nitrogen)	36	24.5	720 (nitrogen)	24	14,4
1000 (nitrogen)	8	4.7	1050 (nitrogen)	45	26.1	1100 (nitrogen)	28,5	15,43

The validity of this conclusion was tested in experiments on the combustion of a granulated mixture of Ti + C, divided in the center of the filling with a layer of Ni + Al granules of a thickness of 2÷3 particles (figure 1).

When a layer of Ni + Al granules was reached, the combustion front continued to spread without delay, but its speed and luminosity sharply decreased. After the end of burning, the luminosity of the entire charge became the same after 4÷5 s, which is significantly less than the cooling time of the sample. Unlike this experiment, in a uniformly mixed granular mixture, the time taken to equalize the temperature of the donor and acceptor mixture granules should be several times less and, most importantly, this time does not depend on the sample size, but is determined only by the characteristics of internal heat transfer. This indicates the prospects of using a granulated mixture during the combustion of thermally conjugated systems.

IOP Conf. Series: Materials Science and Engineering **558** (2019) 012044 doi:10.1088/1757-899X/558/1/012044

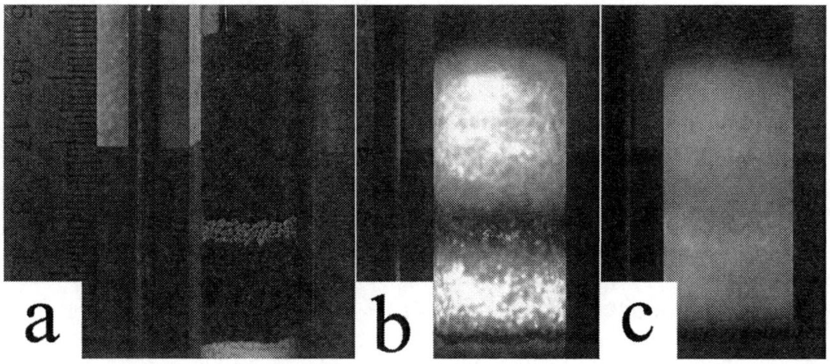

Figure 1. The appearance of the granular layered system: (a) the original sample, (b) immediately after the passage of the combustion front, (c) 5 s after the passage of the combustion front.

It is important to note that the combustion products of the compound mixture of 54% (Ni + Al)/46% (Ti + C) did not sinter and the sample was easily divided into individual granules by color manually (figure 2). Perhaps this is the result of the fact that titanium carbide is not wetted by aluminum nickelide.

Figure 2. The appearance of the separated combustion products of the composite granular mixture. On the left: the products of combustion Ti + C, on the right: the products of combustion Ni + Al.

Next, the combustion parameters of each of the granular mixtures of Ti + C and Ni + Al were determined separately in a co-current stream of argon and nitrogen. The results of measuring the front velocity in a stream of nitrogen are shown in figure 3 (curves *1*, *4*).

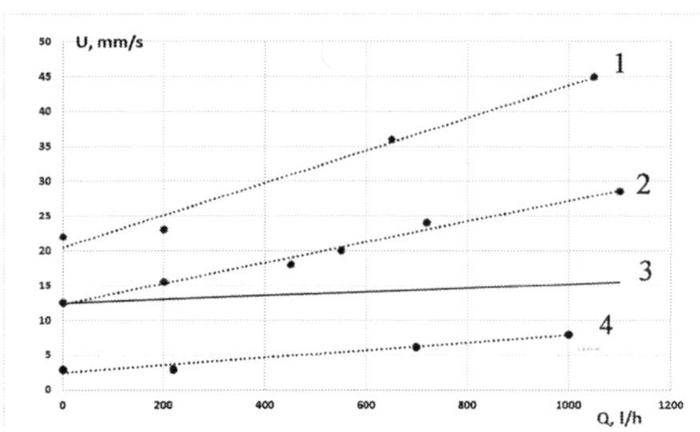

Figure 3. The dependence of the burning rate of granular mixtures on the amount of co-stream of nitrogen: *1* mixture of Ni + Al; *2* mixture of 54% (Ni + Al)/46% (Ti + C); *3* assessment for a mixture of 54% (Ni + Al)/46% (Ti + C) according to the theory of filtration combustion; *4* mixture of Ni + Al.

3. Discussion

In accordance with the theory of filtration combustion, the dependence of the burning velocity U_f on the mass flow rate of gas is linear [10]:

$$U_f = U_0 + c_g \cdot G_m/(c_s \cdot \rho_s) \quad (1)$$

where U_0 is the burning velocity in the absence of gas flow, $G_m = \rho_g \cdot Q/s$ is the mass flow rate of inflowing gas at volumetric flow rate Q through the cross section of the sample (s = 2 cm^2), c_g is the specific gas heat capacity, c_s is the specific heat capacity of porous material, ρ_g is the gas density, ρ_s is the bulk mixture density. When calculating, for Ti + C mixture the density was assumed to be 0.8 g/cm^3, specific heat capacity is 560 J/(kg K), and for Ni + Al mixture the density is 1.7 g/cm^3, specific heat capacity is 590 J/(kg K). As the gas flow increases, the burning velocity growth in the nitrogen flow for the Ti + C mixture is much higher than in the argon flow, and significantly more than theoretical estimates give. This indicates the realization of the convective mode, in which the flow of active gas ignites the granules from the surface and plays a leading role in the mechanism of the combustion front propagation [11]. For the Ni + Al mixture, the calculated increase in the burning velocities agrees well with the experimental data in a flow of argon and nitrogen, which indicates the realization of the conductive mode (see table 2).

When a mixed mixture of 54% (Ni + Al)/46% (Ti + C) was burned in an argon flow, the burning velocity growth was small even at a flow rate of 600–700 l/h in full accordance with the estimates of combustion filtration theory (see table 2). The sample obtained by burning in argon retained the dimensions of the original sample and was fragile. When crushed, it was scattered into individual granules. The latter fact is unusual, since the calculated combustion temperature of the composite mixture (T_{ad} = 2518 K) is significantly higher than the melting point of NiAl (T_{ad} = 1912 K), while it was previously shown that a large amount of liquid phase causes the sample shrinking when burning the granulated mixture [4, 11].

The results of experiments on the combustion of a granulated mixture of 54% (Ni+Al)/46% (Ti + C) in a co-current flow of nitrogen are shown in figure 3, curve *2*. When analyzing the dependence of the burning velocity on the gas flow rate (figure 3, curve *2*), a strong increase in the burning velocity at low nitrogen rates (220 l/h) attracts attention. Such an increase in the burning velocity was observed neither for a mixture of Ti + C or for Ni + Al (see figure 4). This result can be explained as follows: at the contact points of the Ni + Al acceptor granules with hot Ti + C donor granules, the aluminum heats up quickly, melts (melting point 933 K) and begins to interact with nitrogen with the release of a large amount of heat. Since the adiabatic temperature of the interaction of nitrogen with aluminum, calculated on the formation of aluminum nitride of AlN composition (2570 K), is significantly higher than that of nickel with aluminum (1912 K), when burning in a stream of nitrogen, the temperature at the contact points of granules of different composition becomes higher, then in the absence of nitrogen. This development of the combustion process leads to a decrease in heat loss from Ti + C granules in the reaction front, an increase in the combustion temperature and, as a consequence, an increase in the burning velocity both by the conductive and convective mechanisms.

In general, the dependence of the burning velocity of the composite mixture on the nitrogen flow rate in the studied range < 1100 l/h can be approximated by a linear function (figure 3, curve 2), which is steeper than the calculated dependence on the theory of filtration combustion (figure 3, curve *3*). At a high burning velocity corresponding to high gas flow rates, the heat exchange between Ti + C and Ni + Al granules in the combustion wave front could be assumed to have no time to occur, and therefore Ni + Al granules ignite and burn behind the visible combustion front. However, when analyzing the video recording of the combustion process of the granular mixture (Ni + Al)/(Ti + C), no two-front structure was detected, which indicates ignition of the acceptor Ni + Al granules in the combustion front throughout the entire nitrogen flow rate range.

Note that the combustion products of the mixture of 54% (Ni + Al)/46% (Ti + C) in a stream of nitrogen did not sinter and the sample was easily divided into individual granules by hand, as in the absence of a gas flow. This result is a consequence of the convective mode of combustion, when a layer of refractory titanium nitride is formed on the surface of Ti + C granules, which prevents sintering of the granules [6]. Conducted X-ray phase analysis (DRON-3M, Russia) showed that the products consist of phases of TiC and NiAl.

Thus, the experiments confirmed the assumption of the strong influence of the co-current nitrogen flow on the combustion process of the thermally conjugated granular system of 54% (Ni + Al)/46% (Ti + C). The gas flow is a regulating parameter, easily varying in magnitude, duration of exposure and chemical composition. The advantages of carrying out thermally conjugated processes using granulated mixtures are preserved in the co-current gas flow: the heat exchange between donor and acceptor mixtures does not depend on the sample size, but is determined only by the size of the granules, the combustion products consist of separate, easily detachable granules.

4. Conclusions

The regularities of the combustion in a co-current gas flow for the thermally conjugated mixture consisting of granules of different composition, taken in the ratio of 54% (Ni + Al) and 46% (Ti + C), are studied.

It is experimentally proven that the ignition of Ni + Al acceptor granules occurs in the combustion front over the entire range of gas flow rate used.

It is shown that in the absence of a gas flow, the time taken to equalize the temperatures of the donor and acceptor mixtures is substantially less than the cooling time of the entire sample, which allows scaling of the process optimized under laboratory conditions, while retaining the characteristics of the target synthesis product.

A high burning velocity growth of the thermally coupled granulated mixture was detected even at low nitrogen rates, exceeding the burning velocity growth of undiluted Ti + C and Ni + Al mixtures separately at the same flow rates, which is explained by the interaction of aluminum with nitrogen and effective reduction of donor heat loss.

Using the 54% (Ni + Al)/46% (Ti + C) mixture as an example, it is shown that the application of granulated mixtures for conducting conjugated thermal reactions makes it possible to obtain combustion products in the form of an easily destructible sample, even if the melting point of the interaction product of the acceptor mixture is less than the adiabatic temperature of combustion. The possibility of isolating the target product from the burnt sample was demonstrated.

References

[1] Merzhanov A G and Mukasyan A S 2007 *Solid Flame Combustion* (Moscow: Torus Press)

[2] Merzhanov A G 2010 *Dokl. Phys. Chem.* **434** 159 DOI: 10.1134/S0012501610100015

[3] Vadchenko S G, Balikhina N T and Kvanin V L 2004 *Comb. Explos. Shock Waves* **38** 425 DOI: 10.1023/A:1016207115072

[4] Seplyarskii B S, Tarasov A G and Kochetkov R A 2013 *Comb. Explos. Shock Waves* **49** 555 DOI: 10.1134/S0010508213050079

[5] Seplyarskii B S, Kochetkov R A and Vadchenko S G 2016 *Comb. Explos. Shock Waves* **52** 665 DOI: 10.1134/S001050821606006X

[6] Skobeltsov V P, Seplyasckii B S, Lazunin I A, Kharitonov E V and Boldov V V 1988 *Method for producing titanium carbide* Author's certificate 1462702 (USSR)

[7] Seplyarsky B S Kochetkov R A, Lisina T G and Abzalov N I **2017** *Adv. Mater. Tech.* DOI: 10.17277/amt.2017.04.pp.022-028

[8] Braverman B S, Maksimov Y M and Tsybul'nik YV 2012 *Comb. Explos. Shock Waves* **48** 734 DOI: 10.1134/S0010508212060135

[9] Seplyarskii B S and Kochetkov R A 2017 *Int. J. Self-Propag. High-Temp. Synth.* **26** 134 DOI: 10.3103/S106138621702011X

[10] Aldushin A P and Merzhanov A G 1988 *In: Theory of filtration combustion: overview and status of research* (Novosibirsk: Nauka)

[11] Seplyarskii B S and Kochetkov R A 2017 *Russ. J. Phys. Chem.* B **11** 798 DOI: 10.1134/S1990793117050116

The conditions for the implementation of the convective mode of combustion for granular mixtures of Ti + *x*C

B S Seplyarskii, R A Kochetkov, T G Lisina and M I Alymov

Merzhanov Institute of Structural Macrokinetics and Materials Science, Russian Academy of Sciences, Chernogolovka, Russia

E-mail: seplb1@mail.ru

Abstract. An experimental-theoretical method is proposed for calculating the burning velocity of Ti + xC granulated mixtures in a convective mode. The quadratic dependence of the burning velocity on the gas flow rate is shown. The behavior of the burning velocity in the convective mode was studied on the governing parameters and the gas flow rates are calculated, at which the convective mode of combustion replaces the conductive one. The transition between the combustion modes of a granulated mixture of Ti + 0.5C was experimentally realized, and a good agreement between the gas flow rate at the boundary of modes and the calculated value in the transition point was shown. The results can be used to explain and predict fast forest fires, since the fast convective combustion mode is realized with large gas flow rates (high wind speed) and low ignition temperatures typical of forests during the drought period.

1. Introduction

In the experimental works [1, 2], the existence of the convective mode of combustion of fast-burning granulated mixtures Ti + xC ($0.5 \leq x \leq 1$) with pressure drops below 1 ÷ 2 atm was established. The notion "convective mode" is used here for the designation of a combustion process in which the flow of active gas ignites granules from the surface, providing a higher velocity of the combustion front than it follows from estimates on the theory of filtration combustion (conductive mode) [3] and higher velocity than in the inert gas flow. However, the regularities of combustion in this mode were not investigated, and the parametric regions of the implementation of various modes were not established. Known from the literature, numerical calculations have not established the criteria for the transition from the conductive mechanism of combustion to the convective one [4, 5].

In this work, a theoretical answer is derived to the question of the region of the convective combustion mode realization and the nature of the dependence of the burning velocity in this mode on the gas flow rate for Ti + xC mixtures, taking into account the results of work [2]. For a mixture of Ti + 0.5C, experimental confirmation of the transition of the combustion regime from conducting to convective is obtained, and the parameters of this transition are determined.

2. Calculation of combustion velocity dependence on gas flow rate

In [2], for the granular composition of Ti + C, under the assumption of the existence of a convective mode, the heat transfer coefficient of the granules with filtered nitrogen was calculated. In the calculation, we used the values of the effective combustion temperature and the ignition temperature of a mixture of Ti + C and the experimental ignition time of the granule. Then, assuming that the heat transfer coefficient remains the same as for the Ti + C mixture, the combustion temperatures for the Ti

+ 0.5C and Ti + 0.75C mixtures were calculated. The present work used this approach to solve the inverse problem — using the known heat rate coefficients the ignition time for different gas flow rates was calculated, and hence the velocity of the ignition wave was obtained. The velocity of the ignition is considered to be equal to the burning velocity of the mixture in a convective mode. Let us demonstrate the method of calculating the dependence of the burning velocity in a convective mode on the gas flow rate using the example of a mixture of Ti + C.

The velocity of the ignition wave is determined by the formula:

$$u = d/t, \tag{1}$$

where d is the characteristic size of the granules (d = 1 mm), t is the ignition time of the granule.

Taking into account the high burning rate, the time of the granule ignition by the filtering gas t, as in [2], can be determined from the expressions for heating a semi-infinite body under boundary conditions of the third kind [6]:

$$\frac{T_{ig} - T_0}{T_g - T_0} = 1 - \varphi(\omega),$$

where T_g is the combustion temperature of a mixture, T_0 is the initial temperature of the granulated mixture, T_{ig} is the surface temperature of the granule at the instant of ignition, the table additional error function $\varphi(\omega) = \exp(\omega^2)\cdot(1 - erf(\omega))$, and the argument ω of the table function

$$\omega = \alpha^*(Q)\sqrt{at}/\lambda$$

Here $\alpha^*(Q)$ is the coefficient of interphase heat transfer depending on the gas flow rate Q, λ is mean thermal conductivity (λ = 1 W/(m·K) [7]), a is heat diffusivity of granules ($a = 10^{-2}$ sm^2/s [8]). Given (1), the velocity of the ignition wave u is expressed as follows:

$$u(Q) = [\alpha^*(Q)/\omega)\,\lambda]^2\, d/a \tag{2}$$

The value of T_{ig} is equal to the temperature $\alpha \rightarrow \beta$ transition in titanium (1155 K) when burning in a stream of nitrogen and the melting point of titanium (1933 K) when burning in a stream of argon [1,2]. Thus, to calculate the dependence of the burning velocity on the gas flow rate, it is necessary to establish the dependence of the heat transfer coefficient of the granules with the gas α* on the gas flow rate Q.

To calculate the heat transfer coefficient of a porous substance with a gas flow α we will use the formula obtained by transposing the criterion expression from [9]:

$$\begin{gathered} \alpha = \rho_g \cdot Q \cdot c_g \cdot \mathrm{Pr}^{-2/3} \cdot (1-\varepsilon)^{-1} \cdot \Psi(\varepsilon)/4s, \\ \Psi(\varepsilon) = 0.508 - 0.56\cdot(1-\varepsilon),\ \varepsilon < 0.4, \\ \Psi(\varepsilon) = 1 - 1.164\cdot(1-\varepsilon)^{2/3},\ \varepsilon > 0.4. \end{gathered} \tag{3}$$

Here Pr is the Prandtl number, $0 < \varepsilon < 1$ is the open porosity of condensed matter, ρ_g is the gas density (ρ_N nitrogen, ρ_{Ar} argon), c_g is the gas heat capacity (c_N nitrogen, c_{Ar} argon), Q is the gas flow (vol), s = 2 cm^2 is the sample cross section area, d = 1 mm.

We note the directly proportional dependence of α on the gas flow rate, its density and heat capacity in expression (3). To calculate the values of α, ε was taken to be 0.5, which corresponds to the data of [2], Prandtl number for nitrogen was taken to be 0.8. The results of the calculations are given in table 1.

Compare the value of the heat transfer coefficient in the convective mode α = 262 W/(m^{2}K) at Q = 800 l/h, Δp = 1 atm (table 1) and the value calculated in [2] based on experimental data: α* = 2711 W/(m^{2}K). The heat transfer coefficient α* is 10.4 times the theoretical value α calculated by formulas (3). Such a strong difference in the heat transfer rate of gas with granules may be due to the roughness of the surface and the random nature of laying the granules of the mixture, which ensures rapid heating of the surface layer with a hot gas flow and ignition of the granule [10, 11].

Assuming that formulae (3) correctly reflect the dependence of the heat transfer coefficient on the gas flow rate, the values of the effective heat transfer coefficients for other gas flow rates α* were calculated by multiplying α_h from the table 1 to 10.4 The results of calculations α* are given in table 1.

In [2] it was shown that the convective mode is realized by combustion Ti + xC ($0.5 \leq x \leq 1$) mixtures in nitrogen. If we assume that a convective regime is also possible in the argon flow, it is necessary to use the heat transfer coefficients α_{Ar}* for argon in the estimates of the combustion velocity. To obtain them, the values of α^*, calculated for nitrogen, should be multiplied in accordance with the expression (3) by the ratio $\rho_{Ar}c_{Ar}/(\rho_N c_N)$, and the Prandtl number for argon should be used (take 0.6).

Table 1. Heat transfer coefficients of Ti + C granulated mixture for different gas flow rates.

Q, l/h	α, Wm^{-2}K^{-1} nitrogen	α^*, Wm^{-2}K^{-1} nitrogen	α_{Ar}^*, Wm^{-2}K^{-1} argon
50	16	169	125
195	64	661	489
420	138	1423	1053
600	196	2033	1504
800	262	2711	2006
942	308	3192	2362
1100	360	3728	2758

The results of calculating the effective heat transfer coefficients for a mixture of Ti + C in nitrogen and argon are applicable to mixtures of Ti + 0.5C and Ti + 0.75C at the same gas flow rates, since expression (3) does not contain thermal characteristics of mixtures. The values of the temperatures T_g, T_0, T_{ig} used in estimation ω for mixtures Ti + xC in nitrogen and argon gas flows as well as the obtained values of ω are given in table 2. For each meaning of the flow rate Q, formula (2) calculates the burning velocity values in the convective mode in nitrogen and argon. The calculation of the velocities gives quadratic dependences, which is a consequence of the quadratic dependence of the burning velocity on the heat transfer coefficient α^* (expression 2), which in turn is linearly dependent on Q (formula (3)). Table 2 shows expressions for the dependence of the burning velocity of mixtures u on the gas flow rate in the convective mode for mixtures of Ti + xC in a stream of nitrogen and argon.

Table 2. The results of calculations of ω and u for mixtures Ti + xC.

			Ti + C, T_g = 3300K		Ti + 0.75C, T_g = 3100K		Ti + 0.5C, T_g = 2600K	
Gas	T_0	$T(0,t_{ign})$	ω	u, mm/s	ω	u, mm/s	ω	u, mm/s
Nitrogen	300 K	1155 K	0.35	$9\cdot10^{-5}Q^2$	0.38	$8\cdot10^{-5}Q^2$	0.51	$4\cdot10^{-5}Q^2$
Argon	300 K	1933 K	0.90	$8\cdot10^{-6}Q^2$	1.03	$6\cdot10^{-6}Q^2$	1.71	$2\cdot10^{-6}Q^2$

Figure 1 shows the calculated dependences of the burning velocity u on the nitrogen flow rate in the convective mode at a permanent ignition temperature T_{ig} for different values T_g, of the igniting gas temperature, corresponding to the mixtures from table 2 (curves 1, 3, 5). For these mixtures, the figure also shows the dependences of the burning velocity u_f on the nitrogen flow rate (straight lines 2, 4, 6) calculated according to the theory of filtration combustion [3], in which the values of the burning velocity at $Q = 0$ are taken from [2].

From physical reason, it is clear that when the mixture is burned in a stream of nitrogen, the regime which provides a higher velocity of the front propagation is realized. Therefore, the intersection points of curves 1 and 2 (for the mixture of Ti + C), 3 and 4 (for the mixture of Ti + 0.75C), 5 and 6 (for the mixture of Ti + 0.5C) in figure 1 determine the calculated boundary of the transition from conductive to convective mode of combustion in a co-current flow of nitrogen. For the burning velocity of Ti + xC mixtures in argon in the convective mode, the factors at Q^2 are an order of magnitude lower than for combustion in nitrogen (table 2), whereas the calculations according to the theory of filtration combustion give velocities close to velocities of combustion in nitrogen [2]. Consequently, the realization of the convective mode at gas flow rates at the experimental conditions [2] in the argon is impossible. This coincides with the conclusions of [2], made on the basis of a comparison of the ignition time of the granule in the argon flow and the time of its heating.

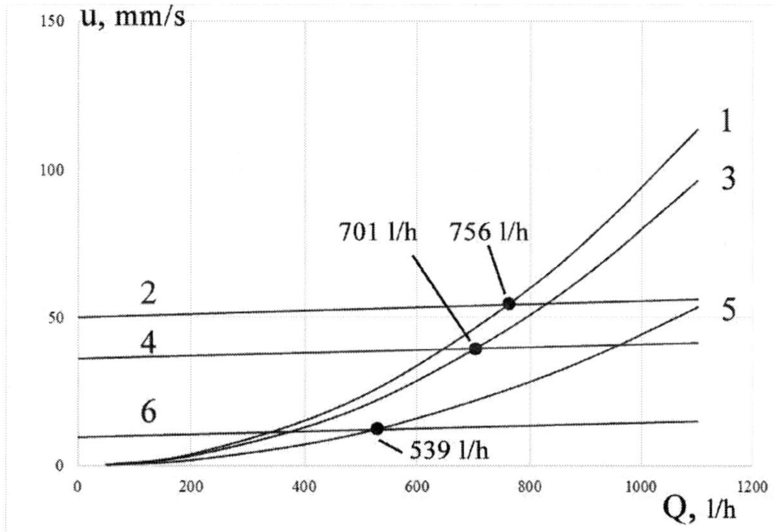

Figure 1. The dependences of the burning velocity u (lines 1, 3, 5) on the nitrogen flow rate Q in the convective mode for mixtures of Ti + C, Ti + 0.75C, Ti + 0.5C (see parameters in table 2) and the velocities u_f calculated for these mixtures according to the theory of filtration combustion (strait lines 2, 4, 6), correspondingly.

3. Discussion of theoretical results and comparison with experiment

The cause of a stronger dependence of the burning velocity in the convective mode on nitrogen flow rates (a larger multiplier at Q^2) with the raising in the soot content in the mixture is an increase in the combustion temperature and, accordingly, the gas temperature T_g, which leads to a decrease in ω in expression (1). Nitrogen flow rates on the boundary between combustion modes (in figure 1, the intersection points of curves 1 and 2, 3 and 4, 5 and 6) increase with raising content of soot x from $x = 0.5$ to $x = 1$, which is a consequence of the growth of the burning velocity of mixtures Ti + xC with increasing x in the absence of a gas stream. Therefore, the region of the conductive regime existence expands with increasing the content of soot x in the mixture.

Let us analyze how the change in the ignition temperature of the granulated mixture T_{ig} affects the velocity of combustion in a convective mode. Figure 2 shows the calculated dependence of the burning velocity u on the nitrogen flow rate at $T_g = 3300$ K for different values of T_{ig}. To demonstrate the effect of T_{ig} on the boundary of the conductive combustion mode in figure 2 (curve 5) we present the results of calculating the dependence of the burning velocity in nitrogen u_f on gas flow rate using the theory of filtration combustion for a mixture of Ti + C. Note that, in accordance with the theory of filtration combustion, the burning velocity does not depend on the ignition temperature T_{ig}, therefore, as T_{ig} varies, the dependence of the burning velocity on the gas flow rates does not change.

As can be seen from figure 2, the higher the ignition temperature of the granule, the higher values of the gas flow rate for the convective combustion mode realization. The dependences of the burning velocity in the convective mode on the nitrogen flow rate at different ignition temperatures $T(0, t_{ig})$ are quadratic, but with increasing T_{ig} the multiplier value at Q^2 decreases (see table 2), which leads to the expansion of the implementation region of the conductive mode. Experimental verification of the proposed method of determining the boundary of the transition between the conducting and convective modes of combustion was carried out for a granular mixture of Ti + 0.5C in nitrogen. The experimental setup, starting materials and the method of preparation of the granules were identical to those described in [2], except that the mixture was made in a "drunk barrel" mixer using steel balls.

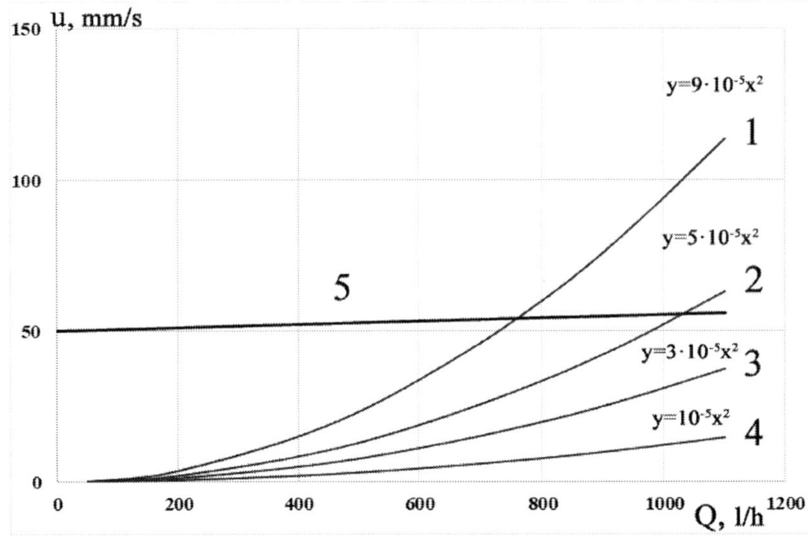

Figure 2. The dependence of the burning velocity on the nitrogen flow rate Q in the convective mode at $T_g = 3300$ K for different values of T_{ig}: 1-1155 K, 2 – 1400 K, 3 - 1600 K, 4 – 2000 K. Straight line 5 shows a dependence of the combustion velocity of a mixture of Ti + C in nitrogen in the conductive mode.

The experimental dependence of the burning velocity of the Ti + 0.5C mixture on the nitrogen flow rate is shown in figure 3. Bold line in figure 3 shows the calculated curve, which consists of two parts: at flow rates below $Q = 779$ l/h, the conductive mode (calculated according to the theory of filtration combustion), above, the convective mode. The experimental value of the mode boundary was estimated from the coordinates of the point of intersection of the straight lines corresponding to a linear extrapolation of the left and right branches of the experimental curve, and was found to be 732 l/h. From figure 3, it can be seen that the proposed calculation method leads to a complete qualitative coincidence of the theoretical and experimental dependences of the burning velocity on nitrogen consumption and a good quantitative coincidence of gas consumption at the mode boundary.

Thus, a simple approximate method is proposed for calculating the burning velocity of granulated mixtures in the convective mode. It is shown that in this mode the dependence of the velocity on the flow rate of the filtering gas is quadratic. The regions of nitrogen flow rates are determined where the convective mode of combustion in granular mixtures Ti + xC ($0.5 \leq x \leq 1$) is realized. The impossibility of the convective mode is shown when burning granulated mixtures Ti + xC in the argon in the range of gas flow rates used in the calculations ($Q \leq 1100$ l/h). It was concluded that the main factors contributing to the transition of combustion of granulated mixtures into a convective mode are an increase in the gas flow rate and a decrease in the ignition temperature of the granules. When burning a granulated mixture of Ti + 0.5C in a co-current stream of nitrogen, the value of the nitrogen flow rate at the boundary of the conductive and convective modes of combustion was experimentally determined, which well coincides with the results of calculations. The results obtained in this work can be used to explain and predict fast forest fires. Indeed, a fast convective mode of combustion occurs at high gas flows (high wind speed) and low ignition temperatures typical of forests during the drought period [12, 13].

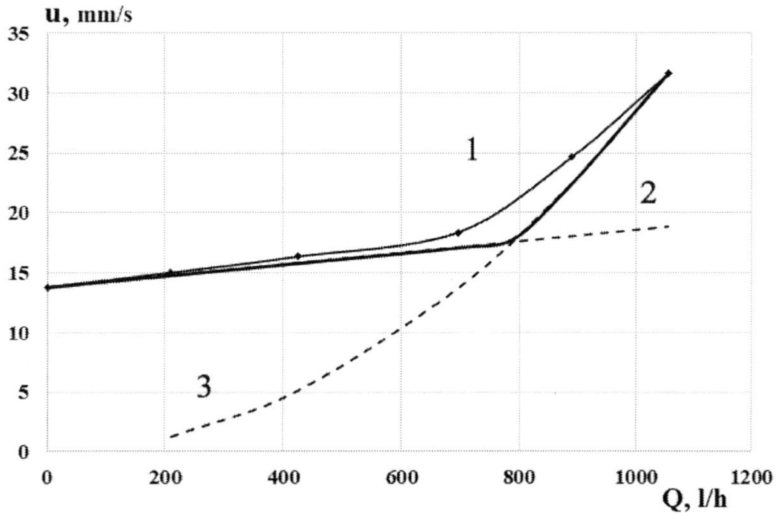

Figure 3. Comparison of the experimental dependence of the burning velocity u of the Ti + 0.5C mixture on the nitrogen flow rates Q (curve 1) with calculations based on the theory of filtration combustion (curve 2) and on the model of the convective mode (curve 3).

References

[1] Seplyarskii B S, Tarasov A G and Kochetkov R A 2013 *Comb. Explos. Shock Waves* **49** 555 DOI: 10.1134/S0010508213050079

[2] Seplyarskii B S and Kochetkov R A 2017 *Russ. J. Phys. Chem.* B **11** 798 DOI: 10.1134/S1990793117050116

[3] Aldushin A P and Merzhanov A G 1988 *In: Theory of filtration combustion: overview and status of research* (Novosibirsk: Nauka)

[4] Aldushin A P 1990 *Comb. Explos. Shock Waves* **26** 180 DOI: 10.1007/BF00742407

[5] Lapshin O V, Prokof'ev V G and Smolyakov V K 2018 *Int. J. Self-Propag. High-Temp Synth.* **27** 14

[6] Lykov A V 1967 *The Theory of Heat Conductivity* (Moscow: Vyssh.Shkola)

[7] Kasatskii N G, Filatov V V and Naiborodenko Yu S 1991 *In: Self-propagating High-Temperature Synthesis* (Tomsk: Tomsk Univ.)

[8] Zenin A A 1981 *Comb. Explos. Shock Waves* **17** 63 DOI: 10.1007/BF00772787

[9] Goldshtik M A 1984 *Transfer processes in granular layer* (Novosibirsk: Institute of Thermophysics of the Siberian Branch of the USSR Acad. Science [in Russian])

[10] Matros Yu Sh 1985 *Unsteady Processes in Catalytic Reactors* (Amsterdam: Elsevier)

[11] Kreith F. and Black W Z 1980 *Basic Heat Transfer* (NewYork: Harper& Row)

[12] Nahmias J, Téphany H, Duarte J et al 2000 *Canadian Journal of Forest Research* **30** 1318 DOI: 10.1139/x00-047

[13] Beer T. 1990 *Comb. Sci. Tech.* **72** 297.

[14] Rothermel R C Predicting Behavior and Size of Crown Fires in the Northern Rocky Mountains, 1991 (Ogden)

IOP Conf. Series: Materials Science and Engineering **558** (2019) 012046 doi:10.1088/1757-899X/558/1/012046

Mathematical modeling of phase separation in three-phase melt formed behind the combustion wave

K G Shkadinsky[1], N I Ozerkovskaya[2] and P M Krishenik[2]

[1]Institute of Problems of Chemical Physics, Russian Academy of Sciences, Chernogolovka 142432, Russia
[2]Merzhanov Institute of Structural Macrokinetics and Materials Science, Russian Academy of Sciences, Chernogolovka 142432, Russia

E-mail: shkad@ficp.ac.ru

Abstract. A mathematical model of the exothermic chemical interaction of a high-calorie solid powder mixture in the presence of centrifugal forces is considered. The process of mathematical modeling is studied for the $3NiO + (2 + \alpha)Al \rightarrow Al_2O_3 + (3 - \alpha)Ni + \alpha NiAl$ thermite composition, as an example. Such models are used to describe the process of high-temperature synthesis for the production of target products in the combustion wave propagating through the initial mixture. This technological process represents a combination of high-temperature combustion with the transformation of the solid porous mixture to liquid products, their separation, and subsequent cooling. A distinctive feature of the proposed model is the use of a "through" description, which makes it possible to examine the mutual influence of the individual stages on the dynamics of the process. This feature provides a more realistic description of the actual process.

1. Introduction

The combustion of high-calorie mixtures is used in high-temperature synthesis [1] to produce desired products. It occurs under extreme conditions: high temperatures and pressures, melt dispersion, and onset of an explosive dynamic frontal mode. The front propagation velocity ranges from several millimeters to several centimeters per second. Typically, three main stages of the process are distinguished: combustion; gravity (or centrifugal) separation of the medium (metal phase, ceramics, oxidizing gas); and cooling and crystallization of products. The process is controlled by changing the temperature of the chemical interaction (the initial mixture composition is changed), pressure or centrifugal forces (melt dispersion is suppressed and phase separation is promoted), and dispersity and quality of mixing (initial mixture density affects the burning rate and the properties of the products). Extreme combustion conditions limit the possibility of performing direct experiments and motivate the use of modeling methods based on the fundamental laws of physics and chemistry. The SHS metallurgy is characterized by the use of models for individual stages of the process with the rest of the factors being treated as perturbations. In what follows, we develop a "through model" that takes into account the mutual influence of the stages on the overall dynamics of high-temperature synthesis. High temperature in the combustion front melts the medium, a process under the action of centrifugal forces, changes the conditions for transport of reagents into the reaction zone. Mass-transfer alters the effect of the force field on the reaction medium and the process of phase separation. In constructing the "through model," we use simplifications that make the problem tractable while preserving the interaction of the stages inherent in the real process.

IOP Conf. Series: Materials Science and Engineering **558** (2019) 012046 doi:10.1088/1757-899X/558/1/012046

2. Mathematical model

The process of exothermic chemical interaction of a mixture of condensed reagents is investigated within the framework of the mechanics of multicomponent chemically active continuous media [2].

2.1 Properties of the Medium

The condensed components of the *i*th medium are assumed to be incompressible. Their volume fraction is designated as ν_i, hence, the density of the *i*th component in the composition $\rho_i = \rho^0_i \nu_i$. For the gas component, $\nu_g = 1 - \Sigma_i \rho_i$. The gas medium is assumed to be an ideal gas obeying the equation $\rho_g = p\mu_g / RT_g$, where T_g is the gas temperature, R is the universal gas constant, p is the pressure, and μ_g is the molecular mass of the gas. A concrete analysis is performed by the example of the interaction 3NiO + (2 + α)Al → Al2O3 + (3 − α)Ni + αNiAl. The macrokinetics of the interaction is rather complicated and is the subject of a separate study. We limited ourselves to an intermediate "truncated" description of the kinetics of the interaction, which can be extended to take into account the details of interest. According to the reference data, above 1503 K, solid β-phase NiO decomposes into metallic nickel and oxygen:

$$2NiO \xrightarrow{W_0(T)} 2Ni + O_2. \tag{1}$$

By this time, the preheating zone of the synthesis front in the porous composition contains molten Al (melting point 933 K, whereas the melting point of the Al_2O_3 film around is ~ 2300 K). In the reactor conditions, this refractory film partially dissolves and collapses under the action of centrifugal forces, being partially restored by the interaction of oxygen with aluminum with heat release Q:

$$4Al + 3O_2 \xrightarrow{W_1(T)} 2Al_2O_3 + Q. \tag{2}$$

A less rapid and less exothermic stage

$$Ni + Al \xrightarrow{W_2(T)} NiAl + Q_1, \tag{3}$$

produces an intermetallic compound. With increasing temperature, the initially strong skeleton of the porous material is converted into a mobile liquid melt, which when being centrifuged is pressed to the initial porous medium. The permeability of the initial medium in the preheating zone changes, and the filtration of oxygen formed into the low-temperature initial medium and into the liquid products is impeded. The concentration of oxidizer in the narrow reaction zone motivates using a simplified kinetic scheme of "gas-free" interaction instead of (1) and (2):

$$3NiO + 2Al \xrightarrow{W(T)} 3Ni + Al_2O_3 \tag{4}$$

2.2 Mass Transfer

Thus, behind the combustion front, a suspension of liquid metal, liquid ceramics, and oxidizer are formed with some characteristic dimensions of the components (for simplicity, we set them identical to the initial ones) and individual velocities of movement. The reactants involved in chemical interaction change their mass, and, consequently, the field of centrifugal forces. Taking into account that, in the condensed state, the diffusion transfer is concentrated in the reaction zone, being partially allowed for in the macrokinetic description of the chemical interaction rate, we confine ourselves to considering convective masstransfer and the mass transfer caused by chemical interaction:

$$\frac{\partial \rho(NiO)}{\partial t} = -2W_0(T)\mu(NiO) - \frac{\partial}{\partial x}(\rho(NiO)\upsilon(NiO)), \tag{5}$$

$$\frac{\partial \rho(Al)}{\partial t} = -[4W_1(T) + W_2(T)]\mu(Al) - \frac{\partial}{\partial x}(\rho(Al)\upsilon(Al)), \tag{6}$$

$$\frac{\partial \rho(NiAl)}{\partial t} = W_2(T)\mu(NiAl) - \frac{\partial}{\partial x}(\rho(NiAl)\upsilon(NiAl)), \tag{7}$$

$$\frac{\partial \rho(Ni)}{\partial t} = [2W_0(T) - W_2(T)]\mu(Ni) - \frac{\partial}{\partial x}(\rho(Ni)\upsilon(Ni)), \tag{8}$$

$$\frac{\partial \rho(AL_2O_3)}{\partial t} = 2W_1(T)\mu(Al_2O_3) - \frac{\partial}{\partial x}(\rho(Al_2O_3)\upsilon(Al_2O_3)), \quad (9)$$

$$\frac{\partial \rho(O_2)}{\partial t} = [W_0(T) - 3W_1(T)]\mu(O_2) - \frac{\partial}{\partial x}(\rho(O_2)\upsilon(O_2)). \quad (10)$$

Here, the density of each reacting element is the product of the true density and its volume fraction. For the incompressible media (all except for the gas), these equations describe the dynamics of the change in the volume fraction of the reagents. The velocity of movement and the temperature of a particular reagent depend on its spatial location and chemical content. Such a multi-velocity and multi-temperature description of mixed continuous media hugely complicates the mathematical model. For simplicity, the property characteristic of solutions is used: typically, the solubility is high when "like dissolves in like" (oxide in oxide, metal in metal). All the components, we categorized into three groups: metal, ceramics, and oxidizing gas. Accordingly, we consider a three-velocity, three-temperature combustion model for a mixture of continuous media, assuming that, in each group, the characteristics of the medium depend on its location. The laws of momentum conservation for the metal, ceramics, and oxidizing gas components moving at velocities much lower than the speed of sound in the test mixture are reduced to the balance of the principal forces acting on the medium. For the metal group,

$$\rho_m\left(\frac{\partial \upsilon_m}{\partial t} + \upsilon_m\frac{\partial \upsilon_m}{\partial x}\right) = 0 = -v_m\frac{\partial p}{\partial x} + v_m\frac{\partial}{\partial x}\left(\eta_m\frac{\partial \upsilon_m}{\partial x}\right) - f(L+x)\omega^2\rho_m - \alpha_{11}v_m\upsilon_m$$
$$+\alpha_{12}(\upsilon_k - \upsilon_m) + \alpha_{13}(\upsilon_g - \upsilon_m) + W_0(T)\mu(Ni)(\upsilon_k - \upsilon_m). \quad (11)$$

For the metal component, the pressure gradient multiplied by the cross-sectional area (proportional to v_m) is balanced, first, by the centrifugal forces acting on the metal components in the emulsion and, secondly, by the forces of viscous interaction with other components of the emulsion in their relative motion, as well as by momentum exchange during chemical mass transfer. Similar equations are valid for the ceramics and gas:

$$\rho_k\left(\frac{\partial \upsilon_k}{\partial t} + \upsilon_k\frac{\partial \upsilon_k}{\partial x}\right) = 0 = -v_k\frac{\partial p}{\partial x} + v_k\frac{\partial}{\partial x}\left(\eta_k\frac{\partial \upsilon_k}{\partial x}\right) - f(L+x)\omega^2\rho_k - \alpha_{22}v_k\upsilon_k$$
$$+\alpha_{21}(\upsilon_m - \upsilon_k) + \alpha_{23}(\upsilon_g - \upsilon_k) + W_1(T)(4\mu(Al)(\upsilon_m - \upsilon_k) + 3\mu(O_2)(\upsilon_g - \upsilon_k)), \quad (12)$$

$$\rho_g\left(\frac{\partial \upsilon_g}{\partial t} + \upsilon_g\frac{\partial \upsilon_g}{\partial x}\right) = 0 = -v_g\frac{\partial p}{\partial x} + v_g\frac{\partial}{\partial x}\left(\eta_g\frac{\partial \upsilon_g}{\partial x}\right) - f(L+x)\omega^2\rho_g - \alpha_{33}v_g\upsilon_g$$
$$+\alpha_{13}(\upsilon_m - \upsilon_g) + \alpha_{23}(\upsilon_k - \upsilon_g) + W_0(T)\mu(O_2)(\upsilon_k - \upsilon_g) \quad (13)$$

Here, L is the distance from the axis of the centrifuge to the open surface of the reactor; f is the multiplier; η is the viscosity; ω is the angular velocity of centrifugal action on the emulsion components; α_{ii}, α_{ij}, and α_{ji} are the coefficients of viscous interaction forces (I and j are in the range of 1–3).

The laws of mass conservation are represented by the heat conduction equations for the three groups (metal, ceramics, and gas), with consideration given to the mobility of the emulsion components and the heat exchange between the groups and the external environment (its temperature is set constant, equal to the initial T_0). The main sources of heat generation are the exothermicity of the chemical interaction in the group and the energy exchange that accompanies the mass transfer between the groups during the chemical reaction. We disregarded less significant sources and confined ourselves to a one-dimensional description in order to simplify and analyze the results obtained. For brevity, only the equation for the group of metals is given here; for the other groups, the equations have a similar structure:

$$c_m\rho_m\left(\frac{\partial T_m}{\partial t} + \upsilon_m\frac{\partial T_m}{\partial x}\right) = \frac{\partial}{\partial x}\left(\lambda_m\frac{\partial T_m}{\partial x}\right) + \overline{(QW)}_m + 2(c\mu)_{Ni}W_0(T)(T_k - T_m) - a_{mk}(T_m - T_k) \quad (14)$$
$$-a_{mg}(T_m - T_{g)} - a_{mw}(T_m - T_0).$$

Here, $\overline{(QW)}_m = Q(\mathrm{Al})W_1(T)\mu(\mathrm{Al}) + Q(\mathrm{NiAl})W_2(T)\mu(\mathrm{NiAl})_l$. The coefficients a_{mk}, a_{mg}, and a_{mw} characterize the heat exchange between the components of the emulsion and wall. The initial conditions read as $T_m = T_k = T_g = T_0$. The boundary conditions at the end faces are written as follows: $T_m(0, t) = T_k(0, t) = T_g(0, t) = TW$ is the condition for initiating the combustion process (they can be pulse type); $\left(\frac{\partial T_m}{\partial x}\right)_{(l,t)} = \left(\frac{\partial T_k}{\partial x}\right)_{(l,t)} = \left(\frac{\partial T_g}{\partial x}\right)_{(l,t)} = 0$ is the condition of adiabaticity at the closed (bottom) end face.

2.3 Quasi-Hydrostatic Approximation

Given that the process of combustion is characterized the velocities of motion of the mixture components much smaller than the speed of sound, a situation arises similar to that described in [3], more specifically, a small parameter appears in the model that leads to its singularity. The liquid combustion products transit into a stressed state, which varies with the composition (density) and the distance from the axis of the centrifuge. Describing the quasi-stationary stress state of the mixture by the equation

$$\frac{\partial p}{\partial x} = f(L+x)\omega^2\left(\nu_g p\mu_g / RT_g + \sum_i \nu_i \rho_i^0\right) \qquad (15)$$

with boundary conditions at the end faces of the reactor, we do not consider the rapid process of onset of hydrostatic equilibrium. The heat conduction equations for the gas, metal, and ceramics determine the temperature fields for the mixture components. Equations (5−9) determine the composition of the incompressible components in the mixed reaction medium. Hydrostatic equilibrium condition (15) determines the evolution of the pressure in the moving combustion products. Equations (11), (12), and (13) with the appropriate boundary conditions determine the velocities of the gas, metal, and ceramics. In the absence of centrifugal forces, equation (15) corresponds to the quasi-isobaric situation considered in [3]. Note also that singularities and degeneracy can arise in analyzing the system of heat conduction equations under the conditions of phase separation caused by centrifugation. An effective algorithm for approximately solving these equations is assumed known.

3. Results

The results of numerical simulations of the dynamics of the combustion process are presented on the space−time plots of temperature, fractions of reagents, and rates of their convective transfer. The process is initiated by a high-temperature source (spiral) at the open end of the reactor ($x = 0$). As a result, the frontal process of exothermic chemical reaction in the mixed composition sets in. Figure 1 displays successive distributions (from left to right) of the temperature fields in the gas, ceramic, and metal at constant time intervals. These distributions show that a high-temperature region propagating to the right is formed. The structure of the right edge of this region (front zone) stabilizes its characteristics. In fact, they change slowly, so the front is called quasi-stationary; the reasons for this change are explained below. The initial stage of the formation (stage of ignition) being a typical mode of induction ignition of the medium with a relatively low-temperature source contrasts against the background of steady combustion. During this period, an excess of enthalpy arises in the front, which moving together with the front, plays an important role in the analysis of its stability and affects the temperature distribution of the medium when the front reaches the reactor (at $x = 20$). The determining role in the combustion process of the mixed medium is played by the exothermic transformation in the metal phase. Through heat exchange, it influences the processes in the ceramic and gas phases, thereby creating a unified structure of the combustion front of the mixed medium. Note that the specific features of the temperature distribution in ceramics and gas phase are determined not only by classic conduction heat transfer, but also by the energy exchange caused by interphase mass transfer. In the conditions of adiabatic steady combustion (in figure 1, the mixed medium burns in the absence of heat exchange with the external medium), the adiabatic combustion temperature is established in the products behind the combustion front. As can be seen, in this case, superadiabatic temperatures arise in the metal phase in the reaction zone. This contributes to an increase in the propagation rate of the front. We can demonstrate the steady propagation of the combustion front due to the influence of the varying field of centrifugal forces. The field is determined by the redistribution of the mass of the reagents in the exothermic chemical interaction, the dynamics of which is shown in figures 2 and 3. Here, like in figure 1, the space–time distributions (at the same instants of time as those for the temperature fields) of the specific volumes of the chemical reagents of the mixed medium during its burning are shown.

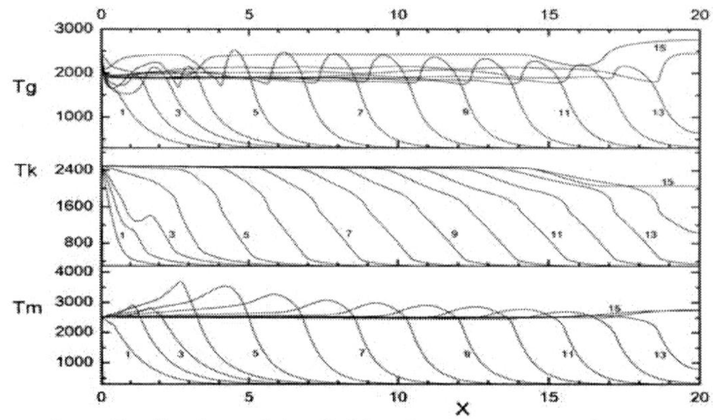

Figure 1. Space–time distribution of the fields of temperature of gas, ceramics and metal.

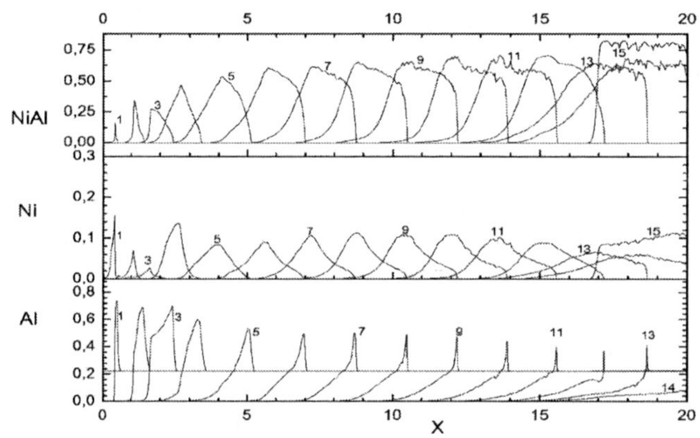

Figure 2. Space–time distribution of specific volumes of metal phase reagents in a mixed medium.

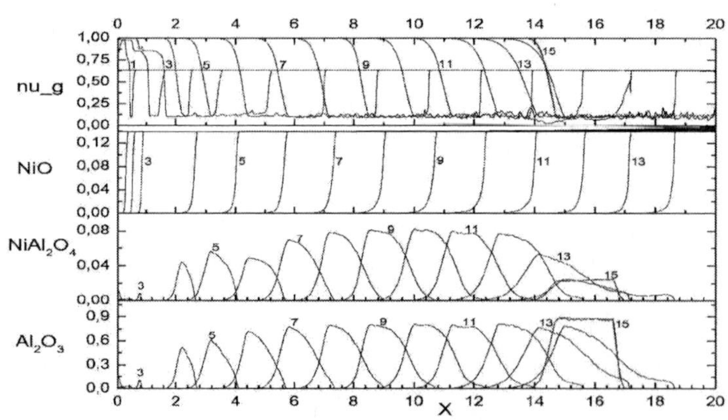

Figure 3. Space–time distribution of specific volumes of ceramics and gas reagents in a mixed medium.

IOP Conf. Series: Materials Science and Engineering **558** (2019) 012046 doi:10.1088/1757-899X/558/1/012046

One can see how the composition of the mixture changes during the propagation of the front due to the chemical interaction and convective transfer of the melt under the action of centrifugal forces. We consider convective transport is represented by the space–time distribution of the velocity fields. At the stage of ignition (small x), negative values of the velocity of the medium are seen, which corresponds to the experimentally observed dispersion of the medium. Later, the melt moves due to the densification of the gas-saturated medium under the action of centrifugal forces. Note the localization of the oxidant in the reaction zone of the combustion front.

4. Conclusions

Thus, we obtained a significant volume of information on the process of high-temperature synthesis. First, its frontal character should be emphasized. Selecting a moment of time and collecting the corresponding spatial distributions of temperatures, chemical composition, and convective motion velocities of the reagents, we obtain information about the front of the exothermic chemical interaction. As the synthesis front moves, the chemical composition of the medium and its density change, which, in turn, causes changes in the distribution of the forces acting on the multicomponent medium. The conditions of chemical interaction change, and the synthesis front becomes quasi-stationary. The separation of the products occurs according to the evolution of their temperature. Under adiabatic conditions, stratification occurs. The heavy metal phase is concentrated in the bottom part of the reactor. Above is a ceramic that protects the metal phase from contact with the gaseous medium. Being the lightest gas phase is displaced from the melt. These processes are described by the presented model and are realized during their characteristic times.

References

[1] Yukhvid V I 2002 Combustion Science and Technology Series **5** Technology of SHS Casting Self-Propagating High-Temperature Synthesis of Materials ed. Anatoli A B, Luigi D L, Merzhanov A Taylor & Francis, pp. 238.

[2] Nigmatulin R I 1987 *Dynamics of Multiphase Media* (Fizmatlit, Moscow) [in Russian].

[3] Shkadinskii K G 2014 *Russ. J. Phys. Chem.* B **8**, 356.

Nanostructured Cu–Cr–W pseudoalloys by combined use of high-energy ball milling and spark plasma sintering

N F Shkodich[1], K V Kuskov[2], I D Kovalev[1] and Yu B Scheck[1]

[1]Merzhanov Institute of Structural Macrokinetics and Materials Science, Russian Academy of Sciences, Chernogolovka, 142432 Russia
[2]National University of Science and Technology MISIS, Moscow, 119049 Russia

Abstract. Nanostructured pseudoalloys of immiscible Cu, Cr and W metals were prepared by combined use of high-energy ball milling (HEBM) and spark plasma sintering (SPS). Mechanical alloying in an Activator-2S planetary ball mill for 150 min provides a uniform distribution of refractory Cr and W particles in a moldable Cu matrix on a submicron and nanoscale level. Activated W–Cu–Cr powders were SPS-consolidated at 800–1000°C and 50 MPa for 10 min. The effect of preparation conditions on the microstructure, crystal structure, and properties (density, hardness, electric resistivity) of synthesized nanostructured W–Cu–Cr pseudoalloys was studied. The consolidated W–Cu–Cr composites showed a hardness of up to 9.5 GPa and an electric conductivity of 27–28% IACS, which makes them promising for use as an interrupter contact material.

1. Introduction

Over the past years, an increasingly growing attention has been given to tailoring the pseudoalloys of immiscible metals with strongly different melting points, which exhibit unique electric and thermophysical properties [1–3]. Combination of a refractory metal with a low-melting electro/thermo-conducting one opens up a way to designing novel materials for various applications in modern engineering. So copper-based pseudoalloys find their application in interrupter contacts, arc-resistant electrodes, heat-removing elements in microelectronic devices, etc. [4].

Among other metals, tungsten seems most attractive as a basis of items and materials for high temperature/pressure applications due to its exceedingly high refractoriness and mechanical strength. In addition, heavy-weight tungsten (19.3 g/cm^3) exhibit a high absorption coefficient for γ-radiation, which makes this metal promising for use in protection against ionizing radiation. Despite the fact that the melting point of chromium (2180 K) is lower than that of tungsten (3695 K), Cr is rather promising for use in electrical engineering in view of its high-temperature strength, corrosion resistance, chemical inertness, and hardness.

Despite a wide range of existing materials suitable for use in various electrical contacts, the problem of highly reliable electrical contacts for use as current interrupters in electric power networks still remains far from its resolution [5–7]. As for Cu–Cr composite materials, these turned out to be especially suitable for vacuum current interrupters in the range of 12–30 kV and higher; within the range below 1 kV, Cu–W contactors are used [8, 9]. Due to incongruous properties of pure refractory components, W and Cr, contact materials suitable for use in the remaining gap between 1 and 12 kV are very difficult to fabricate and thus meet ever growing requirements to vacuum contacts. To ensure reliable contact, such materials must have specific structure, properties, and durability.

In order to further improve the performance of Cu–Cr alloys, it was suggested to add still another metal X and thus obtain ternary Cu–Cr–X alloy [10–14]. The addition of 2.0 wt % W was found to improve the voltage resistance of contacts; adding 0.05–4.00 wt % Te improved the fusion-welding resistance of contact surface; adding 3.00–4.00 wt % Fe increased the solubility of Cr in Cu; and the addition of Ni improved the dielectric strength.

To ensure better mechanical strength of tailored alloys, the latter ones must be uniform and fine-grained. Such materials can in principle be prepared by high-energy ball milling (HEBM) and subsequent spark plasma sintering (SPS). In conventional powder metallurgy (sintering, hot pressing) this is hard to achieve since prolonged holding at high temperature leads to the loss of nano-crystallinity because of melting. In particular, this technique was recently used to fabricate nanostructured Cu–Cr pseudoalloy with a high Cr content [15].

In this communication, we report on fabrication of fine-grained Cu–Cr–W pseudoalloys by combined use of high-energy ball milling and spark plasma sintering.

2. Materials and Methods

Commercial powders of Cu (99.97% pure, particle size $d < 60$ μm), Cr (99.7%, $d = 10$–30 μm), and W (99.98%, $d < 6$ μm) were mixed in Cu/Cr/W ratios (by weight) of 55–40–5 (mix **1**) and 20–10–70. (mix **2**). The blends were processed in a water-cooled planetary ball mill Activator-2S using stainless-steel cylindrical vials and balls 7–9 mm in diameter. In all cases, the ball/mill ratio was 20 : 1. The drum was evacuated and then filled with Ar gas at 4 bar. The HEBM was carried out at a rotating speed of sun wheel/grinding chamber of 694 (900)/1388 rpm. Milling time (τ) was varied between 5 and 150 min. The milled powders were then consolidated by spark plasma sintering (SPS) in vacuum (Labox 650, Sinter Land, Japan). The activated powder was placed into a cylindrical graphite die (inner diameter 15 mm) and uniaxially loaded at 50 MPa. Then the sample was heated at a rate of 500°C/min up to some preset sintering temperature in the range 800–1000°C by passing rectangular pulses of electric current through the sample and the die. The dwell time at sintering temperature was 5–10 min. The consolidated samples had a shape of thin (2–3 mm) disks 15 mm in diameter.

Crystal structure, local composition of initial, milled powders and consolidated samples were characterized by XRD (DRON-3M diffractometer, Cu-K_α radiation, 2θ =20–80°), SEM (Zeiss Ultra+, Carl Zeiss) and EDS (Oxford Inca spectrometer).

Microhardness H_V of the composites was measured with a Vickers indenter. Electrical resistivity of the materials was measured by four-point probe method using a Cryotel apparatus (Cryotechnics & Electronics Ltd., Russia).

3. Results and Discussion

Figure 1 presents the XRD patterns of Cu–Cr–W mixtures taken before and after HEBM for different milling time τ. Gradual broadening of diffraction peaks of activated powders is seen to be accompanied by a decrease in their intensity. This can be caused by a decrease in crystallite size, increase in defects density, and lattice strains. As for milled Cu–Cr–W powder mixture with a high tungsten content (70 wt %), an increase in milling time (to 30–120 min) leads to vanishing the (111), (200) and (220) peaks of Cu and the (110) and (220) peaks of Cr. Furthermore, during HEBM the W peaks shifted toward larger angles, thus indicating a change in the W lattice. Such a behavior of the XRD peaks coincides with the formation of nanoscale W–Cu–Cr structure. In a softer mode of mechanical treatment (694 rpm), the structural changes proceed slower than at 900 rpm of the sun wheel (figure 1b).

As for milled Cu–Cr–W (5 wt % W) powder mixture, an increase in HEBM time up to 30 min leads to a decrease in the intensity of W, Cu, Cr diffraction peaks, while the main peaks of Cu (111) and Cr (110) partially overlap. Starting from $\tau = 60$ min, a common diffraction peak appears at $2\theta = 41$–$42°$, while the other reflexes become strongly broadened.

The activated powders were consolidated by SPS method at 800–1000°C.

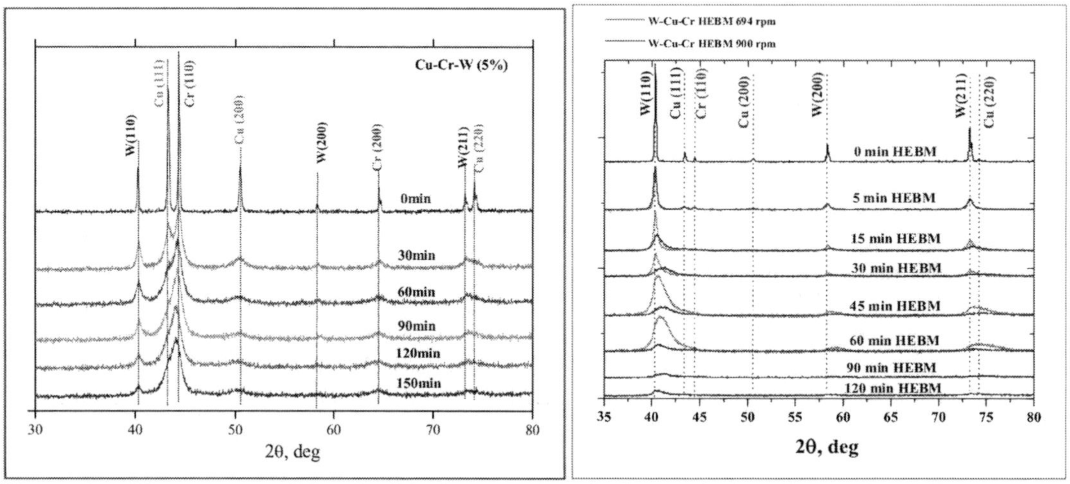

Figure 1. XRD patterns of Cu–Cr–W mixtures taken before and after HEBM for different time τ (indicated): (a) mix **2** (70% W) and (b) mix **1** (5% W).

The SEM results for mix **1** in figure 2 suggest that the composite Cu–Cr–W derived from non-activated powder (figure 2a) is comprised of Cr particles (dark grey areas), rounded W particulates (white) embedded in a Cu matrix (grey). Copper particles are seen to be partially sintered while the chromium and tungsten ones exhibit no sintering with copper.

The composite derived from the mix **1** activated for 150 min (figures 2b, 2c) represents the Cu matrix uniformly doped with nanograins of tungsten (20–100 nm) and chromium (20–50 nm). It is worth noting that the Cu–Cr–W structures formed during HEBM are retained after SPS-consolidation.

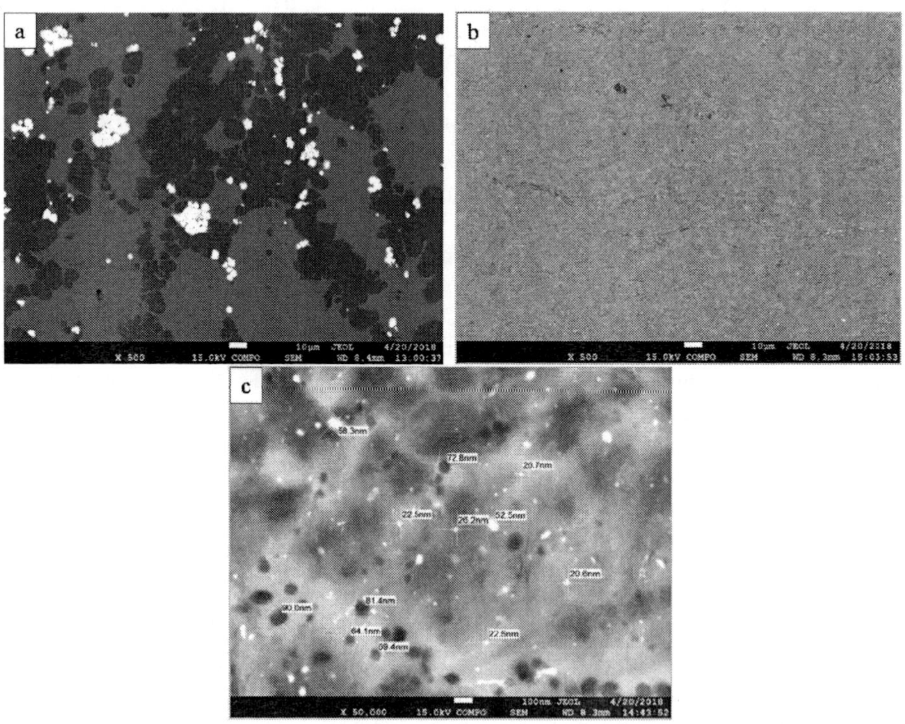

Figure 2. SEM images of Cu–Cr–W composites derived from staring (a) and activated mix **1** (τ = 150 min) SPS-consolidated at 800°C.

The Vickers hardness and relative density of W–Cu–Cr pseudoalloys derived from mix **1** and mix **2** are presented in table 1.

Table 1. Mechanical properties of fabricated W–Cu–Cr pseudoalloys

Composition and milling conditions	H_V, GPa	ρ_{rel}
Mix **2** 60 min, 694 rpm, 900°C	3.14±0.27	0.75
Mix **2** 60 min, 694 rpm, 950°C	6.35±0.12	0.91
Mix **2** 60 min, 900 rpm, 900°C	5.72±0.83	0.7
Mix **2** 60 min, 900rpm, 1000°C	9.5 ±0.72	0.82
Mix **1** 0 min, 694 rpm, 800°C	1.02±0.05	0.62
Mix **1** 150 min, 694 rpm, 800°C	5.5±0.09	0.79
Tungsten	3.43*	
Copper	0.369*	
Chromium	1.06*	

As follows from table 1, the Vickers microhardness (H_V) of produced Cu–Cr–W pseudoalloys exceeds that of Cu, Cr, and W by a factor of 30, 9 and 3, respectively.

HEBM leads to a fivefold increase in the microhardness of consolidated Cu–Cr–W pseudoalloys (mix **1**) compared to the alloys derived from non-activated powders. The maximum value of H_v = 9.5±0,72 GPa was attained for consolidated composites derived from mix **2** at 900 rpm and 1000°C. Also it was demonstrated that energy-intensive mechanical treatment (900 rpm) is a more effective way to increasing the microhardness of Cu–Cr–W materials.

As is seen in table 1, relative density ρ_{rel} attains its maximum (0.91) in case of mix **2** sintered at 950°C. Further rise in sintering temperature up to 1000°C leads to a decrease in H_V. This can be explained by a competition of two factors: material porosity and grain size. The initial increase in H_V can be explained by a decrease in overall sample porosity. But since at low porosities the Hall–Petch effect is prevailing, the hardness of polycrystalline material can be expected to decrease with an increase in the grain size. This can be associated with the process of grain growth at this temperature.

Electrical conductivity of our Cu–Cr–W composite is higher than 25% of International Annealed Copper Standard, which is tolerable for its use as an interrupter contact material.

Our results suggest that the combined use of high-energy ball milling (HEBM) and spark plasma sintering (SPS) can be readily recommended as a technological basis for fabrication of nanocrystalline Cu–Cr–W pseudoalloys.

Acknowledgments

This work was financially supported by the Russian Foundation for Basic Research (project no. 18-38-00843mol_a).

References

[1] Myshkin N K, Konchits V V and Braunovich M 1984 *Elektricheskie kontakty* (Electric Contacts) (Intellekt, Moscow) in Russian
[2] Slade P 1984 *IEEE Trans.* **7** 25
[3] *Binary Alloy Phase Diagrams* ed Massalski T B, Okamoto H, Subramanian P R 1990 (ASM International Materials Park, Ohio)
[4] Avramov Yu S and Shlyapin A B 1999 *New Composites of Immissible Components: Preparation, Structure, Properties* (Moscow) in Russian
[5] Davis J R *Copper and copper alloys* Ohio: ASM International materials Park 2001
[6] Garay J F, Glade S C and Anselmi-Tamburini U 2004 *Appl. Phys. Lett.* **85** 573
[7] Gordeev Yu I, Zeer G M and Zelenkova A K 2010 *Izv Vyssh. Uchebn. Zaved. Poroshk. Metall. Funkts. Pokr.* **4** 47

[8] Mula S, Sahani P, Pratihar S K, Mal S and Koch C C 2011 *Mater. Sci. Eng. A* **528** 4348
[9] Maneshian M N and Simchi F 2008 *J. Alloys Comp.* **463** 153
[10] Zhang F B, Wang Z D and Chen Q 2012 *Plasma Sci. Technol.* **14** 71
[11] Turchanin M A, Bondar A A, Dreval L A, Abdulov A R and Agraval P G 2014 *Powder Metall. Met. Ceram.* **53** 70
[12] Xiu S H, Yang R, Xue J and Wang J X 2009 *Trans. Nonferr. Metal. Soc.* **19**(S2) 444
[13] Espevik S, Rapp R A, Daniel P L and Hirth J P 1980 *Oxid. Met.* **14** 85
[14] Wei X, Yu D M, Sun Z B, Yang Z M, Song X P and Ding B J 2014 *Vacuum* **109** 162
[15] Shkodich N F, Rogachev A S and Vadchenko S G 2014 *J. Alloys Comp.* **617** 39

IOP Conf. Series: Materials Science and Engineering **558** (2019) 012048 doi:10.1088/1757-899X/558/1/012048

Titanium carbide based metal-ceramic material with porous layered structure

V S Shustov

Baikov Institute of Metallurgy and Materials Science of the Russian Academy of Sciences, Leninskii 49, Moscow, Russia

Abstract. The method of obtaining a porous material with a layered structure is considered. This method makes it possible to create a finely porous layer on a highly porous base. The method was used to obtain a material based on titanium carbide powders. In the structure of flat samples of such material there were areas with different characteristics of porosity, such as volume and pore size. Creating a layered (or gradient) structure allows to increase the efficiency of membrane filters based on such material. The characteristics of the porous structure are determined. Filters based on titanium carbide powders were synthesized, the isotropic porous structure of which has the following characteristics: filtration fineness is about 460 nm with a total porosity of 39%, gas permeability of $3.3 \cdot 10^{-19}$ kg/(m$^2 \cdot$Pa$\cdot$s) at a pressure drop of 0.05 MPa. Filters with a gradient structure had a filtration fineness of 320 nm and a gas permeability of $2.4 \cdot 10^{-19}$ kg/(m$^2 \cdot$Pa$\cdot$s) at a pressure drop of 0.05 MPa.

1. Introduction.

Porous materials occupy one of the most significant places in modern life, due to their wide use in various fields of human activity. Obtaining porous materials in the form of final products with the necessary characteristics and physicomechanical, chemical properties is the key to the development of the economy, technology and science, and would also contribute to the solution of environmental problems.

Membranes are one of the main areas of research and the use of porous materials. The current state of the technology of ceramics and metals allows the manufacture of porous samples from any inorganic raw materials. The composition of the starting material, the porosity and the structure of the material determine the properties of the porous product. The characteristics of the structure are total, open and closed porosity, permeability, shape and size of pores, their size distribution, specific surface area. Of these indicators, the most important structures are porosity and pore size [1]. Depending on the tasks, for which the membrane will be used, it becomes necessary to obtain a porous material with a different value of porosity and pore size, which will be the basis of the membrane. The main task of membrane filters is that it is necessary to make the separation of the components of a liquid or gaseous medium with lower energy costs [2].

It should be noted that currently produced organic polymer filters have significant drawbacks: low strength, corrosion resistance, heat resistance, lack of sorption properties, etc. Most of these shortcomings are absent in filters made of ceramic materials, which are used, for example, when filtration processes, gas separation, various electrochemical processes are carried out at high pressure drops or high temperatures, in aggressive environment.

The membrane consisting of several layers of different materials, each of which has its own structural organization at the micro and nano levels, is a modern and competitive product with a range of technological characteristics, such as high transport and separation properties, the ability to regenerate in the process of pollution [2]. In the case of manufacturing porous materials, characterized by a homogeneous and isotropic in volume distribution of porosity and pore size, a rapid decrease in permeability due to clogged pores with colloidal and suspended particles contained in filtered substances is observed [1]. The creation of a fine-pore structure on a highly porous base allows one to increase the cleaning efficiency of technological environment from submicron dispersed contaminants with an insignificant increase in the resistance to the flow of the technological environment in comparison with a single-layer highly porous material. In addition, the advantage of such membrane filters, in addition to the possibility of regeneration by heat treatment, washing with acids or other solvents, is the possibility of regeneration by means of a reverse flow of liquid or gas.

It is known that the performance of a membrane filter increases with decreasing thickness of the membrane layer [3]. However, in the production of membranes of small thickness, difficulties arise in obtaining a defect-free membrane layer.

In this paper, the study was carried out to obtain a metal-ceramic material with a gradient porous structure on the basis of titanium carbide powders with the consolidation of pressing powders and their subsequent sintering. In this case, a material production method was used, including the manufacture of a porous carrier substrate that provides mechanical strength, and the formation of a thin membrane layer on it that determines the filtration fineness.

2. Used materials and research methods

In this work, we used titanium carbide powders obtained by the following methods: calcium-hydride method and a melting method in an electric arc furnace of a consumable electrode in a graphite crucible with subsequent crushing and sieving of the melted ingot. The average particle size of the powder of titanium carbide obtained by the calcium-hydride method was 130 nm. Fraction of crushed titanium carbide powders was less than 56 μm. Commercial highly dispersed powders of carbonyl nickel and molybdenum were also used.

Using scanning electron microscopes LEO 430i and JSM 6480, particles of titanium carbide powders were examined (Figures 1, 2). Sintering was carried out in a shaft electric furnace in a vacuum of 10^{-2} Pa. The porosity of the material was determined by hydrostatic weighing. The determination of the pore size in the samples was carried out by the "bubble" method (State standard of Russia R 50516-93. Polymer membranes, the method for determining the bubble point of flat membranes). The method applies to membranes with a maximum pore size of 0.1 to 15 μm and consists in determining the minimum gas pressure required to force a gas bubble through the pores of a flat hydrophilic membrane impregnated with water or through the pores of a flat hydrophobic membrane impregnated with alcohol.

The permeability coefficient K was calculated by the following equation:

$$K = \frac{2\mu hVP}{S\tau(P_1^2 - P_2^2)}$$

where V is the volume of gas passing through the test sample in cm^3; h is the height of the sample in cm; S is the sample area in cm^2; μ is the gas viscosity in Pa·s; τ is the time in seconds; P is the pressure at which the volume of gas is determined in dyn/cm^2; P_1 and P_2 are the gas pressures before the sample and after the sample in dyn/cm^2, respectively [4].

3. Experiment and results

In the production of such a filter with a gradient (layered) structure, a porous material with a pore size of about 1–10 μm was used as a substrate obtained by pressing of a mixture of crushed titanium carbide powders (Fig. 1), nickel and molybdenum in a cylindrical matrix under a pressure of 75 MPa. A binder, polyvinyl alcohol, was added to the powders, and the compacts were then sintered at a temperature of 1250°C in vacuum. The crushed titanium carbide powder was mixed with 7.5 mass % nickel and 2.5 wt % molybdenum. The good wettability of titanium carbide with Ni–Mo alloy ensures the formation of the fine-grained structure of alloys, the possibility of reducing the sintering temperature, thereby reducing the degree of overgrowth of pores [5, 6].

Selective filtering layer with smaller pores was obtained by sequential deposition of fine-pores layers and heat treatment. This allows creating a structure with high efficiency of trapping highly dispersed pollutants. The front surface of the sample was covered with a thin uniform layer of a suspension of ultrafine powder of titanium carbide, obtained by the calcium hydride method (figure 2), and a solution of a binder with the addition of 50 wt % nickel and 25 mass % molybdenum. Depending on the required efficiency and material resistance to the flow of the substance being filtered, a different number of such layers can be applied. The sintering temperature was chosen to be 1000°C. After sintering the sample on the selective layer, there were no macrocracks and exfoliation.

In figure 3, we can see the presence in the sample volume of a region with a coarse-porous (pore size ~ 10 μm) structure and a surface layer with smaller pores.

Figure 1. SEM-photograph of crushed titanium carbide powder.

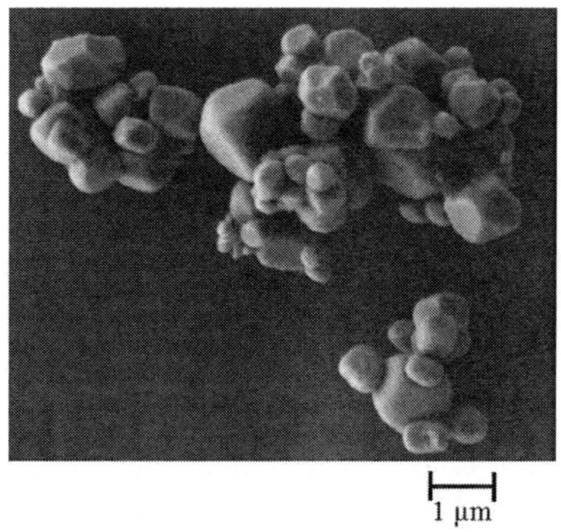

Figure 2. SEM-photograph of titanium carbide powder obtained by the calcium hydride method.

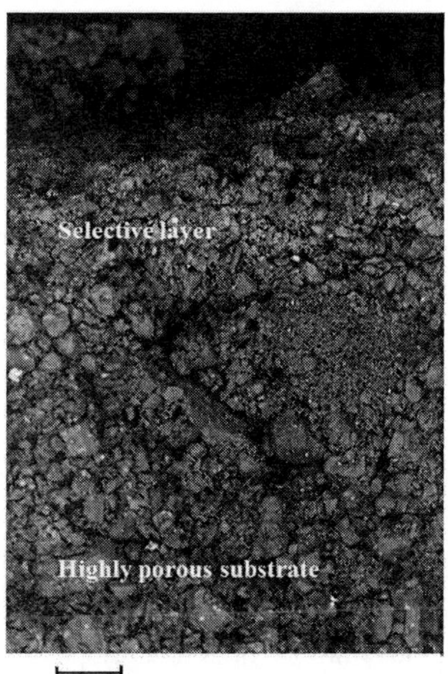

Figure 3. Micrograph of the surface region of the sample.

The results of the permeability study of the obtained porous samples are presented in table 1. The processing of the experimental data made it possible to determine such characteristics of a sintered sample as porosity, permeability, and filtration fineness. The results of the calculations are presented in table 2.

Table 1. The results of the study of porous samples on permeability.

Sample	Air flow through the sample at a pressure drop of 0.05 MPa, 10^{-3} $m^3/(min\times cm^2)$	Air flow through the sample at a pressure drop of 0.1 MPa, 10^{-3} $m^3/(min\times cm^2)$	Water flow through the sample at a pressure drop of 0.05 MPa, 10^{-3} $m^3/(min\times cm^2)$	Water flow through the sample at a pressure drop of 0.1 MPa, 10^{-3} $m^3/(min\times cm^2)$
before applying the selective layer	26.2	32.0	0.09	0.18
with selective layer	18.9	26.0	0.05	0.1

Table 2. Characteristics of the obtained porous material.

Sample	Total porosity, %	Filtration fineness, μm	Permeability at a pressure drop of 0.05 MPa, 10^{-19} $kg/(m^2\cdot Pa\cdot s)$
before applying the selective layer	39	0.46	3.3
with selective layer	39	0.32	2.4

As you can see, the application of the selective layer and its sintering at a temperature of 1000°C reduced the filter throughput by almost one and a half times as compared with the sample without such a layer. Filtration fineness also increased by approximately one and a half times. The decrease in the flow of air or liquid indicates that the large pores of the substrate after application of the selective layer were blocked by a fine-pore structure. The application of additional layers will increase and regulate the degree of purification of such a material.

Thus, a porous material based on titanium carbide is obtained, the isotropic porous structure of which has the following characteristics: filtration fineness is about 460 nm with a total porosity of 39%, gas permeability of $3.3\cdot 10^{-19}$ $kg/(m^2\cdot Pa\cdot s)$ at a pressure drop of 0.05 MPa. Filters were also synthesized with a layered (gradient) structure with a filtration fineness of 320 nm and a gas permeability of $2.4\cdot 10^{-19}$ $kg/(m^2\cdot Pa\cdot s)$ at a pressure drop of 0.05 MPa.

Acknowledgments
This work was supported by the Russian Foundation for Basic Research (grant no. 17-03-00867). Also investigations were carried out with the involvement of government funds no. 007-00129-18-00.

References

[1] Alymov M I, Uvarov V I and Shustov V S 2015 *Inorg. Mat.* **51** 984–990
[2] Volkov V V, Mchedlishvili B V, Roldugin V I et al. 2008 *Russ. Nanotech.* **3** 67–69
[3] Mulder M 1996 *Basic Principles Of Membrane Technology* (Dordrecht: Kluwer Academic Publishers)
[4] Bakunov V S and Balkevich V D 1972 *Practical Course in the Technology of Ceramics and Refractories* (Moscow: Literatura po Stroitel'stvu).
[5] Yanyushkin A S, Yakimov S A, Petrov N P and Arkhipov P V 2009 *Syst. Meth. Tech.* **2** 70–77
[6] Rotagha R, Langer R, El-Sherik A M, et al. 1991 *Scripta Metall. Mater.* **25** 2867–72

IOP Conf. Series: Materials Science and Engineering **558** (2019) 012049 doi:10.1088/1757-899X/558/1/012049

Morphology and phase composition of hybrid powders based on aluminum, synthesized for a gas-dynamic spraying of coatings by milling

A V Sobol`kov and A V Aborkin

Vladimir State University named after Alexander and Nikolay Stoletovs, Vladimir, Russia

E-mail: aborkin@vlsu.ru

Abstract. Hybrid powders of the $AlMg_2$ + 1 wt % TiC/MWNT + 10–50 wt % Al_2O_3 system were obtained using the method of mechanical processing in a planetary ball mill. The powders were characterized using the methods of scanning electron microscopy, particle size distribution testing and X-ray analysis. It is shown that an increase in the content of Al_2O_3 particles from 10 to 50 wt % leads to a decrease in the average particle size of the synthesized powder from 49 to 16 μm, respectively. The proportion of particles smaller than 5 μm, depending on the concentration of Al_2O_3, varies in the range of 1.2–15.1%. At the same time, the size of the matrix material crystallites decreases from 58 and 52 nm, respectively. The resulting hybrid powder can be used to create functional coatings using cold gas-dynamic spraying.

1. Introduction

At present, both in the Russian Federation and abroad, intensive studies of the structure-phase composition and functional properties of coatings obtained by gas-dynamic spraying are carried out. The diversity of the research and the results obtained can be judged at least by detailed reviews [1–4], most of which is occupied by the development of new powder materials for the synthesis of coatings with an increased complex of properties.

Currently, the possibilities of improving the properties of coatings due to the introduction of ceramic particles are limited or even exhausted. Therefore, it is important to conduct research on the synthesis and study of the properties of hybrid coatings with two or more types of reinforcing additives. In this case, the most promising is the creation of coatings with the nanocrystalline structure of a matrix material that simultaneously contains nano- and micro-sized reinforcing particles.

The creation of hybrid powders is possible through the introduction of reinforcing particles into the matrix material using the method of mechanical synthesis in high-energy ball mills. In [5], a significant increase (more than two times) in the tribological properties of coatings obtained by surfacing babbit powders modified with carbon nanostructures additives was shown.

At the same time, promising carbon nanostructures that can be used as reinforcing nanoparticles should include multiwalled carbon nanotubes (MWCNTs) decorated with TiC coating. Confirmation of the promise of using TiC is the positive experience of creating bulk aluminum matrix composites strengthened with nano- and micro-sized TiC additives, which made it possible to significantly increase tribological properties [6].

However, the gas-dynamic spraying of powders with a particle size of less than 2 μm is difficult due to their inhibition in the compressed layer, which occurs when an ultrasonic gas stream flows onto an obstacle. In addition, when accelerating particles of a powder in a gas stream, there is a strong variation in velocity for particles of different sizes, while the nanoscale fraction can scatter without falling into the coating. Therefore, the use of polydispersed powder, which is a mechanical mixture of nano- and micro-sized particles, for gas-dynamic spraying seems to be unpromising. For the formation of hybrid coatings containing, in addition to micro-sized nanoscale particles, it is necessary to use powder in the form of a mechanical mixture consisting of agglomerates of complex composition and ceramic particles. Use for gas-dynamic spraying of powders having such a structure, allows to obtain hybrid coatings with a uniform structure and a high level of mechanical properties [7, 8]. Since the presence of nanoscale particles increases the mechanical properties of the matrix material, increasing the strength of fixing in the matrix of micro-sized ceramic particles. At the same time, the plasticity of the particles of the matrix material remains sufficient for the effective formation of a coating with a low porosity [9].

2. Research methods and equipment

Mechanical synthesis of the powder was carried out in two steps. As the matrix material, granules of aluminum alloy $AlMg_2$ obtained by spraying the melt were used. As reinforcing particles, we use multi-walled carbon nanotubes decorated with TiC particles. The weight fraction of TiC/MWCNT was 1 wt %. The mechanical treatment of the raw materials was carried out in a FRITSCH PULVERISETTE 6 planetary ball mill with the use of surfactants [10]. Next, α-Al_2O_3 (n = 10, 30, and 50 wt %) with an average particle size of 18 μm was added to the obtained nanocomposite powder and mechanical processing was continued.

The study of the morphology of the composite powder particles was performed using a Quanta 200-3D scanning electron microscope. The granulometric composition of the obtained powders was determined on a Microsizer-201C instrument. Studies of the structural and phase composition were performed using a D8 ADVANCE X-ray diffractometer. The size of the matrix material crystallite was calculated using the Selyakov-Scherrer formula. The volume-averaged crystallite sizes were calculated on the assumption of their spherical shape.

3. Results

Figure 1 shows the SEM images of composite powders based on the $AlMg_2$ alloy obtained after the first step. From figure 1 it can be seen that the particles of the obtained powders have sizes varying in a wide range of 5–100 μm with an average particle size of d_{50} = 53 μm. Particles have an irregular shape with a developed surface.

Figure 1. SEM images characterizing the morphology of the powder particles.

The study of the surface of the powder particles using a scanning electron microscope shows a small amount of reinforcing particles on their surface. The length of the reinforcing particles visible on the powder surface was ~ 600 nm (see figure 1, indicated by the arrow). A small amount of reinforcing particles on the surface is due to the peculiarities of the formation of powder particles under the conditions of mechanical processing in a ball mill. In the process of mechanical synthesis, a significant part of the reinforcement may not be on the surface, but inside the agglomerates based on the matrix material. This can protect them from destruction under high-energy effects of grinding bodies during mechanical processing.

The powder obtained after the second step was a mechanical mixture consisting of agglomerates of complex composition and micro-sized ceramic particles. Agglomerates are particles of a nanocrystalline matrix material containing TiC/MWNTs with micro- and nano-sized Al_2O_3 particles embedded in them, as well as on the surface. The study of the particle size distribution of mechanically synthesized powders of different composition shows that an increase in the content of ceramic particles from 10 to 50 wt % leads to a decrease in the average particle size of the synthesized powder from 49 to 16 μm, respectively. At the same time, the proportion of particles with a size of less than 5 μm, gas-dynamic spraying of whose is difficult, varies depending on the concentration of Al_2O_3 in the range of 1.2–15.1%.

The results of X-ray phase analysis of the original matrix material show the presence of only peaks of aluminum. The X-ray diffraction patterns of the reinforcement were recorded with Bragg diffraction peaks corresponding to MWCNT and TiC. According to X-ray diffraction data, the phase composition of the composite powder is similar to the original $AlMg_2$ matrix alloy. Also noted the absence of peaks corresponding to the reinforcement. The broadening of aluminum peaks is associated with a decrease in the size of the structure of the material due to severe plastic deformation of particles when exposed to grinding bodies. The results of the X-ray phase analysis of the powder obtained at the second step are shown in figure 2. Analysis of the data obtained shows that diffraction patterns have a qualitatively similar character. The presence of peaks corresponding to aluminum and Al_2O_3 is noted. Moreover, the intensity of Al_2O_3 peaks increases with an increase in its share in the charge. The calculation of the crystallite size, performed for the powders obtained both in the first and second steps using the Selyakov-Scherrer dependence, shows that the nanocrystalline structure corresponds to the matrix material. At the first step, the crystallite size of the $AlMg_2$ + 1 wt % TiC/MWCNT composite powder is 68 nm. The introduction of 10–50 wt % Al_2O_3 into the powder mixture and the subsequent processing in a planetary mill lead to a decrease in the crystallite size to 58 and 52 nm, respectively.

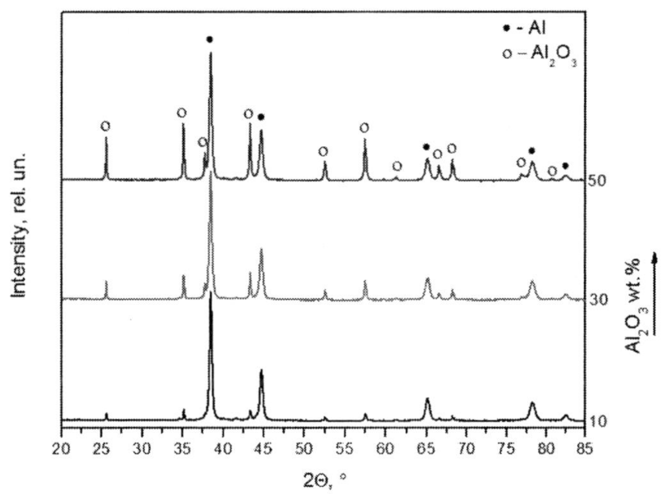

Figure 2. X-ray diffractometry results of mechanically synthesized powders.

Thus, powder mixtures $AlMg_2$ + TiC/MWNT + Al_2O_3 were synthesized by the method of mechanical processing in a spherical planetary mill. The resulting composite powder will be used to create coatings using cold gas-dynamic spraying.

Acknowledgements
The study was carried out with the financial support of the Russian Foundation for Basic Research in the framework of the scientific project no. 18-38-00590 mol_a.

IOP Conf. Series: Materials Science and Engineering **558** (2019) 012049 doi:10.1088/1757-899X/558/1/012049

References

[1] Raoelison R N, Verdy C and Liao H 2017 *Mater. Design* **133** 266
[2] Raoelison R N, Xiea Y, Sapanathan T, Planche M P, Kromer R, Costil S and Langlade C 2018 *Additive Manufacturing* **19** 134
[3] Moridi A, Hassani-Gangaraj S M, Guagliano M and Dao M 2014 *Surf. Eng.* **36** 369
[4] Li W, Assadi H, Gaertner F and Yin S 2018 *Critical Rev. Solid State Mater. Sci.* **43** 1
[5] Aleshin N P, Kobernik N V, Mikheev R S, Vaganov V E, Reshetnyak V V and Aborkin A V 2016 *Russ. Eng. Res.* **36** 46
[6] Mikheev R S and Chernyshova T A 2013 *Aluminum-matrix composite materials with carbide reinforced to solve the problems of new technology* (Moscow: Maska) p 356
[7] Aborkin A V, Sobol'kov A V, Elkin A I and Arkhipov V E 2018 *J. Phys.: Conf. Ser.* **951** 012010
[8] Aborkin A V, Alymov M I, Arkhipov V E and Khrenov D 2018 *Dokl. Phys.* **63** 50
[9] Aborkin A V, Alymov M I, Kireev A V, Sobol'kov A V and Arkhipov V E 2018 *Metall.* **62** 809
[10] Aborkin A V, Alymov M I, Kireev A V, Elkin A I and Sobol'kov A V 2017 *Nanotech. Russ.* **12** 395

IOP Conf. Series: Materials Science and Engineering **558** (2019) 012050 doi:10.1088/1757-899X/558/1/012050

Arctic protective materials based on complex oxides of titanium

I A Sologubova, M K Kotvanova and S S Pavlova

Yugra State University, Federal State Budgetary Educational Institution of Higher Education, Khanty-Mansiysk, 628012 Russia

E-mail: i.a_sologubova@mail.ru

Abstract. This paper presents the components and method for obtaining an effective protective coating. The coating has chemical and thermal stability in a wide range of temperatures, and has good mechanical characteristics. It was found that one of the important factors determining the properties of the coating is the degree of dispersion of powders of the starting materials. The optimal average particle size is 60 nm.

1. Introduction

Corrosion of metals causes significant economic and ecological damage to the national economy. The issue is getting worse while operation of metal structures and items in Arctic regions. In this regard the requirements to the maintenance engineering reliability and corrosion resistance of structural materials and metal structures as well as methods of their protection have been improved.

There are many ways to protect steel from corrosion, however, as a rule, each of them is aimed at protecting from any specific effects, without providing comprehensive protection. Also often, many methods are expensive or intractable. Thus, the using of composite aluminium-oxide coating is complicated by the requirement of the control its porosity [1]; diffusion zinc coating as a method of protecting a metal surface requires use of large amounts of expensive zinc [2].

In this article an attempt to obtain a protective coating meeting the requirements of the Arctic materials resistant to low temperatures, intense wind and wave and ice dynamic loads, corrosion and erosion impact has been made.

It is known that complex oxides of d-elements have a high chemical resistance to aggressive environments. Complex oxides of titanium (oxide bronze) of K_xTiO_2 composition, where x = 0.06–0.12 were used as the basis of the protective coating in this study.

2. Experimental

Mechanochemical synthesis was performed in a planetary mill AGO-3 with 1200 rotations of the shell at min. Steel balls with a diameter of 8 mm were used as grinding media. The ratio "reaction mixture:grinding body" was 10:220 by weight. The charge mixture was calculated according to the equation $TiO_2 + xKI = K_xTiO_2 + x/2I_2$. The duration of the synthesis was 400 s. The product of the synthesis was purified with concentrated nitric acid from impurities of iron which gets into the reaction mixture from the surface of the grinding bodies.

For coating steel base were cleaned with sandpaper M40 and degreased with acetone. The suspension was prepared in a weight ratio of $H_20:Na_2SiO_3$ (in $NaOH_{water}$):K_xTiO_2 = 1:0.05:0.05. The resulting suspension was poured evenly on the steel base and dried during the day. Heat treatment was carried out using a gas burner (temperature range 1100–1200°C).

3. Results and discussion

It was found that one of the important factors determining the properties of the coating is the degree of dispersion of powders of the starting materials.

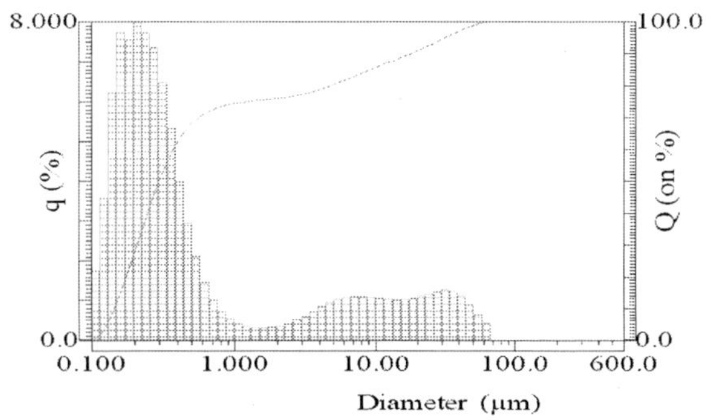

Figure 1. Diagram of distribution of particles K_xTiO_2 by sizes (LA-300 Horiba).

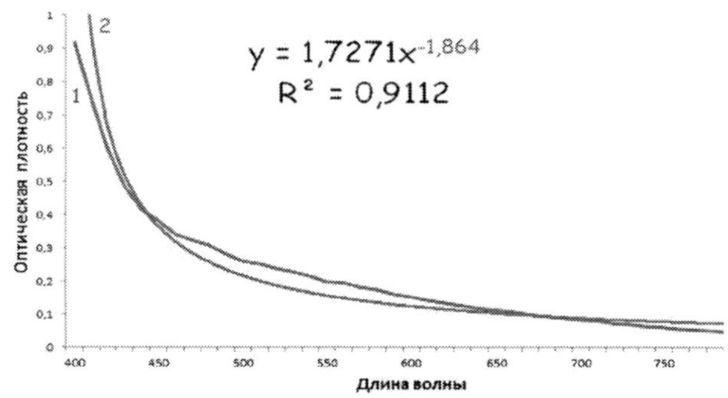

Figure 2. Absorption spectrum of an aqueous suspension K_xTiO_2 (Promekolab PE5100).

The coating prepared from powders with an average particle size of 60 nm, has a smaller thickness and porosity, and more continuity. The small thickness of the coating contributes to its adhesion to the steel core, thus reducing the consumption of coating material.

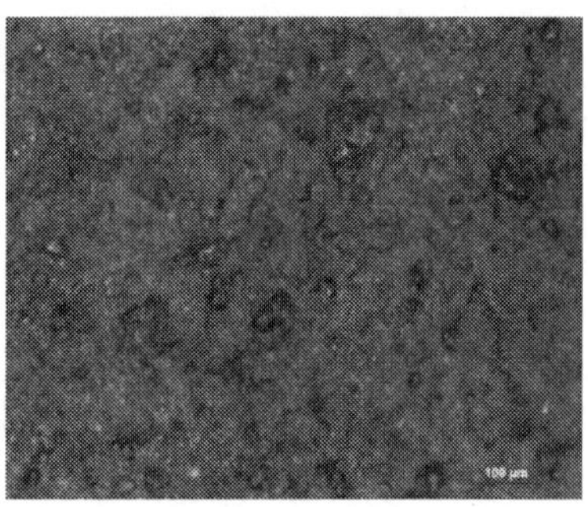

Figure 3. SEM image of the surface of the protective coating (average particle size 60 nm).

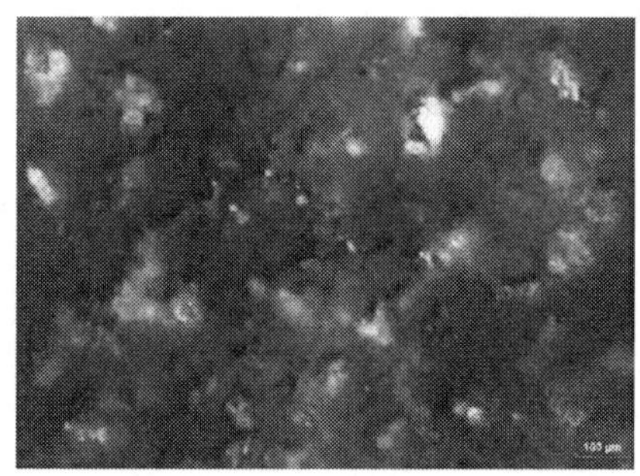

Figure 4. SEM image of the surface of the protective coating (average particle size 400 nm).

The obtained protective coating was subjected to test for adhesion by the method of separation σ = 120 H/м²; the peeling of the coating occurred when bending of the metal substrate by 20°. The following indicators of coatings were obtained: the force of static friction F_{sf} = 0.70 N; the force of sliding friction F_{sf} = 0.59 N.

In addition, coatings were tested for resistance to high and low temperatures. Exposure to high temperatures was carried out using a gas burner (the temperature range 1100–1200°C) – coating was destroyed due to the deformation of the steel plate. Cracking of the coating at low temperatures (–70°C) occurred with prolonged exposure (3 months).

Chemical resistance of the obtained coatings was evaluated by the corrosion rate of the samples. The results are presented in table 1.

Table 1. Chemical properties samples of the resulting coatings.

Reactive atmosphere	Corrosion velocity of the sample, g/min	
	Without coating	With coating
HNO_3, $\rho = 1.48$ g/cm^3	0.037	0.015
H_2SO_4, $\rho = 1.83$ g/cm^3	0.044	0.018
HCl, $\rho = 1.19$ g/cm^3	0.236	0.113

Thus, we succeeded in obtaining a protective coating having chemical and thermal stability in a wide temperature range, and having high mechanical characteristics. On the basis of conducted research it can be concluded that the presented method of protecting steel from corrosion can be used in hard environmental conditions, including the Arctic regions.

References

[1] Pat. 90440 Russia Suminov I V *Composite aluminium-oxide coating to protect steel from corrosion and wear* applicant and patentee OOO "Innovative technological center "NANOMER" No RU2009135311U; Appl. 23.09.09; publ. 01.10.10

[2] Pat. 2147046 Russia *Method of thermal diffusion galvanizing* applicant and patentee Institute of high temperature electrochemistry, Ural branch, RAS No. 98115707/02; Appl. 17.08.98; publ. 27.03.00

IOP Conf. Series: Materials Science and Engineering **558** (2019) 012051 doi:10.1088/1757-899X/558/1/012051

The use of impulse electric and magnetic fields in obtaining permanent joints

E L Strizhakov, S V Nescoromniy, S O Ageev and V G Vinogradov

Don State Technical University, Rostov-on-don, 344000 Russia

E-mail: strizhakov@inbox.ru

Abstract. The article discusses the possibility of using electromagnetic influence in the production of permanent compounds such as metal–ceramics, metal–glass, as well as composite compounds from powder materials. It is proposed to use magnetic pulse action to form such joints in the solid phase: in the atmosphere with the position of parts at an angle; in vacuum with preheating and subsequent radial action. When consolidating powder materials (sintering with simultaneous pressing), a scheme of synchronous action of electromagnetic energy is proposed: thermal and mechanical.

In the manufacturing of products in electronics, instrument engineering there is a necessity to connect dissimilar materials such as metal–ceramics, metal–glass, or dissimilar alloys together. Joints of the tips with wires, various composite insulating metal–ceramic components of microwave devices, metal-glass fuses, resistors and other electronic products are obtained by magnetic impulse crimping [1].

The greatest interest is the consolidation of materials in the solid phase by magnetic impulse welding. The scheme of magnetic impulse welding with oblique collision (Magnetic Impulse Welding MIW) is shown in figure 1 [1].

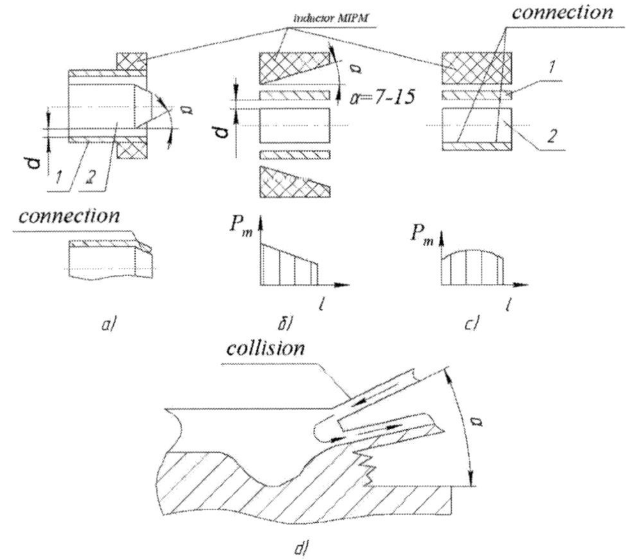

Figure 1. Technological schemes of magnetic impulse welding. (a) the location of parts at an angle; (b, c) with a changing length zones of processing by magnetic pressure; (d) formation of cumulative jet; α, δ-initial angle and gap; P_m is the magnetic pressure.

Parts 1 and 2 are installed at an acute angle α overlap to each other and with a gap δ between them. Opposite the welded surface inductor 3 is set.

Part 1 was fixed to prevent movement during welding rigidly. The workpiece 2 must provide the movement of a welded tip in the direction of the workpiece 1. The inductor 3 is connected by its terminals to the impulse current generator. In the gap between the inductor and the workpiece during the discharge of the capacitor bank, there is a strong magnetic field inducting current in the workpiece. The repulsion forces between the inductor and part 2 arise from the interaction of the inductor current with the induced current in the workpiece, as a result part 2 instantly with high speed moves from the inductor in the direction of fixed part 1; in the contact area at collision, high pressure develops, and the welded joint is formed. In fact, the scheme used in explosion welding is implemented [2, 3]. The replacement of the energy carrier is implemented.

Overlapping joints of single- and multithickness fine-walled tubes of dissimilar materials can be obtained by magnetic impulse welding. For example, copper and aluminum.

Devices of magnetic impulse processing in vacuum can be successfully used for combined process of pressing and welding with the emission of the powder compositions to cathode cores [4, 5]. The scheme of magnetic-impulse pressing in vacuum is shown in figure 2.

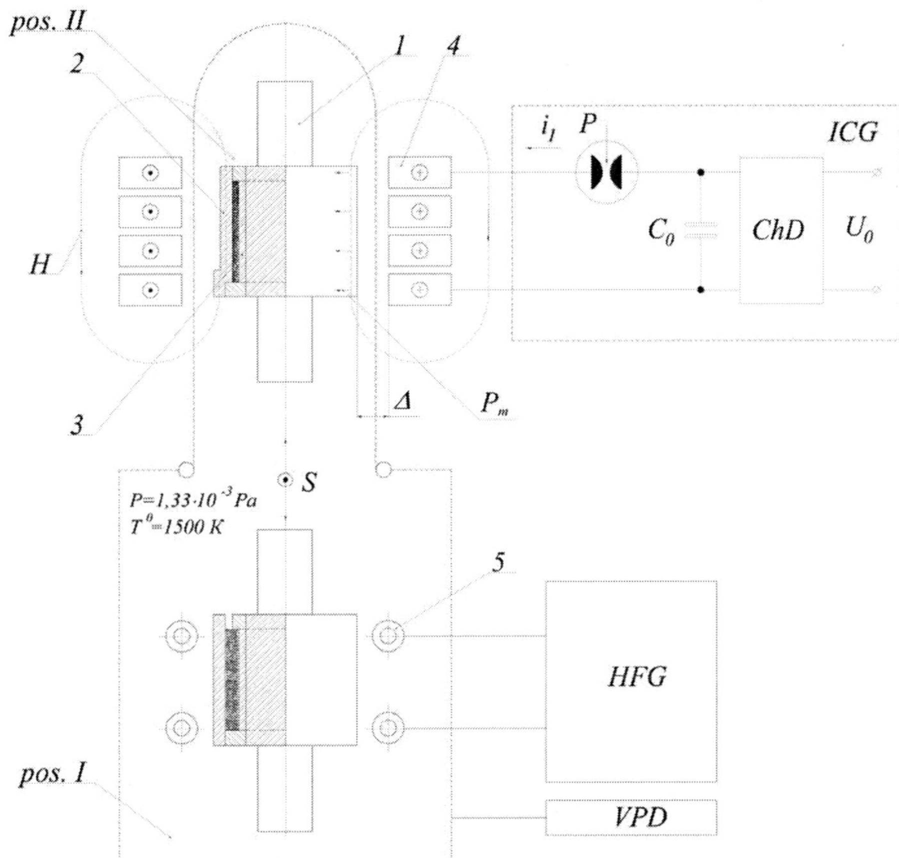

Figure 2. Scheme of magnetic impulse pressing in vacuum: *1* base; *2* cover; *3* powder coating; *4* MIPM (magnetic impulse processing of materials) inductor; *5* high frequency inductor; U_0 is the line voltage; ChD – charger device; C_0 is the capacity drive MII (magnetic-impulse installation); *P* is the commuting device; i_1 is the discharge current; P_m is the magnetic pressure; *H* is the magnetic flux; IGG – impulse current generator; VPD – vacuum pumping device; Δ is the gap between the inductor and the cover; pos. I is the preliminary heating in vacuum; pos. II is the magnetic impulse welding (pressing) in vacuum; *S* is the movement of the workpiece.

IOP Conf. Series: Materials Science and Engineering **558** (2019) 012051 doi:10.1088/1757-899X/558/1/012051

The process is carried out as follows. Base with coating and the satellite is attached to the stem of electromechanical drive of the vacuum chamber and is inserted into the area of the high-frequency inductor heating. Using vacuum system, the rarefaction is created in the working volume. After that, the HF generator is switched on. Reaching the required temperature node is transferred by electromechanical drive to the MIP (magnetic-impulse processing) zone. Radial crimping is carried out.

Magnetic Impulse Pressing Technology (MIPT) allows to combine outgassing operations, compaction of powder materials, sintering of powder particles and welding with cathode core in one technological operation proceeding in a vacuum chamber without intermediate technological transitions, and exclude the operation of subsequent sintering. It significantly (2–3 times) reduces the complexity of manufacturing cathode nodes ultra-high frequency EVD (electric vacuum device) [1].

Consolidation of dissimilar materials by using powders can be reached by using electroimpulse pressing sintering (EIPS). Schematic diagram process EIPS is presented in figure 3.

The device is assembled so that the thermal (electric current) and mechanical (dynamic loading) impact occurred simultaneously [6, 7].

When discharging capacitor batteries, high frequency current consistently passes through the electrode carried thermal effects on the powder, which leads to intense local heating in the contact area and through the spirals of the inductor, which form a force magnetic field. Magnetic pressure is converted into mechanical force and gives extra compaction and is also a way of activation of interparticle contacts. In this way copper wire is connected to steel basis.

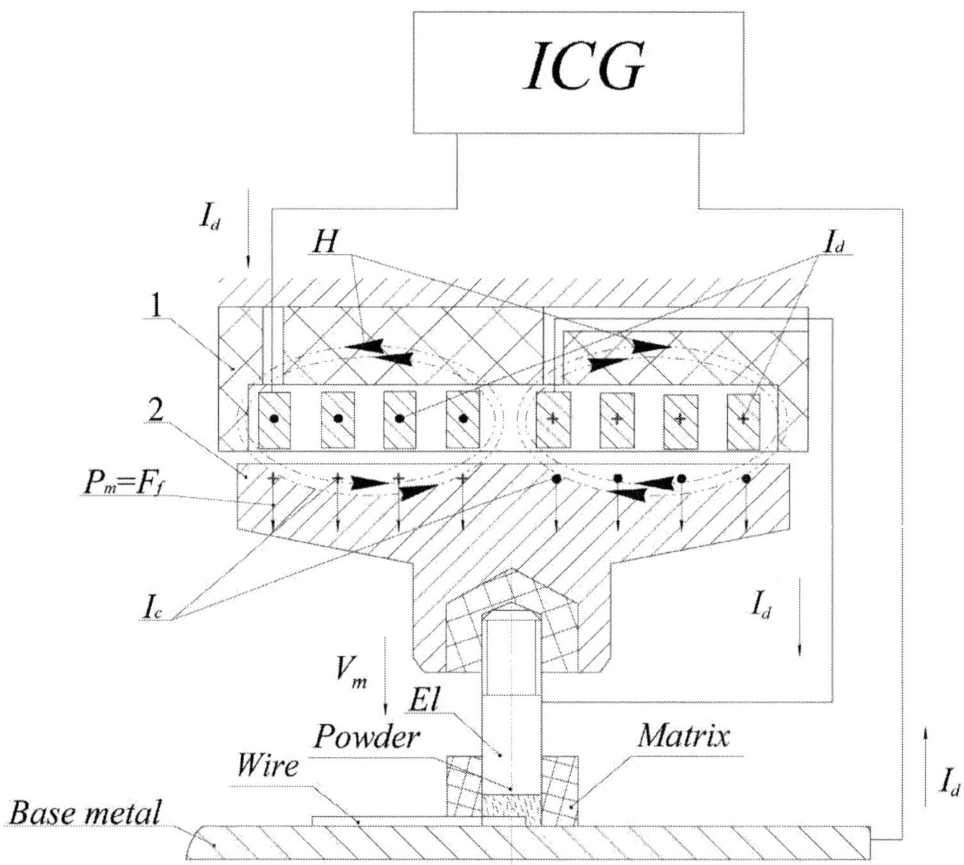

Figure 3. Schematic diagram of the EIPS process: *1* inductor; *2* pusher; I_d is the discharge current; I_c is the induced current; V_m is the speed of movement of the pusher; H is the magnetic flow; P_m is the magnetic pressure; F_f is the forging force; ICG – impulse current generator; El – electrode.

References

[1] Strizhakov E L, Nescoromniy S V and Minko D V 2016 *Discharge-impulse processing of materials* (Rostov n/D: Publishing center DSTU)
[2] Glushchenkov A V and Karpuhin V F 2014 *The technology of magnetic impulse processing of materials* (Samara: Publishing house «Fedorov»)
[3] Konyushkov G V and Musin R A 2009 *Special methods of pressure welding* (Saratov: Publ. AI PI Er Media)
[4] Grigoriev E G 2013 *Regularities of high-voltage electric impulse consolidation of composite materials* (Moscow: MEPhI)
[5] Strizhakov E L, Nescoromniy S V and Ageev S O 2018 *Welding production* **5**
[6] Pat. 2551329 Russian Federation, IPC51 B23K 11/02, B22F 7/08. Method for obtaining joints of dissimilar conductive materials and device for its implementation / Nescoromniy S V et al. 2013158412/02; appl. 27.12.2013: publ. 20.05.2015 Bul. No. 14-4 p.
[7] Lemeshev S V, Nescoromniy S V, Ageev S O and Strizhakov E L 2015 (Don State Technical University Herald) **15** 76

IOP Conf. Series: Materials Science and Engineering **558** (2019) 012052 doi:10.1088/1757-899X/558/1/012052

Influence of the binder composition on the properties of the silicon carbide green compacts and sintered parts prepared from the powders produced by SHS

K S Torosyan[1,2] **and V G Pak**[1]

[1]Penza State University, Krasnya 40, Penza, Russia
[2]National University of Science and Technology "MISIS", Leninskii 4, Moscow, Russia

Abstract. In this work, the possibility of the application of binders based on the polyvinyl alcohol and paraffin instead of environmentally-dangerous phenol-formaldehyde resin-based binders in the production of silicon carbide ceramics are investigated. Introduction of citric acid and carbamide in the binder based on water solution of polyvinyl alcohol ensures the high wettability of the powders of non-oxide ceramics by the binder. The addition of castor oil and oleic acid in the solution of paraffine in petroleum leads to a decrease in the inter-particles frictions and an increase in the density of green compacts. The developed binders allow to reduce compaction pressure by 12–15%.

1. Introduction

In this work, SiC powders produced by self-propagating high-temperature synthesis (SHS) were used. SHS powders are peculiar due to the fine particles size (≤ 1 μm), which allows to produce parts with fine and uniform microstructure and increased mechanical, physical and electromagnetic properties. However, the application of fine powders leads to problems in the pressing of green compacts due to the higher friction between the particles and the agglomeration.

In the production of silicon carbide ceramics, phenol-formaldehyde resin-based binders and complex plasticizers are conventionally used to increase the powders pressability and decrease the sintering temperature. [1]. However, this binder does not provide a decrease in the inter-particular friction. Therefore, the development of surfactants which will allow to decrease the friction and to increase the wettability of powders by the binder is of current interest. Replacement of phenolformaldehyde resin-based binders by environmentally-friendly water soluble binders is an another important task addressed in this study [2].

2. Materials and Methods

SiC specimens were produced by conventional powder metallurgy route, which includes the following operations: mixing of silicon and graphite powders, introduction of 20 wt % binder, granulation by rubbing through the mesh with 0.5 mm cells, pressing of $20 \times 12 \times 15$ rings under a pressure of 150–200 MPa, vacuum sintering in the furnace shown in figure 1. SiC powders produced by SHS were characterized by median particles size of 0.7 μm, whereas the powders synthesized by conventional Acheson process had an average size of 63 μm. To produce the specimens with various grains size the sintering temperature was varied from 1700 to 1800°C.

Figure 1. Vacuum furnace ZGS-200.

To produce the paraffin-based binders, the calculated amount of molten paraffine (grade P-1, GOST 23683-89) and petroleum were mixed until the full dissolution of paraffine; afterwards, oleic acid (GOST 3652 - 69 "Ch") and castor oil (GOST 6691-77 "Ch") were added during the mixing. The produced binder was introduced into the mixture of powders of SiC (particles size of 0.5–0.8 μm) and carbon black. Binder ratio was equal to 20 wt %. The mixtures were then additionally mixed in the pan mixer with subsequent rubbing through the 0.5 mm mesh. The mixtures were dried until the residual petroleum content was 0.5–2.7 %.

Water-soluble binder for granulation was produced by the dissolution of polyvinyl alcohol (grade 16/1 GOST 10779-79), citric acid ("Ch" grade, GOST 3652-69) and urea ("Ch" grade, GOST 6691-77). 0.1% zinc stearate was introduced into the granulated powder before the pressing as the "surface binder" to decrease the pressure loss due to the friction of granulated powder against the die wall and therefore to decrease the wear of the dies and increase the density uniformity of the compacts. For comparison, press-powder was prepared using the phenol-formaldehyde resin. Influence of the binder on the rate of wear of dies was estimated as the decrease of the mass of the pressing insert after the pressing of 20× 10 × 16 ring-shaped compacts at 100 MPa.

3. Results and Discussion

Characteristics of powder compacts and wear rate of the die pressing inserts are given in tables 1 and 2. Data are averaged from 10 measurements. Compositions of complex binders are also given in the tables 1, 2. Binder based on the solution of paraffine in petroleum and containing the castor oil and oleic acid allows to increase the density of compacts and decrease the wear of dies in the silicon carbide production. Addition of castor oil and oleic acid in a particular proportion ensure the simultaneous increase in the powder wettability, decrease in the pressure loss due to friction against the die walls, retention of high mechanical properties of the compacts. Paraffin glues the powder particles, providing the necessary compacts strength.

Binder based on the water solution of polyvinyl alcohol and containing the citric acid and urea also allows the increase on the compacts density and decrease the die wear in the production of silicon carbide ceramics. Addition of citric acid and urea in a particular proportion ensure the simultaneous increase in the powder wettability, decrease in the pressure loss due to friction against the die walls, retention of high mechanical properties of the compacts. Polyvinyl alcohol glues the powder particles, providing the necessary compact strength.

Application of the binder based on the solution of paraffin in petroleum led to the residual petroleum content of 1–2 wt %. Application of the binder based on the solution of polyvinyl alcohol in water leads to the residual water content of 0.5–2 wt %. Comparison of the data given in table 1 demonstrates that the application of binders based on the solution of paraffine in petroleum and containing castor oil and oleic acid leads to an increase in the relative density of the compacts up to 2.05–2.22 g/cm^3 and reduction of the wear of the die inserts to 19.1–20.2 μm/cycle.

Table 1. Influence of the composition of binder based on a solution of paraffine in petroleum on the density of compacts and wear resistance of the dies.

Technological properties of the press-powders and compacts	Residual petroleum content in the press-powder, wt %	Binder components ratio, wt %			
		Paraffin	7.0	8.5	10.0
		Oleic acid	0.2	0.6	1.0
		Castor oil	2.0	5.0	8.0
		Petroleum	90.8	85.9	81.0
Compacts density, g/cm^3	1.1		2.05	2.15	2.16
	2.0		2.16	2.20	2.22
Die insert wear rate, μm/cycle	1.1		20.2	20.1	20.0
	2.0		19.8	19.2	19.1

Table 2. Influence of the composition of binder based on water solution of polyvinyl alcohol and phenol-formaldehyde resin on the density of compacts and wear resistance of the dies.

Technological properties of the press-powders and compacts	The residual water content in the press-powder, wt %	Binder components ratio, wt %			
		Polyvinyl alcohol	5.0	5.0	
		Water	95.0	90.5	
		Citric acid		0.5	
		Urea		4.0	
		Phenol-formaldehyde resin			100.0
Compacts density, g/cm^3	0.0				2.02
	1.1		1.96	2.08	
	2.0		2.06	2.12	
Die insert wear rate, μm/cycle	0.0				31.9
	1.1		26.5	21.1	
	2.0		24.8	19.9	

As the tables 2 and 3 demonstrate, the binder based on water solution of polyvinyl alcohol without the addition of urea and citric acid does not provide the required quality of press-powders and compacts. Addition of urea and citric acid increases the wettability of the particles by the binder during the granulation, decreases the dust content and increases the density and strength of compacts.

Table 3. Influence of the binder composition on the technological properties of press-powders and compacts

Technological properties of press-powders and compacts	The water content in the press-powder, wt %	Binder components ratio, wt %		
		PVA - 5.0 Water - 95.0	PVA - 5.0, Citric acid - 0.5, Urea - 4.0, Water - 90.5	Phenolformaldehyde resin - 100
The density of the compacts at pressure $P = 100$ MPa, kg/m³	0	1960	2160	2140
	2.7	2130	2220	
The tensile strength of the ring specimens, kPa	0	0.45	0.57	0.59
	2.7	0.40	0.38	
Dust content in the press-powder, wt %	0	13.8	3.0	7.6
	2.7	7.8	1.5	
Wetting angle, °		47	22	25
The density of the sintered specimens, kg/m³	0	2630	3120	3085
	2.7	2870	3180	

The developed binders with the addition of urea and citric acid allow to decrease the binder consumption by 10–15%, increase the density of compacts by 10–12% and increase the density of sintered parts by 5–7% as compared to the parts prepared on the base of polyvinyl alcohol without the additions. Results suggest that the used water-soluble binder can replace the phenol-formaldehyde resin-based binders.

Mechanical tests revealed that the structural ceramic produced using the powders synthesized by SHS surpass the analogs produced from conventional Acheson process (table 4).

Table 4. Influence of the composition of binder on the strength and density of silicon carbide ceramics produced from powders synthesized by SHS and Acheson process.

	Synthesis route	Binder composition, wt %		
		Paraffin – 8.5 Castor oil – 0.6 Oleic acid – 5.0 Petroleum – 85.9	PVA – 5.0 Citric acid – 0.5 Urea – 4.0 Water – 90.5	Phenolformaldehyde resin – 100
The tensile strength of the sintered rings, MPa	SHS	135	132	123
	Acheson	108	104	97
The density of the sintered rings, kg/m³	SHS	3184	3180	3085
	Acheson	3118	3107	3004

Application of the developed binder compositions for the production of parts from non-plastic powders allows to reduce the production cost due to an increase in the wear resistance of the press-forms and due to an increase in the prime yield.

4. Conclusions

Based on the results of the research, the following conclusions could be made:

1. Binders based on the solution of the paraffin in petroleum and of PVA in water were used for powders produced by the SHS and Acheson process. Developed binders are advantageous as compared to phenol-formaldehyde resin due to being non-toxic. Water-soluble binders are non-flammable, providing the possibility to the increase in the working conditions during the production of silicon carbide ceramics.

2. Introduction of the solution of paraffine in petroleum allows to increase the density and strength of ceramics and increase the wear resistance of the dies. The inclusion of the citric acid and urea increases the wettability of the silicon carbide powder by the binder and increase the density and strength of the ceramics.

References

[1] Henit G and Rol R 1972 *Silicon Carbide* (M.: Mir) 349 p

[2] Belyaev A E and Konakova R V 2010 *Silicon Carbide: technology, properties, application* (Kharkov, «ISMA») 532 p

SHS membranes based on materials of mica-like structure

V I Uvarov[1], M I Alymov[1], A R Kachin[1], V E Loryan[1], V S Shustov[1], A S Fedotov and M V Tsodikov[2]

[1]Merzhanov Institute of Structural Macrokinetics and Materials Science, RAS, Chernogolovka, 142432 Russia
[2]Topchiev Institute of Petrochemical Synthesis, RAS, Moscow, 119991 Russia

E-mail: uvar@ism.ac.ru, kachin@ism.ac.ru

Abstract. The possibility of synthesizing a functional porous material of a mica-like structure to manufacture filter membranes for purification of liquids and gases, in particular tap water, was investigated. Mixtures of available powders based on mineral raw materials — quartz sand (SiO_2) and Karelian schungite with reducing agents and secondary cryolite, a product of processing of fluorine-containing wastes of aluminum production, were used for the synthesis. The synthesis was carried out in the furnace at $T = 930°C$ in the atmosphere of air at $P = 1$ atm. A ceramic filter membrane 40 mm in diameter and ~ 5 mm in height was synthesized. The size of the open pores in the synthesized sample, which is measured according to GOST 26849-86, is 1–3 μm. The bending strength is 5.5–6 MPa. The product has a porosity of ~ 60% and an open porosity of up to 98%. The synthesized sample efficiently purifies the tap water from the high content of heavy metal ions and reduces the hardness of the water by 30%.

1. Introduction

Man-made burden on the environment can lead to irreversible changes in the composition of natural water. In particular, surface and underground water sources are contaminated by substances of anthropogenic origin - heavy metal ions, nitrates, volatile organochlorine compounds, herbicides, pesticides, radionuclides, etc. [1, 2]. Therefore the research aimed at purification of drinking water, in particular development of filter technologies using various materials, is relevant.

A comparative analysis of characteristics of filters manufactured in different countries [3] shows the expected advantages of SHS filters, especially multiple application after regeneration. It is impossible for the filters produced by traditional methods.

Fluoroflogopite is known to have a high ion-exchange capacity and can be used for heterovalent replacement of Na+ ions with metal ions that determine the water hardness (Ca2+, Mg2+, Sr2+, Ba2+) [4–7]. The authors show the possibility of SHS of a porous material based on fluoroplogopite [8, 9].

Introduction of up to 20% of CuO and Cr_2O_3 into the porous sample with the formation of spinels based on them will allow us to obtain catalytically active membranes for dehydrogenation of hydrocarbons [10] and catalytic oxidation of carbon monoxide [11].

The aim of this study is to create highly efficient, corrosion-resistant SHS membranes for purification of liquids and gases in both industrial enterprises and small plants. They allow adjusting the salt composition of water and ensuring the active human activity at its consumption.

IOP Conf. Series: Materials Science and Engineering **558** (2019) 012053 doi:10.1088/1757-899X/558/1/012053

2. Experimental methods

Quartz sand (SiO_2), Karelian schungite (Schung, table 1), aluminum PAD-1 (GOST 6058-73), and magnesium MPF-3 (GOST 6001-79) were used as initial elements in the experiments. Recycled cryolite Na_3AlF_6 was used as a source of fluorine, potassium perchlorate $KClO_4$ (TU 6-09-3801-76) – as an oxygen source. Schungite is used due to the high content of silicon oxide, as the main reagent, and carbon, which forms an oxide loosening the product. Some amounts (up to 5%) of iron and chromium oxides were introduced into the mixture as catalytically active additives. The quartz sand was crushed into powder of less than 150 and 50 μm in ball drums. The particle size was measured using a MicroSizer 201 analyzer.

Table 1. Schungite chemical composition.

1	Silicon (Si, weight %)	25
2	Oxygen (O, weight %)	30
3	Carbon (C, weight %)	32

The mixtures which can react in the SHS mode were used. The initial mixture composition was estimated by the following chemical scheme:

$$SiO_2 + Shung + Al + Mg + Na_3AlF_6 + KClO_4 \rightarrow NaMg_3Si_3AlO_{10}F_2 + KCl\uparrow + CO\uparrow \quad (1)$$

The initial billets 40 mm in diameter and ~5 mm in height were sintered in the furnace at T = 930°C in the air of atmospheric pressure. The temperature was being raised during 1 h. The exposure time at 930°C was 10 min. The phase composition of the product was determined using a DRON-3M diffractometer with Cu K_α radiation. The quantitative elemental composition of phases was carried out using a Zeiss Ultra Plus high-resolution SEM with INCA 350 Oxford Instruments X-ray microanalysis system.

The porosity of the synthesized material was determined in accordance with GOST 2409-80. The size of the open pores of the synthesized material was determined according to GOST 26849-86. The method consists in determining the minimum gas pressure required for forcing a gas bubble through the pores of a flat hydrophilic membrane impregnated with water or alcohol. The bending strength was measured on the installation Instron. The concentration of chemical elements in tap water before and after filtration was determined by the atomic absorption method. In order to reuse the synthesized membranes, they were regenerated by a reverse pulsed flow of the purified water.

3. Results

The overall view of the synthesized sample is shown in figure 1. The main phases of the synthesized material are those of spinel and fluorophlogopite in approximately equal mass fractions (figure 2). At the same time, the presence of unidentified lines indicates a more complex composition of the material.

Figure 1. Overall view of the synthesized sample.

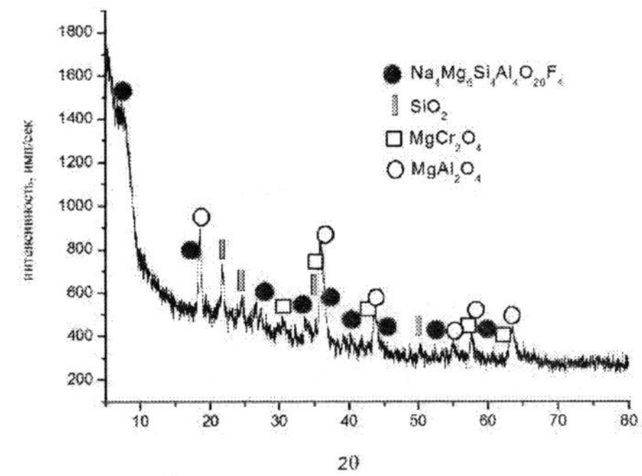

Figure 2. Phase composition of the synthesized material.

IOP Conf. Series: Materials Science and Engineering **558** (2019) 012053 doi:10.1088/1757-899X/558/1/012053

Figure 3 demonstrates the porous structure of the synthesized material. According to the analysis, the material porosity is 60%; the open porosity is 98%. The measured value of open pores (according to GOST 26849-86) in the synthesized material is 1–3 μm. The bending strength of the sample is 5.5–6 MPa.

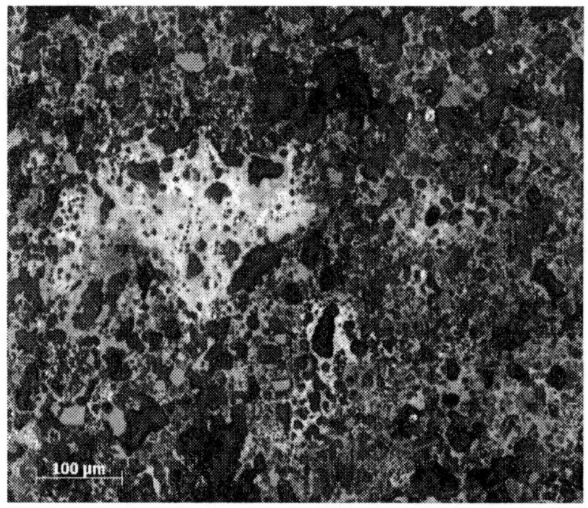

Figure 3. Synthesized material structure.

The elemental composition of the material is shown in figures 4 and 5. The analysis confirms the formation of spinel based on Cr, Mg and Al (figure 4) and the presence of fluorophlogopite (figure 5) in the material composition.

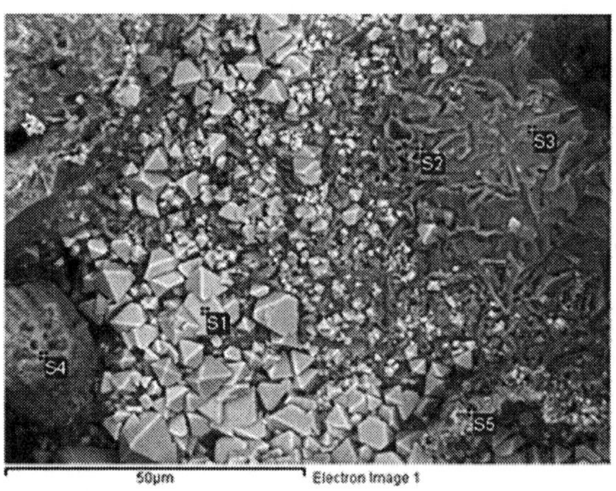

Spectrum	O	Na	Mg	Al	Si	K	Ca	Cr
S1	32.40		6.67	0.71				60.22
S2	39.09	0.74	3.03	3.61	15.88	1.36	0.72	35.56
S3	45.08	1.13	3.48	12.68	16.69	1.75		19.20
S4	49.47		3.93		4.60		25.85	16.15
S5	37.38	0.51	4.64	4.79	12.40	2.04	0.62	37.63

Figure 4. Spinel elemental composition.

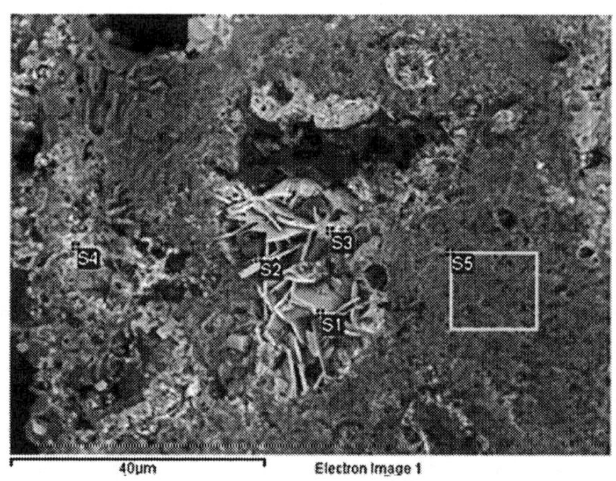

Spectrum	O	F	Na	Mg	Al	Si	Cl	K	Cr
S1	37.83			0.72	0.59	1.46			59.40
S2	35.85	18.78	1.00	4.86	1.31	6.51			31.69
S3	39.54			4.75	3.71	1.01	0.37		50.62
S4	41.75		2.07	1.21	5.47	1.82		0.40	47.27
S5	39.14	18.95	0.47	10.78	4.82	15.16	0.51		10.17

Figure 5. Elemental composition of mica-like matrix.

As a result of the work, laboratory devices were developed and experiments in water purification were carried out. Table 2 demonstrates the content of metal ions in the water before and after its filtration through the synthesized membrane. The results prove that the obtained sample of the filter purifies the tap water rather efficiently from the high content of metal ions and reduces the water hardness by 30%.

Table 2. Content of Mg and Ca ions in the water before and after filtration.

Chemical element	Concentration of chemical elements in the tap water, mg/l	
	Tap water before filtration	Tap water after filtration through the synthesized filter
Mg	23.706	16.856 16.343 15.963
Ca	81.350	63.276 62.735 63.186

4. Conclusions

Membranes with mica-like structure were obtained by the direct synthesis. Their pore size was 1–3.5 μm, porosity was ~ 60%, open porosity was up to 98%. It was established that the bending strength of the obtained samples was 5.5–6 MPa, the specific surface area was 1.203 m^2/g.

Multiple regeneration of filters can be carried out by a reverse pulsed flow of the purified water, washing with solvent or high-temperature annealing.

The investigation results can be used for the development of production technologies of porous mica-crystalline filters for fine purification of water and technological gases from dispersed micro-impurities.

Acknowledgments

This work was performed by using the set of modern scientific instruments available for multiple accesses at the ISMAN Center of Shared Services.

References

[1] Mironenko V I, Mironenko K V, Reznikov K L, Kofanov S P, Krotov A P and Goncharov V D 1996 *The impact of man-made factors on the environment of the Altai Territory* (Barnaul: Altai) pp 25–28

[2] Varshavsky V Ya, Nikoladze G I, Rakhmanin Yu A, Skvortsov L S and Tseskis A B 1996 *Installations (Filters) for Drinking Water Purification* ed L Skvortsov (Moscow: Moskva)

[3] Borovinskaya I P, Merzhanov A G and Uvarov V I 2001 *Nauka proizvodstvu* **10** 28–32

[4] Gordienko P S, Yarusova S B, Bulanova S B, Shabalin I A and Kudryavy V G 2013 *Khimicheskaya tehnologiya* **14** 185–192

[5] Gordienko P S, Yarusova S B, Shabalin I A, Zheleznova V V, Zarubina N V and Bulanova S B 2014 *Radiokhimiya* **56** 518–52

[6] Tresvyatski S G, Parkhomenko M A and Kondratenko A D 1965 *Neorg. Materialy* **1** 449–59

[7] Gordienko P S, Yarusova S B, Shabalin I A and Somova S N 2015 *Proc. IV Int. Conf. on Chemical Technology* (Yerevan) pp 31–34

[8] Loryan V E, Kachin A R and Borovinskaya I P 2014 *Proc. Int. Conf. on Modern Methods and Technologies for Material Development and Processing* (Minsk: PTI NAS) pp 130–136

[9] Loryan V E, Kachin A R and Uvarov V I 2017 *Perspektivnye materialy* **2** 72–77

[10] Gates B, Katzer J and Schuit G 1981 *Chemistry of catalytic processes* (Moscow)

[11] Novikov V A, Komzolov A V and Zhadyaev A A 2017 *Vestnik of Samara State Technical University/Tech. Sci.* **2** 182–190

IOP Conf. Series: Materials Science and Engineering **558** (2019) 012054 doi:10.1088/1757-899X/558/1/012054

SHS in the Ta–Zr–Si–B system and properties of the produced ceramics

S Vorotilo* and E A Levashov

National University of Science and Technology "MISIS", Leninskii 4, Moscow, 119049 Russia

*E-mail: s.vorotilo@misis.com

Abstract. In this work, we investigated the macrokinetic parameters of combustion of the Ta–Zr–Si–B mixtures, the mechanisms of phase and structure formation in the combustion wave, and the structure and properties of the consolidated combustion products. During the combustion, primary grains of ZrB_2 and TaB_2 are formed in the preheating zone due to the gas-phase mass transfer mechanism, which involves the evaporation of boron oxide and chemisorption of boron onto the surface of metals. Upon reaching the combustion zone, the Si–B eutectic and Zr particles melt, and crystals of ZrB2, TaB_2, $TaSi_2$ and $ZrSi_2$ precipitate from the melt. In the secondary structure formation zone, unreacted boron interacts with the ZrSi2 and produces ZrB_2. TaB_2 and ZrB_2 partially interact and form solid solution $(Ta,Zr)B_2$. Ceramics, produced by hot pressing of SHS, posses relative density 95–98% and a specific microgradient grain structure.

1. Introduction

Transitional metals borides (TMB) are prospective for the development of advanced ultra-high temperature ceramics (UHTCs) for aerospace applications. Metallic bonding in the TM layers provides a high electrical and thermal conductivity, whereas the strong covalent bonding in the B layers is responsible for the high hardness, elastic modulus and relatively low coefficient of thermal expansion of TMBs. This combination of mechanical and thermophysical properties provides the high thermal shock resistance of zirconium diboride [1–4].

Oxidation resistance of TMBs is governed by the formation of a protective B_2O_3 layer, which hinders the diffusion of oxygen in the material [1, 2, 5–7]. However, at $T \geq 1100$ °C, B_2O_3 becomes volatile and evaporates, leading to the destruction of the protective oxide layer. Decomposition of the protective layer can be inhibited by alloying of ZrB_2 with silicon-containing compounds, such as SiC, $MoSi_2$, $ZrSi_2$, and $TaSi_2$ [2, 5, 6, 8]. During the oxidation, silicon yields either a SiO_2 layer or a borosilicate glass layer, which remains stable up to 1600°C [1, 4, 5]. Alloying of ZrB_2 with TaB_2 to produce $(Zr,Ta)B_2$ solid solution is another viable way to increase its oxidation resistance. [9, 10]. Another approach is adding $TaSi_2$ to the ZrB_2-based ceramics [11–20]. In previous works, $TaSi_2$ was introduced into the ceramic mixtures as a powder.

Meanwhile, the simultaneous in situ co-synthesis of both refractory boride and disilicide is of a great interest. Phase and structure formation mechanisms during the combustion of Ta–Zr–Si–B mixture were investigated in the previous work [20]. The aim of this work was to investigate the influence of $TaSi_2$ addition on the microstructure of the ZrB_2–TaB_2 ceramics, produced by elemental SHS.

2. Materials and Methods

TaPM grade Ta, PTsrK-1 grade Zr, KEF-4.5 grade Si and B-99A boron powders were used in this work. The reaction mixtures were mixed in a 3 L ball mill for 4 h using hard-alloy balls at a 1:8 mixture to ball weight ratio. The mixing duration was optimized to achieve homogeneous distribution of the reagents in the reaction charge mixture. The SHS was carried out in a sand mold. The combustion products were ground in a ball mill for 6 h until the micron-sized ZrB_2–TaB_2–$TaSi_2$ composite powder was obtained. Ceramic powders were hot-pressed (HP) on a DSP-515 SA Direct Hot Pressing apparatus (Dr. Fritsch Sondermaschinen GmbH) under the following conditions: temperature 1700°C, pressure 30 MPa, exposure time 10 min. The density of consolidated samples was determined by hydrostatic weighing on an AND1 GR-202 analytical balance (A&D, Japan); true density was measured using an AccuPyc 1340 helium pycnometer (Micromeritics, USA). The hardness of the dense ceramics after hot pressing was measured using a Vickers hardness testing machine under a load of 10 N. The SEM and EDS of the sintered samples were conducted using an S-3400N scanning electron microscope (Hitachi) equipped with a NORAN energy-dispersive X-ray spectrometer.

3. Results and Discussion

Figure 1 demonstrates the microstructure of the ZrB_2–20%TaB_2 and ZrB_2–20%TaB_2–35%$TaSi_2$ samples, produced by the hot pressing of SHS products.

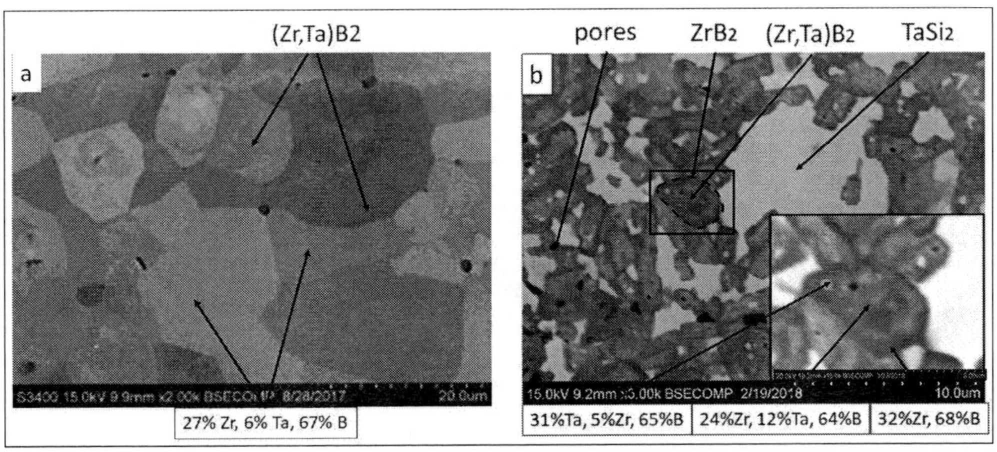

Figure 1. SEM and EDS of ceramics (a) ZrB_2-20%TaB_2 and (b) ZrB_2–20%TaB_2–35%$TaSi_2$

As is evident from the figure 1, the ZrB_2–20%TaB_2 and ZrB_2–20%TaB_2–35%$TaSi_2$ samples have drastically different microstructures. Single-phase solid solution ZrB_2–20%TaB_2 is comprised of equiaxial grains 4–20 μm in diameter, with no noticeable compositional gradients between or within the grains. In the case of ZrB_2–20%TaB_2–35%$TaSi_2$ ceramic, the diboride matrix had a microgradient structure. Tantalum content was maximal in the center of grains and decreases towards the grain boundaries; vice versa, zirconium content increased from the center of grain towards the grain boundaries. To better estimate the grain size, a fracture of ZrB_2–20%TaB_2–35%$TaSi_2$ was investigated (figure 2).

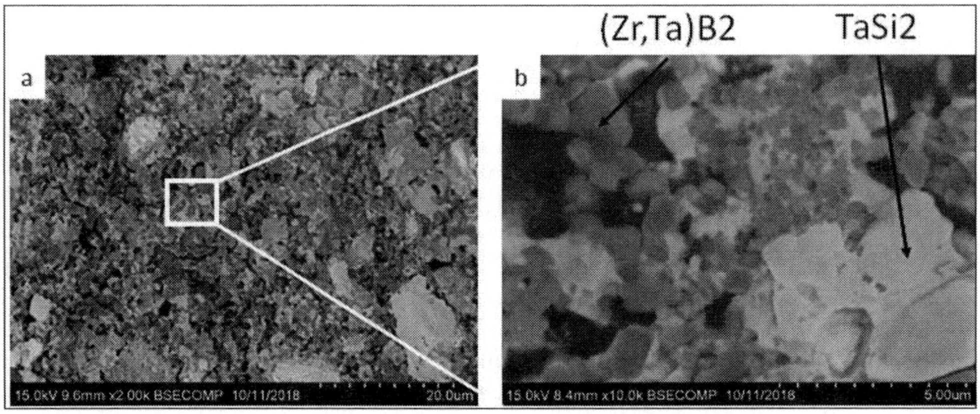

Figure 2. SEM of the fracture of ZrB_2–20%TaB_2–35%$TaSi_2$

As can be seen at figure 2, relatively large $TaSi_2$ particles (3–5 μm) are surrounded by submicron boride grains (0.2–0.7 μm), embedded in the disilicide matrix. Apparently, such structure is the result of the formation and decomposition of the solid solutions $(Zr,Ta,Si)B_2$ during the SHS and hot pressing of the ceramics. This effect provides the opportunity for optimization of the microstructure of boride-silicide ceramics, including the production of nanostructured materials.

The fabrication scheme and microstructure of ZrB_2–TaB_2–$TaSi_2$ composite are summarized in the figure 3.

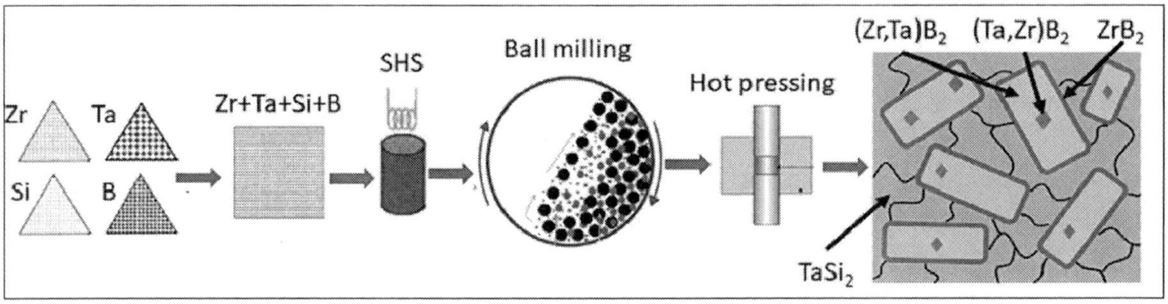

Figure 3. Fabrication scheme and microstructure of ZrB_2–TaB_2–$TaSi_2$ composite

Acknowledgment

This work was conducted with the financial support from the Ministry of Science and Higher Education of Russian Federation in the framework of state assignment № 11.1207.2017/ПЧ.

References

[1] Low I M, Sakka H and Hu Ch 2013 *MAX phases and ultra-high temperature ceramics for extreme environments* ed J Gamon (Hershey: IGI Global) p. 649
[2] Opeka M M, Talmy I G and Zaykoski J A 2004 *J. Mater. Sci.* **39** 5887–5904
[3] Zhang Z, Nan C W, Xu J J, Gao Z H, Li M and Wang J 2014 *J. Mater. Sci. Tech.* **30** 1223–29
[4] Guo S Q 2009 *J. Eur. Ceram. Soc.* **29** 995–1011
[5] Fahrenholtz W G, Hilmas G E, Talmy I G and Zaykoski J A 2007 *J. Am. Ceram. Soc.* **90** 1347–64
[6] Wuchina E, Opila E, Opeka M, Fahrenholtz W and Talmy I 2007 *Interface* **16** 30–36

[7] Silvestroni L, Kleebe H J, Lauterbach S, Muller M and Sciti D 2010 J. *Mater. Res.* **25** 828–834
[8] Rodríguez-Sánchez J, Sánchez-González E, Guiberteau F and Ortiz A L 2015 *J. Eur. Ceram. Soc.* **35** 3179–85
[9] Opila E and Levine S 2004 *J. Mater. Sci.* **39** 5969–77
[10] Levine S R and Opila E J 2003 Tantalum Addition to Zirconium Diboride for Improved Oxidation Resistance, NASA/TM-2003-212483
[11] Sciti D, Silvestroni L, Guicciardi S, Fabbriche D D and Bellosi A 2009 *J. Mater. Res.* **24** 2056–65
[12] Talmy I G, Zaykoski J A and Opeka M M 2008 *J. Am. Ceram. Soc.* **91** 2250–57
[13] Ghaffari S A, Faghihi-Sani M A, Golestani-Fard F and Mandal H 2013 *J. Am. Ceram. Soc.* **33** 1479–84
[14] *Ultra-High Temperature Ceramics: Materials for Extreme Environment Applications* 2014 ed W G Fahrenholtz, E J Wuchina, B Lee and Y Zhou (Wiley) p. 465
[15] Silvestroni L and Sciti D 2011 *J. Am. Ceram. Soc.* **94** 1920–30
[16] Silvestroni L and Sciti D 2101 *Characterization of Ultra High Temperature Ceramics via Transmission Electron Microscopy. Part II: UHTCs sintered with addition of* $TaSi_2$ (technical report) ISTEC-CNR (Institute for Science and Technology for Ceramics)
[17] Justin J F and Jankowiak A 2011 *AerospaceLab* 1–11.
[18] Opila E J, Smith J, Levine S R, Lorincz J and Reigel M 2010 *Open Aero Eng J* **3** 41–51
[19] Wang S, Xu C, Ding Y and Zhang X 2013 *Int. J. Refract. Met. H.* **41** 507–516
[20] Vorotilo S, Levashov E A, Petrzhik M I and Kovalev D Yu 2018 *Ceram. Int.* doi:10.1016/j.ceramint.2018.10.020

IOP Conf. Series: Materials Science and Engineering **558** (2019) 012055 doi:10.1088/1757-899X/558/1/012055

Application of nickel nanopowders of different morphology for the synthesis of highly porous materials by powder metallurgy

V A Zelensky[1], A B Ankudinov[1], M I Alymov[1,2], N M Rubtsov[2] and I V Tregubova[1]

[1]Baikov Institute of Metallurgy and Materials Science, RAS, Moscow, Russia
[2]Merzhanov Institute of Structural Macrokinetics and Materials Science, RAS, Chernogolovka, Russia

Abstract. In this work, we studied the morphology of nickel nanopowders synthesized in various ways: by chemical-metallurgical method, by decomposition of nickel formate and by electroexplosive technology. A comparative analysis of the characteristics of these nanopowders has been carried out. A sample carrier was made from the powder synthesized by electroexplosive technology. A porous material with a gradient structure was created by modifying the pore surface of a sample carrier with catalytically active centers of copper oxide.

1. Introduction

At present, of interest are the studies of gas-phase heterogeneous catalytic reactions in membrane reactors, in which, compared with traditional reactors having a bulk catalyst layer, mass and heat transfer has been significantly improved. To intensify endothermic catalytic reactions (for example, steam reforming of methane), accelerated heat transfer from an external source to the reaction zone is necessary. Therefore, the use of catalysts on porous metal supports is promising.

Methods are known for producing highly porous metal and ceramic materials by self-propagating high-temperature synthesis, casting, gas melting, rolling, electrochemical processing, deposition, and others. Methods of producing porous metal materials using powder metallurgy methods using temporary pore-forming agent are being actively developed [1–3]. Research is being conducted in the field of creating highly porous metallic materials (they are also called metal foams, sponges, or cellular materials).

Materials having a bimodal pore size distribution, otherwise materials with a gradient porosity [4, 5] are very interesting. Improved transport properties in such materials due to the presence of large pores (macropores). Pores of small sizes (micro- and nanopores) together with the micro-relief provide a large specific surface, which is important for practical use, since adsorption, desorption and chemical transformations of reagents occur on the surface. The technology of powder metallurgy allows to obtain macropores due to large particles of a pore-forming agent, while micro- and nanopores are created through the use of highly dispersed metal powders.

Another interesting direction is the modification of the surface of the pore space with catalytically active oxide or metal centers of nanometer size [6, 7]. Such centers can be created by the methods of precipitation from solutions, by the sol-gel method, by recondensation, etc.

In the present work, nickel nanopowders were used in the manufacture of porous materials. It is known that the size of powder particles, their morphology and tendency to form agglomerates depend on the method of preparation.

The purpose of this work is the study of nickel powders of various origin: obtained by chemical-metallurgical method [8], by decomposition of nickel-containing salts [9] and by the method of wire explosion. On the basis of a comparative analysis of powders for the manufacture of porous carrier samples, preference was given to a powder produced by an electroexplosive method. Additionally, copper oxide particles, which are catalytically active centers, were deposited on the surface of the porous carrier.

2. Synthesis of nickel nanopowders and their characterization

2.1. Chemical-metallurgical method of obtaining nanopowders

Production of powders by chemical-metallurgical method is carried out in two stages. First, low-water nickel hydroxide is synthesized. For the synthesis, a portion of the solid salt of nickel chloride was treated with 6 mol/l alkali solution of NaOH at a constant pH = 9 and kept for 45 min. The alkali solution was taken 2.5 times the amount required by stoichiometry. The washed powders were dried in air at room temperature for a day before dusting. The nickel hydroxide obtained as a result of precipitation has the structural formula $Ni(OH)_2 \cdot H_2O$. The specific surface of the hydroxide obtained is 154 m^2/g.

The resulting hydroxide was reduced in 60 min in hydrogen at 180°C to the metal. The recovery mode was selected on the basis of the balance: complete recovery of the hydroxide to the metal while maintaining the minimum particle size. The crystallite size of the obtained nickel nanopowder calculated by the Selyakov-Scherer formula is about 30 nm. The particle size determined by the BET method is about 70 nm. Electron microscopy of the reduced nickel nanopowder is shown in figure 1. It is seen that the powder consists of dense agglomerates, in which the particles are strongly burned together.

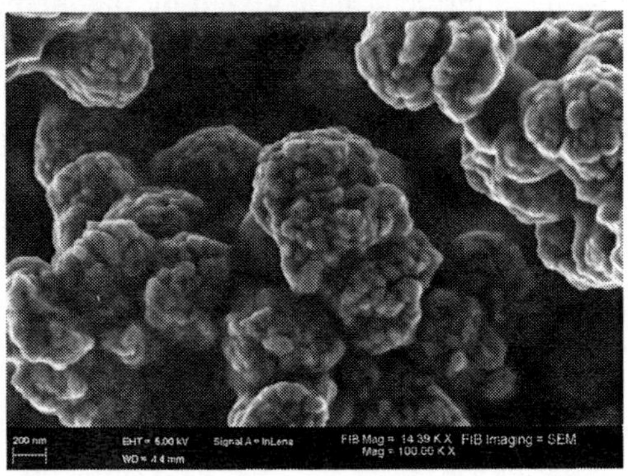

Figure 1. Electron microscopy of reduced nickel nanopowder.

2.2. Production of nanopowders by decomposing unstable compounds

This method is one of the easily implemented and low-cost. To obtain nickel nanopowder, nickel formate Ni $(HCOO)_2 \cdot 2H_2O$ was used, which tends to decompose when heated to pure nickel and exhaust gases. The decomposition reaction was carried out in a quartz reactor in a stream of argon at a temperature of 210°C for 70 min. Electron microscopy of the obtained nanopowder is shown in figure 2. The size of the powder particles is about 70 nm. Particles form large-scale agglomerates.

2.3. Electroexplosive technology of nanopowders

When an electrical explosion of conductors occurs, the wire material, when a powerful current pulse passes through it, is destroyed to form particles from 5 nm to 1 mm, as a result, the wire material partially evaporates or scatters as liquid droplets. From the evaporated phase, due to the condensation process, very small particles are formed in a stream of rapidly expanding vapor. The technology is quite productive, although not without flaws. The main disadvantage is the presence in the nanopowders of this type of a small amount of powder of a large fraction. Figure 3 presents electron microscopy of a commercial nickel powder with an average particle size of 68 nm obtained by this technology. It can be seen that along with the smallest nanoparticles, it contains rather large round particles of submicron size.

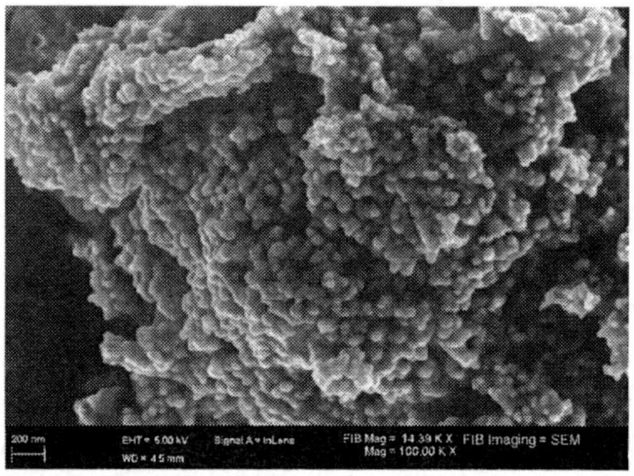

Figure 2. Electron microscopy of nickel nanopowder obtained by decomposition of nickel formate.

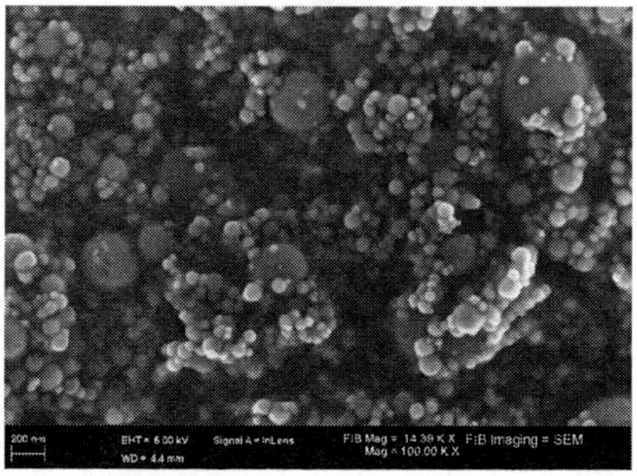

Figure 3. Electron microscopy of nickel nanopowder obtained by electroexplosive technology.

3. Fabrication of highly porous nanopowder samples

Porous metal samples of the carrier for deposition on the surface of the through channels of catalytically active nanostructured copper oxide layers were made using SDP (sintering-dissolution process). When conducting preliminary experiments, it was found that the use of an electroexplosive nickel nanopowder was preferred. The use of an electroexplosive powder is explained by the fact that, as compared with powders obtained by a chemical-metallurgical method and by the method of decomposition of nickel formate, the highest porosity of carrier samples is achieved. Cylindrical samples were prepared using sodium chloride as a pore former. The ratio of Ni/NaCl by volume in the compacts was 30/70. Samples were sintered for 120 min at a temperature of 800°C in a stream of hydrogen. Their diameter is 26 mm, height 8 mm; porosity of about 70%.

4. Copper oxide precipitation

The deposition of copper oxide was as follows:

1. Samples were soaked with 0.02 M aqueous solution of copper sulfate hydrate $CuSO_4 \cdot 5H_2O$. Samples were pre-evacuated.
2. Samples were dried.
3. Samples were treated with 7M aqueous NaOH solution for 20 min. Samples were pre-evacuated. The treatment with alkali is necessary in order for the product to transform to the hydroxide Cu $(OH)_2$: $CuSO_4 + 2NaOH = Cu(OH)_2 + Na_2SO_4$. $Cu(OH)_2$ precipitates; Na_2SO_4 remains in solution and is removed during subsequent washing.
4. Samples were thoroughly washed by pumping water through the sample.
5. Samples were calcined at a temperature of 150–180°C for 30 min. When heated, the hydroxide goes into oxide: $Cu(OH)_2 \Rightarrow CuO + H_2O$.

On several samples, the salt deposition procedure (operations 1 and 2) was repeated three times. After treatment with an aqueous solution of alkali, washing and calcining (steps 3–5) with an initial mass of samples of 10.3–10.6 g, the mass of precipitated copper oxide was 0.8–1.0 g.

An energy dispersive analysis (EDA) was performed on a TESCAN VEGA II SBU scanning electron microscope. The surface-normalized distribution of Ni/Cu/O elements was 80.60/12.75/6.65%. The results of EDA indicate a relatively uniform deposition of a layer of copper oxide on the pore walls.

The appearance of a porous nickel tablet modified with copper oxide, micrographs of the surface area of its cleavage and the results of EDA are presented in figures 4–6. In contrast to existing industrial technologies, nickel nanopowders were used as a starting material for creating highly porous samples. This allowed, along with micron-sized macropores created by the pore-forming agent, to obtain a developed surface of the pore walls formed by agglomerates of Ni nanoparticles and having their own nanoporosity.

Figure 4. Porous nickel tablet modified with copper oxide.

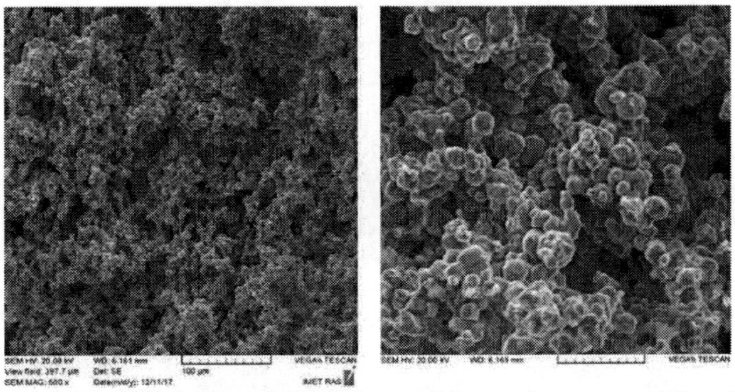

Figure 5. SEM images at different magnifications of the surface area of the cleaved porous Ni modified with copper oxide.

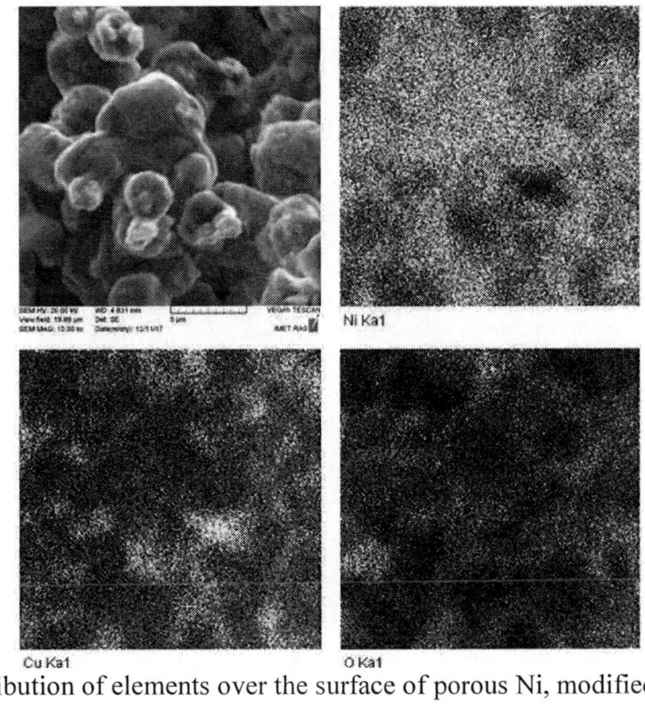

Figure 6. The distribution of elements over the surface of porous Ni, modified with copper oxide.

Thus, it was shown that powder metallurgy can create volumetric highly porous nickel catalyst carriers with a developed surface and a gradient structure of the pore space based on preliminary pressing of mixtures of metal nanopowders and blowing agents, followed by sintering metal particles and removing the blowing agent. The experiments on the deposition of catalytically active centers (oxides) on the pore walls allow the creation of a porous membrane-catalytic converter based on copper oxide powders.

Acknowledgements
This work was supported by the Russian Foundation for Basic Research, grant no. 17-03-00867, and the Presidium of the Russian Academy of Sciences, the Program of Basic Research of the Russian Academy of Sciences 34P. Investigations by electron microscopy and X-ray phase analysis were carried out with the involvement of government funds no. 007-00129-18-00.

References

[1] Kränzlin N and Niederberger M 2015 *Mater. Horiz.* **2** 359–377
[2] Stanev L, Kolev M, Drenchev B and Drenchev L 2017 *Rev. J. Manuf. Sci. Eng.* **139** 050801
[3] Stanev L, Kolev M, Drenchev B and Drenchev L 2017 *Rev. J. Manuf. Sci. Eng.* **139** 050802
[4] Singh S and Bhatnagar N 2017 *J. Porous Mater.* **24** 1–18
[5] Izaak T I and Vodyankina O V 2009 *Uspekhi khimii* **78** 80–92
[6] Zhang F, Yuan C, Lu X, Zhang L, Che Q and Zhang X 2012 *J. Power Sour.* **203** 250–256
[7] Danilova M M, Fedorova Z A, Zaikovskii V I, Porsin A V, Kirillov V A and Krieger T A 2014 *Appl. Catal. B: Environ.* **147** 858–863
[8] Tregubova I V, Alymov M I, Ankudinov A B, Zelenskiy V A and Tikhomirov S A 2011 *Tsvetnyye metally* **5** 47–52
[9] Alymov M I, Rubtsov N M, Seplyarskii B S, Kochetkov R A, Zelensky V A and Ankudinov A B 2017 *Mendeleev Commun.* **27** 631–633

Institute of Physics
Dirac House, Temple Back
Bristol BS1 6BE UK

ISSN: 1757-8981
ISBN 978-1-5108-9092-3

9 781510 890923

CURRAN ASSOCIATES INC.
proceedings.com